Arithmetic, Geometry, Cryptography and Coding Theory

CONTEMPORARY
MATHEMATICS

832

Arithmetic, Geometry, Cryptography and Coding Theory

19th International Conference
Arithmetic, Geometry,
Cryptography and Coding Theory
June 5–9, 2023
Centre International de Rencontres Mathématiques,
Marseille, France

Nils Bruin
David Kohel
Chloe Martindale
Editors

AMS AMERICAN MATHEMATICAL SOCIETY

2020 *Mathematics Subject Classification.* Primary 11G05, 11G10, 11G20, 11G25, 11G30, 11G35, 11T71, 14G10, 14H40, 14Q05.

Library of Congress Cataloging-in-Publication Data

Cataloging-in-Publication Data has been applied for by the AMS.
See http://www.loc.gov/cip/.

Contemporary Mathematics ISSN: 0271-4132 (print); ISSN: 1098-3627 (online)
DOI: https://doi.org/10.1090/conm/832

---

10 9 8 7 6 5 4 3 2 1   31 30 29 28 27 26

Contents

Contents

Preface

The 19th edition of the conference Arithmetic, Geometry, Cryptography and Coding Theory (AGC^{2}T) took place in June 2023 at the *Centre International de Rencontres Mathématiques* (CIRM). This conference is a biennial event going back to 1987, and has become a significant forum for researchers in arithmetic geometry and number theory, and its applications to cryptography and coding theory. This particular edition served an additional role as the capstone event for the Jean-Morlet Chair thematic semester on Arithmetic Statistics, organized by Peter Stevenhagen (Chair) and Samuele Anni (Chair host).

Some 95 researchers participated in the 2023 edition of this conference. In particular, we would like to thank the plenary speakers for accepting an invitation to speak about their work: Wouter Castryck, Carel Faber, Elena Kirshanova, Rachel Newton, Padmavathi Srinivasan, Mima Stanojkovski, Yan Bo Ti, and Tony Várily-Alvarado. We also thank the other speakers for highlighting current developments in the field: Yves Aubry, Elena Berardini, Jeremy Booher, Irene Bouw, Thomas Bouchet, Mar Curcó Iranzo, Martin Djukanovic, Damián Gvirtz-Chen, Tomoyoshi Ibukiyama, Kiran Kedlaya, Harun Kır, Eda Kırımlı, Dimitri Koshelev, Avinash Kulkarni, Boris Kunyavskii, Christophe Levrat, Luciano Maino, Stefano Marseglia, Travis Morrison, Steffen Müller, Jade Nardi, Ciaran Schembri, Peter Stevenhagen, Michael Tsfasman, and Yuri Zarhin.

We thank all the participants for actively contributing to the success of the meeting, as well as the steering committee for their support. We also thank the contributing authors for their submissions as well as the many reviewers who carefully evaluated them.

We are also grateful to the staff of CIRM for their help in organizing the conference and in providing such a welcoming and mathematically stimulating environment, as well as the American Mathematical Society and the editors at the editorial office of *Contemporary Mathematics*.

Last but not least we express our gratitude to the sponsors of AGC^{2}T: CIRM, Institute Archiméde Mathématiques-Informatique, Centre National de la Recherche Scientifique, Institut de Mathématiques de Marseille, Fondation AMIDEX, Foundation Compositio Mathematica, Fonds d'Intervention Recherche, and JC^{2}A.

Nils Bruin
David Kohel
Chloe Martindale

Contemporary Mathematics
Volume **832**, 2026
https://doi.org/10.1090/conm/832/16604

Closed points on curves over finite fields

Yves Aubry, Fabien Herbaut, and Julien Monaldi

ABSTRACT. We are interested in the quantity $\rho(q,g)$ defined as the smallest positive integer such that $r \geq \rho(q,g)$ implies that any absolutely irreducible smooth projective algebraic curve defined over $\mathbb{F}_q$ of genus g has a closed point of degree r. We provide general upper bounds for this number and its exact value for $g = 1, 2$ and 3. We also improve the known upper bounds on the number of closed points of degree 2 on a curve.

1. Introduction

In the whole paper we consider a power q of a prime, $\mathbb{F}_q$ the finite field with q elements and $\overline{\mathbb{F}}_q$ its algebraic closure. Let X be an absolutely irreducible smooth projective algebraic curve (just called curve from now on) defined over $\mathbb{F}_q$ of genus g. This article deals with the notion of closed point, that is an orbit under the action of $\mathrm{Gal}\left(\overline{\mathbb{F}}_q/\mathbb{F}_q\right)$ on $X(\overline{\mathbb{F}}_q)$, or equivalently a place of the corresponding function field. If P is a closed point of X, we define its degree as the cardinality of the orbit, or equivalently as the dimension over $\mathbb{F}_q$ of the residue field, that is the quotient of the local ring at P by its maximal ideal.

Clearly the numbers $B_r(X)$ (B_r for short) of closed points of degree r are related to the numbers $N_r(X) = \sharp X(\mathbb{F}_{q^r})$ (N_r for short) of rational points over $\mathbb{F}_{q^r}$ by the following formula:

$$N_r(X) = \sum_{d|r} dB_d(X). \tag{1}$$

The aim of the article is to study the quantity $\rho(q,g)$ introduced in Problem 3.2.13 of [**12**] by Tsfasman, Vlăduţ and Nogin and defined as the smallest positive integer such that $r \geq \rho(q,g)$ implies that $B_r(X) \geq 1$ for any genus g curve X defined over $\mathbb{F}_q$.

They ask to find its exact value. In this paper we contribute to this issue by determining the exact value of $\rho(q,g)$ for $g = 1, 2, 3$. We also establish a new upper bound for the number $B_2(X)$ of closed points of degree 2. We summarize our contributions in the following theorem.

2020 *Mathematics Subject Classification.* Primary 14H25; Secondary 11G20.

Key words and phrases. Algebraic curves, finite fields, closed points, diophantine stability.

This work is partially supported by the French Agence Nationale de la Recherche through the Barracuda project under Contract ANR-21-CE39-0009-BARRACUDA.

THEOREM (Proposition 3.1, Theorem 4.1, Theorem 5.2, Theorem 6.1). *Let g be a positive integer and X be a genus g curve defined over $\mathbb{F}_q$.*

(i) *An upper bound on $B_2(X)$ is given by*

$$B_2(X) \leq \begin{cases} \frac{q^2+1+2gq}{2} - \frac{(q+1)^2}{2g} & \text{if } g \geq 2q+2 \\ \frac{q^2+1+2gq}{2} - \frac{4(q+1)-g}{8} & \text{otherwise.} \end{cases}$$

(ii) *The values of $\rho(q,g)$ for $g = 1, 2$ and 3 are as follows*

TABLE 1. Values of $\rho(q,g)$ for $g = 1, 2$ and 3.

g \ q	2	3	4	5	7	8	9	11	13	16	17	19	23	25	27	29	31	32	≥ 37
1	5	3	3	1	$\cdots$													$\cdots$	1
2	4	4	2	3	2	2	2	2	1	$\cdots$								$\cdots$	1
3	7	5	3	3	2	2	3	2	2	2	2	2	2	2	1	2	1	2	1

Let us stress the connection between the topic and the notion of Diophantine stability introduced (over number fields) by Mazur and Rubin in [**8**]: a variety V defined over a field K is said to be diophantine-stable for the field extension L/K if $V(K) = V(L)$. When r is a prime number and X is a curve defined over $\mathbb{F}_q$ which is diophantine-stable for the extension $\mathbb{F}_{q^r}/\mathbb{F}_q$ then Formula (1) implies $\rho(q,g) > r$. Curves with Diophantine stability (also called DS-curves) over finite fields have been studied by Lario who provides in [**6**] the complete list of isomorphism classes of DS-curves up to genus 3. Vrioni has studied in [**13**] curves and surfaces with Diophantine stability over finite fields.

2. Known bounds on B_r and $\rho(q,g)$

We collect in this section the known general bounds on B_r and $\rho(q,g)$ as well as existence results for points of given degree. In this direction we also propose one contribution, namely Proposition 2.3, which sometimes slighty improves the known bounds and which helps reduce the number of cases to consider in the last three sections.

The zeta function Z_X of a curve X of genus g defined over $\mathbb{F}_q$ is defined by

$$Z_X(T) = \exp\Big(\sum_{n=1}^{\infty} \sharp X(\mathbb{F}_{q^n})\frac{T^n}{n}\Big).$$

It is well-known that it is a rational function of the form

$$Z_X(T) = \frac{\prod_{j=1}^{g}(1-\omega_j T)(1-\overline{\omega_j}T)}{(1-T)(1-qT)}$$

for some complex numbers $\omega_1, \ldots, \omega_g$, and the Riemann Hypothesis, proved by Hasse for elliptic curves and by Weil for curves of any genus, says that the ω_j have absolute value $\sqrt{q}$. We obtain from the two previous expressions of the zeta function

$$N_n(X) = \sharp X(\mathbb{F}_{q^n}) = q^n + 1 - \sum_{j=1}^{g}(\omega_j^n + \overline{\omega_j}^n) \tag{2}$$

for any $n \geq 1$, and the Riemann Hypothesis implies that:

$$q^n + 1 - 2gq^{n/2} \leq N_n(X) \leq q^n + 1 + 2gq^{n/2}. \tag{3}$$

When r is a prime, one can combine the equality $B_r = (N_r - N_1)/r$ with the Weil bounds (3) to obtain the following Lemma.

LEMMA 2.1. *Let X be an absolutely irreducible smooth projective algebraic curve of genus g defined over $\mathbb{F}_q$ and let r be a prime number. If $g < \frac{\sqrt{q^r}-\sqrt{q}}{2}$ then $B_r(X) > 0$.*

More broadly, for any integer $r \geq 1$, the Möbius inversion formula, and the Weil bounds lead to the following inequality (see for instance Proposition 3.2.10 in [**12**])

$$\left| B_r(X) - \frac{q^r}{r} \right| \leq \left(\frac{q}{q-1} + 2g\frac{\sqrt{q}}{\sqrt{q}-1} \right) \frac{q^{r/2}-1}{r} < (2+7g)\frac{q^{r/2}}{r}. \tag{4}$$

The asymptotic expansion $B_r = q^r/r + O(\sqrt{q^r}/r)$ follows, for a fixed genus g and for large values of q. As a second consequence of (4), we can deduce that if $g = 0$ then $B_r(X) > 0$ for any $r \geq 1$. Therefore for any q we have

$$\rho(q, 0) = 1.$$

The case of genus zero curves is thus solved and we will consider curves of positive genus from now on.

As a third consequence of (4), one can deduce the Corollary 3.2.11 in [**12**] which reads

$$\rho(q, g) \leq \left\lceil 2\log_q\left(\frac{2g+1}{\sqrt{q}-1} \right) + 1 \right\rceil \tag{5}$$

where $\lceil x \rceil$ stands for the smallest integer greater than or equal to a real number x. We can therefore deduce a bound which is rough when compared to (5) but which has the advantage to be uniform with respect to q

$$\rho(q, g) \leq 4g + 3. \tag{6}$$

Combining the Möbius inversion formula and the Weil bounds is also the point of departure of inequalities obtained by Elkies et al. in [**3**]. For instance, Lemma 2.1 in their paper states for every $r > 0$ the lower bound

$$B_r(X) > \frac{q^r - (6g+3)q^{r/2}}{r}. \tag{7}$$

This inequality together with Weil bounds again enable them to improve the uniform bound (6) the following way when $g \geq 2$

$$\rho(q, g) \leq 2g + 1. \tag{8}$$

But dividing by $q^{r/2}$ in (7) also leads to the asymptotically better bound

$$\rho(q, g) \leq \lceil 2\log_q(6g+3) \rceil. \tag{9}$$

When $r \geq 2$ the first author with Haloui and Lachaud manage to improve Bound (7). In Proposition 3.7. of [**1**] they study similar bounds in the context of abelian varieties, but the proof adapts *mutatis mutandis* if we consider a curve rather than an abelian variety. More precisely, they start from the same Möbius inversion formula $B_r = \frac{1}{r}\sum_{d|r} \mu(r/d)N_d$ where μ is the Möbius function and thus replace the numbers N_d by their expression in function of the roots ω_j's. They prove $rB_r > G(q^{r/4})$ where $G(x) = (x+1)^2\left((x-1)^2 - 2g\right)$ and so the second point of Proposition 3.7 in [**1**] remains true when the inequality becomes strict, that is

$$B_r > \frac{(q^{r/4}+1)^2\Big((q^{r/4}-1)^2 - 2g\Big)}{r}. \tag{10}$$

A consequence is that if $g \geq 1$, $r \geq 2$ and $(q^{r/4}-1)^2 \geq 2g$ then $B_r > 0$. We have thus checked that point (i) of Proposition 3.8 in [**1**] is still valid even if $g \geq 1$ and we can state:

PROPOSITION 2.2 (Aubry, Haloui and Lachaud, Proposition 3.8 in [**1**]). *If $g \geq 1$ then*

$$\rho(q,g) \leq \text{Max}\left(2, \left\lceil 4\log_q(1+\sqrt{2g})\right\rceil\right). \tag{11}$$

One can also notice that Bound (4) involves a second degree polynomial in $q^{r/2}$ whose study leads to a quite efficient bound on $\rho(q,g)$.

PROPOSITION 2.3. *If $g \geq 2$ or if $g = 1$ and $q \leq 9$, then:*

$$\rho(q,g) \leq \left\lfloor 2\log_q\Big(\frac{A+\sqrt{A(A-4)}}{2}\Big) + 1\right\rfloor \tag{12}$$

where $A := \frac{q}{q-1} + 2g\frac{\sqrt{q}}{\sqrt{q}-1}$ and $\lfloor x \rfloor$ stands for the integer part of a real number x.

In particular, with the same conditions on g and q, if $g < \frac{q}{2(\sqrt{q}+1)}$ then $\rho(q,g) = 1$.

PROOF. When setting $X = q^{r/2}$, the lower bound of Inequality (4) yields $rB_r \geq X^2 - AX + A$ where A is as in the statement. The discriminant of the rigth hand side is $A(A-4)$ which is positive if $g \geq 2$, or if $g = 1$ and $q \leq 9$. In these cases we see that if $r > 2\log_q\left(\frac{A+\sqrt{A(A-4)}}{2}\right)$ then $B_r \geq 1$ and the bound follows.

An easy computation shows that $2\log_q\left(\frac{A+\sqrt{A(A-4)}}{2}\right) < 1$ if and only if $A < \frac{q}{\sqrt{q}-1}$, which is equivalent to $g < \frac{\sqrt{q}}{2}\left(1 - \frac{\sqrt{q}-1}{q-1}\right) = \frac{q}{2(\sqrt{q}+1)}$. Hence the previous condition on g implies by Inequality (12) that $\rho(q,g) = 1$. □

Now one can deduce bounds on $\rho(q,g)$ for small values of g and large values of q as stated in the following corollary. These bounds will prove useful to reduce the work of Sections 4, 5 and 6 to a finite number of cases.

COROLLARY 2.4. *We have $\rho(q,1) = 1$ for any $q \geq 8$, $\rho(q,2) \leq 2$ for any $q \geq 7$ and $\rho(q,3) \leq 2$ for any $q \geq 37$.*

PROOF. If $q \in \{8,9\}$ then Proposition 2.3 asserts that $\rho(q,1) = 1$. Moreover, if $q \geq 11$ then $A(A-4) < 0$ and thus $B_r \geq 1$ for any r, which implies that $\rho(q,1) = 1$. Finally, the proof of the cases of genus 2 and 3 is straightforward by Proposition 2.3. □

To conclude this section one should also mention that the Stöhr-Voloch theory leads to specific upper and lower bounds on B_r for irreducible plane curves. See for instance Theorem 9.62 and Theorem 9.63 in [**5**].

3. Upper bound for B_2

In this section we focus on the number of closed points of degree 2 for which we improve the known upper bounds.

We begin by summarizing the upper bounds obtained in the most natural ways. The first idea is to specialize Inequality (4) for $r = 2$ to obtain

$$B_2 \leq \frac{q^2+1+2gq}{2} + \frac{q-1+2g\sqrt{q}}{2}.$$

For comparisons it is interesting to adopt the asymptotic viewpoint for a fixed genus g and for large values of q. This way the asymptotic expansion of the upper bound reads

$$B_2 \leq \frac{q^2}{2} + (g+\frac{1}{2})q + O(\sqrt{q}). \tag{13}$$

Another natural idea is to start from Formula (1), namely $N_2 = B_1 + 2B_2$. Then the Weil upper bound for N_2 yields to the following bound which involves N_1:

$$B_2 = \frac{N_2 - N_1}{2} \leq \frac{q^2+1+2gq}{2} - \frac{N_1}{2}. \tag{14}$$

Again from the Weil lower bound for N_1, one deduces the following bound which only depends on q:

$$B_2 \leq \begin{cases} \frac{q^2+1+2gq}{2} & \text{if } g \geq \frac{q+1}{2\sqrt{q}} \\ \frac{q^2-q+2g(q+\sqrt{q})}{2} & \text{otherwise.} \end{cases} \tag{15}$$

If we adopt the same asymptotic point of view for large values of q we get

$$B_2 \leq \frac{q^2}{2} + (g-\frac{1}{2})q + O(\sqrt{q}). \tag{16}$$

Let us now present our new upper bound.

PROPOSITION 3.1. *Let X be a curve of genus $g > 0$ defined over $\mathbb{F}_q$. We have:*

$$B_2(X) \leq \begin{cases} \frac{q^2+1+2gq}{2} - \frac{(q+1)^2}{2g} & \textit{if } g \geq 2q+2 \\ \frac{q^2+1+2gq}{2} - \frac{4(q+1)-g}{8} & \textit{otherwise.} \end{cases} \tag{17}$$

The following recent result is the point of departure to prove Proposition 3.1. It has been interpreted by Hallouin and Perret as a consequence of an inequality of Euclidean geometry in the numerical space $Num(X \times X)_{\mathbb{R}}$.

THEOREM (Hallouin and Perret, Proposition 12 in [**4**]). *Let X be a curve of genus g defined over $\mathbb{F}_q$. Then*

$$\sharp X(\mathbb{F}_{q^2}) - (q^2+1) \leq 2gq - \frac{1}{g}\left(\sharp X(\mathbb{F}_q) - (q+1)\right)^2. \tag{18}$$

PROOF OF PROPOSITION 3.1. The previous theorem gives immediately (see also Proposition 3.1. in [**2**] where it has already been noticed):

$$B_2 \leq \frac{q^2+1+2gq}{2} - \frac{N_1}{2} - \frac{(N_1-(q+1))^2}{2g} \tag{19}$$

which clearly improves Bound (14). Now we consider the function $x \mapsto -\frac{1}{2g}(x-(q+1))^2 - \frac{x}{2}$. If $g \geq 2q+2$ this function reaches its maximum for $x=0$ which implies $B_2 \leq \frac{q^2+1+2gq}{2} - \frac{(q+1)^2}{2g}$. If $g \leq 2q+2$ it reaches its maximum for $x = \frac{2q+2-g}{2}$ and thus we deduce $B_2 \leq \frac{q^2+1+2gq}{2} - \frac{4(q+1)-g}{8}$. □

Straightforward computations in the three intervals $[1, \frac{q+1}{2\sqrt{q}}]$, $[\frac{q+1}{2\sqrt{q}}, 2q+2]$, and $[2q+2, \infty)$ enable us to check that the new bounds of Proposition 3.1 are always better than Bound (15) deduced from the Weil bounds.

Let us come back to the asymptotic viewpoint for a fixed genus g and for large values of q to compare Bound (15) with our new bound (17) which reads

$$B_2 \leq \frac{q^2}{2} + (g - \frac{1}{2})q + O(1). \tag{20}$$

In the asymptotic expansion of the upper bound we have managed to replace the $O(\sqrt{q})$ in the previous bounds with an $O(1)$.

We now provide examples to show that the integer part of the new bound (17) of Proposition 3.1 is reached for different values of g and q. The third and fourth columns of the following table enable us to compare the bound (15) deduced from the Weil bounds to our new bound (17) for some values of q and g. The fifth column provides equations of curves which attain our new bound. To describe the curves we sometimes need to introduce a generator a of the multiplicative group $\mathbb{F}_q^*$. We also give the numbers of rational points N_1 and N_2 over $\mathbb{F}_q$ and $\mathbb{F}_{q^2}$ of the curves.

TABLE 2. Ex. of curves which attain the bound of Proposition 3.1. In column 5 we denote by a a generator of $\mathbb{F}_q^*$.

g	q	Bound (15)	Bound (17)	Curves reaching bound (17) of Proposition 3.1	N_1	N_2
1	2	4	3	$y^2 + xy = x^3 + x^2 + 1$	2	8
1	3	7	6	$y^2 = x^3 + x^2 - 1$	3	15
1	4	12	10	$y^2 + xy = x^3 + a$	4	24
2	3	11	9	$y^2 = 2x^6 + x^4 + 2x^3 + x^2 + 2$	2	20
2	4	16	14	$y^2 + (x^2 + x)y = x^5 + x^3 + x^2 + x$	3	31
2	5	23	20	$y^2 = 4x^6 + x^5 + x^4 + x^3 + x^2 + x + 4$	4	44
2	7	39	35	$y^2 = 3x^6 + 3x^3 + 3$	6	76
2	8	48	44	$y^2 + (x^2 + x)y = a^2x^5 + a^2x^3 + ax^2 + ax$	7	95
2	9	59	54	$y^2 = 2ax^6 + ax^5 + 2ax^4 + ax^2 + ax + a$	8	116
2	11	83	77	$y^2 = 7x^6 + 5x^5 + 9x^4 + 8x^3 + 5x^2 + 6x + 7$	10	164
3	2	8	7	$x^4 + x^2y^2 + x^2yz + x^2z^2 + xy^2z + xyz^2 + y^4$ $+ y^2z^2 + z^4 = 0$	0	14

4. Elliptic curves

In this section we consider an elliptic curve X defined over $\mathbb{F}_q$.

THEOREM 4.1. *The values of $\rho(q,1)$ are as follows.*

(i) $\rho(2,1) = 5$ *(ii)* $\rho(3,1) = 3$

(iii) $\rho(4,1) = 3$ *(iv)* $\rho(q,1) = 1$ *if* $q \geq 5$

PROOF. (i) First we consider the case where $q = 2$. Proposition 2.2 leads to $\rho(2,1) \leq 6$. Lemma 2.1 also applies and ensures the existence of a degree 5 point, so $\rho(2,1) \leq 5$. Note that a curve X defined over $\mathbb{F}_2$ satisfies $B_4(X) = 0$ if and only if it is a DS-curve for the extension $\mathbb{F}_{2^4}/\mathbb{F}_{2^2}$. As Lario indicates in [**6**] the elliptic curve of equation $y^2 + y = x^3$ is an example of such a curve and we can conclude that $\rho(2,1) = 5$.

(ii) For $q = 3$ we use Proposition 2.3 to get $\rho(3,1) \leq 3$. Lario gives in [**6**] a DS-curve of genus 1 for the extension $\mathbb{F}_9/\mathbb{F}_3$, namely the curve of equation $y^2 = x^3 + 2x + 1$. So there exists an elliptic curve defined over $\mathbb{F}_3$ with no closed point of degree 2, and thus we have that $\rho(3,1) \geq 3$ which gives the result.

(iii) For $q = 4$ Proposition 2.3 gives $\rho(4,1) \leq 3$. Since the elliptic curve of equation $y^2 + y = x^3$ defined in [**6**] has no closed point of degree 2 we can conclude that $\rho(4,1) = 3$.

(iv) We use Proposition 2.3 to get $\rho(5,1) \leq 2$ and $\rho(7,1) \leq 2$. But any elliptic curve over a finite field has a rational point, so $\rho(5,1) = 1$ and $\rho(7,1) = 1$. At last, Corollary 2.4 shows that for any $q \geq 8$ we have $\rho(q,1) = 1$, which concludes the proof. □

5. Genus 2 curves

In this section we consider a curve X of genus 2 defined over $\mathbb{F}_q$. To make the proof of Theorem 5.2 more readable we first establish the following lemma.

LEMMA 5.1. *Any curve X of genus 2 defined over $\mathbb{F}_2$ (respectively over $\mathbb{F}_4$) admits at least one closed point of degree 4 (respectively of degree 2 and 3).*

PROOF. We can relate the issue to the existence of a DS-curve and thus conclude with [**6**]. □

As an important ingredient of the following theorem we will also make use of the classification of genus 2 curves over $\mathbb{F}_q$ up to $\mathbb{F}_q$-isomorphism and quadratic twist provided by Maisner and Nart in [**7**].

THEOREM 5.2. *The values of $\rho(q,2)$ are as follows.*

(i) $\rho(2,2) = 4$	*(ii)* $\rho(4,2) = 2$	*(iii)* $\rho(3,2) = 4$
(iv) $\rho(5,2) = 3$	*(v)* $\rho(q,2) = 2$ *for* $7 \leq q \leq 11$	*(vi)* $\rho(q,2) = 1$ *if* $q \geq 13$

PROOF. As a consequence of Theorem 2 in [**10**] Stark has proved that there is no pointless curve of genus 2 over $\mathbb{F}_{13}$. Moreover Weil bounds (3) imply that any curve of genus 2 defined over $\mathbb{F}_q$ has at least one point when $q > 13$. Hence we conclude that $\rho(q,2) = 1$ for any $q \geq 13$ and point (vi) follows.

Corollary 2.4 implies that for $q \geq 7$ we have $\rho(q,2) \leq 2$ whereas Theorem 3.2. in [**7**] provides an example of a pointless genus 2 curve defined over $\mathbb{F}_q$ for any $q \leq 11$. Thus point (v) is proved.

Now we consider the case where $q = 2$. The genus 2 curve X described in Table 2 of [**7**] by the equation $y^2 + y = x^5 + x^2$ satisfies $N_1(X) = N_3(X) = 5$, which means that $B_3(X) = 0$, and thus $\rho(2,2) \geq 4$. Lemma 5.1 ensures the existence of a closed point of degree 4 on any curve whereas (8) gives $\rho(2,2) \leq 5$. We can conclude that $\rho(2,2) = 4$.

When $q = 3$, Proposition 2.2 asserts that $\rho(3,2) \leq 4$. The genus 2 curve X defined in Table 3 in [**7**] and also in [**6**] by the equation $y^2 = (1+x^2)(-1+x+x^2)(-1-x+x^2)$ is such that $N_3(X) = N_1(X) = 8$, and so $B_3(X) = 0$ which implies $\rho(3,2) = 4$.

If $q = 4$ we use Proposition 2.2 to get $\rho(4,2) \leq 4$ and Lemma 5.1 to conclude.

Now suppose that $q = 5$. Proposition 2.2 gives $\rho(5,2) \leq 3$ but the genus 2 curve X defined in [**6**] by the equation $y^2 = x^5 + 4x$ satisfies $B_2(X) = 0$, and the other inequality follows. □

6. Genus 3 curves

We would like to emphasize two of the main ingredients of the proof of the main theorem of this section. First we will make use of the *L-functions and modular forms database* (sometimes refered as LMFDB, see [**11**]). Second, we will exploit Theorem 1.1. proved by Howe, Lauter and Top in [**5**] which ensures that there exists a pointless genus 3 curve defined over $\mathbb{F}_q$ if and only if either $q = 32$, $q = 29$ or $q \leq 25$. Hence for such values of q we know that $\rho(q,3) \geq 2$.

THEOREM 6.1. *The values of* $\rho(q,3)$ *are as follows.*

(i)	$\rho(2,3) = 7$	*(ii)*	$\rho(3,3) = 5$	*(iii)*	$\rho(4,3) = 3$
(iv)	$\rho(5,3) = 3$	*(v)*	$\rho(7,3) = 2$	*(vi)*	$\rho(8,3) = 2$
(vii)	$\rho(9,3) = 3$	*(viii)*	$\rho(q,3) = 2$ *if* $11 \leq q \leq 25$	*(ix)*	$\rho(27,3) = 1$
(x)	$\rho(29,3) = 2$	*(xi)*	$\rho(31,3) = 1$	*(xii)*	$\rho(32,3) = 2$
(xiii)	$\rho(q,3) = 1$ *if* $q \geq 37$.				

PROOF. As already mentioned, Theorem 1.1 of [**5**] implies that $\rho(q,3) \geq 2$ for $q \leq 25$, $q = 29$ and $q = 32$.

(i) The inequality (8) yields $\rho(2,3) \leq 7$. Moreover, LMFDB provides the curve X of equation $x^4 + x^2y^2 + x^2yz + x^2z^2 + xy^2z + xyz^2 + y^4 + y^2z^2 + z^4 = 0$ which satisfies $B_6(X) = 0$, and thus $\rho(2,3) = 7$.

(ii) By Proposition 2.2 we have $\rho(3,3) \leq 5$. But the curve $x^3z + xz^3 + y^4 = 0$ given in [**11**] is a curve of genus 3 over $\mathbb{F}_3$ without points of degree 4, so $\rho(3,3) > 4$ and we are done.

(iii) Proposition 2.2 gives $\rho(4,3) \leq 4$ whereas the curve $x^3y + x^3z + x^2y^2 + xz^3 + y^3z + y^2z^2 = 0$ defined over $\mathbb{F}_4$ and provided by [**11**] has genus 3 and no points of degree 2. So $\rho(4,3) \geq 3$. Furthermore, the Weil bounds (3) yield $N_1 \leq 17 \leq N_3$. But, there does not exist a curve over $\mathbb{F}_4$ of genus 3 with 17 rational points since $N_4(3) = 14$ (see Table 1 in [**9**]). Now use $B_3 = (N_3 - N_1)/3$ to get $B_3 > 0$ and the desired result.

(iv) Here Proposition 2.3 improves Proposition 2.2 and we get $\rho(5,3) \leq 3$. Moreover, the curve $y^2 = x^7 + x^5 + 3x^3 + x$ found in [**11**] has no points of degree 2, so $\rho(5,3) \geq 3$.

(v) Proposition 2.2 gives $\rho(7,3) \leq 3$. Moreover, by Theorem 1 of [**13**], there does not exist a DS-curve for the extension $\mathbb{F}_{7^2}/\mathbb{F}_7$. So one can conclude that $\rho(7,3) = 2$.

(vi) Proposition 2.2 leads to $\rho(8,3) \leq 3$. Furthermore, by [**11**] there is no genus 3 curve X defined over $\mathbb{F}_8$ such that $B_2(X) = 0$. One can see the coherence with the results provided in [**6**]: Lario states that there does not exist a DS-curves for the field extension $\mathbb{F}_{64}/\mathbb{F}_8$.

(vii) By Proposition 2.2 again we have $\rho(9,3) \leq 3$. Moreover, the genus 3 curve X of equation $x^4 + y^4 + z^4 = 0$ defined over $\mathbb{F}_9$ proposed in [**6**] has no degree 2 points since $N_1(X) = N_2(X) = 28$. This implies $\rho(9,3) = 3$.

(viii) We use the bound of Proposition 2.2 to find $\rho(q,3) \leq 2$ for $11 \leq q \leq 25$. But for such values of q, Theorem 1.1. of [**5**] states that there exists a pointless curve of genus 3.

(ix), (x), (xi) and (xii) We first use Proposition 2.2 to learn that $\rho(q,3) \leq 2$ for $q \in \{27, 29, 31, 32\}$. Moreover, by Theorem 1.1. of [**5**], there exists a pointless curve defined over $\mathbb{F}_{29}$ and $\mathbb{F}_{32}$ whereas there does not exist such a curve neither over $\mathbb{F}_{27}$ nor $\mathbb{F}_{31}$.

(xiii) Corollary 2.4 enables us to check that for any $q \geq 37$ we have $\rho(q,3) \leq 2$. But by Theorem 1.1. of [**5**], there does not exist a genus 3 pointless curve for such values of q, so $\rho(q,3) = 1$. □

Acknowledgment

The authors would like to thank Joan-C. Lario for a useful discussion on diophantine stability.

References

[1] Yves Aubry, Safia Haloui, and Gilles Lachaud, *On the number of points on abelian and Jacobian varieties over finite fields*, Acta Arith. **160** (2013), no. 3, 201–241, DOI 10.4064/aa160-3-1. MR3106095

[2] Yves Aubry and Annamaria Iezzi, *Optimal and maximal singular curves*, Arithmetic, geometry, cryptography and coding theory, Contemp. Math., vol. 686, Amer. Math. Soc., Providence, RI, 2017, pp. 31–43, DOI 10.1090/conm/686. MR3630608

[3] Noam D. Elkies, Everett W. Howe, Andrew Kresch, Bjorn Poonen, Joseph L. Wetherell, and Michael E. Zieve, *Curves of every genus with many points. II. Asymptotically good families*, Duke Math. J. **122** (2004), no. 2, 399–422, DOI 10.1215/S0012-7094-04-12224-9. MR2053756

[4] Emmanuel Hallouin and Marc Perret, *A unified viewpoint for upper bounds for the number of points of curves over finite fields via Euclidean geometry and semi-definite symmetric Toeplitz matrices*, Trans. Amer. Math. Soc. **372** (2019), no. 8, 5409–5451, DOI 10.1090/tran/7813. MR4014282

[5] Everett W. Howe, Kristin E. Lauter, and Jaap Top, *Pointless curves of genus three and four* (English, with English and French summaries), Arithmetic, geometry and coding theory (AGCT 2003), Sémin. Congr., vol. 11, Soc. Math. France, Paris, 2005, pp. 125–141. MR2182840

[6] J.-C. Lario, *Curves over finite fields with Diophantine stability*, `https://web.mat.upc.edu/joan.carles.lario/DS.html`.

[7] Daniel Maisner and Enric Nart, *Abelian surfaces over finite fields as Jacobians*, Experiment. Math. **11** (2002), no. 3, 321–337. With an appendix by Everett W. Howe. MR1959745

[8] Barry Mazur and Karl Rubin, *Diophantine stability*, Amer. J. Math. **140** (2018), no. 3, 571–616, DOI 10.1353/ajm.2018.0014. With an appendix by Michael Larsen. MR3805014

[9] Jean-Pierre Serre, *Rational points on curves over finite fields*, Documents Mathématiques (Paris) [Mathematical Documents (Paris)], vol. 18, Société Mathématique de France, Paris, 2020. With contributions by Everett Howe, Joseph Oesterlé and Christophe Ritzenthaler; Edited by Alp Bassa, Elisa Lorenzo García, Christophe Ritzenthaler and René Schoof. MR4242817

[10] H. M. Stark, *On the Riemann hypothesis in hyperelliptic function fields*, Analytic number theory (Proc. Sympos. Pure Math., Vol. XXIV, St. Louis Univ., St. Louis, Mo., 1972), Proc. Sympos. Pure Math., Vol. XXIV, Amer. Math. Soc., Providence, RI, 1973, pp. 285–302. MR332793

[11] The LMFDB Collaboration, *The L-functions and modular forms database*, https://www.lmfdb.org, 2024.

[12] Michael Tsfasman, Serge Vlăduţ, and Dmitry Nogin, *Algebraic geometric codes: basic notions*, Mathematical Surveys and Monographs, vol. 139, American Mathematical Society, Providence, RI, 2007, DOI 10.1090/surv/139. MR2339649

[13] B. Vrioni, *A census for curves and surfaces with Diophantine stability over finite fields*, PhD thesis. Polytechnic University of Catalonia, School of Mathematics and Statistics, 2021.

Institut de Mathématiques de Toulon - IMATH, Université de Toulon, France: and Institut de Mathématiques de Marseille - I2M, Aix Marseille Univ, UMR 7373 CNRS, France

Email address: yves.aubry@univ-tln.fr

INSPE Nice-Toulon, Université Côte d'Azur, France and Institut de Mathématiques de Toulon - IMATH, Université de Toulon, France

Email address: fabien.herbaut@univ-cotedazur.fr

Institut de Mathématiques de Toulon - IMATH, Université de Toulon, France

Email address: julien.monaldi@ac-nice.fr

Contemporary Mathematics
Volume **832**, 2026
https://doi.org/10.1090/conm/832/16605

Multiplication in finite fields with Chudnovsky-type algorithms over the projective line

Stéphane Ballet, Alexis Bonnecaze, and Bastien Pacifico

Abstract. We propose a recursive polynomial generic construction (RPGC) of multiplication algorithms in any finite field $\mathbb{F}_{q^n}$ based on the method of D.V. and G.V. Chudnovsky (1987) specialized to the projective line. These algorithms improve those of Lempel-Seroussi-Winograd (1983) and the Karatsuba algorithm can be seen as a particular case of this construction. We give upper bounds for the bilinear complexity of these algorithms, and show that it is uniformly in q quasi-linear with respect to the extension degree n. We also prove that the construction of these algorithms is deterministic and can be done in polynomial time. We give an asymptotic bound for the complexity of their construction.

1. Introduction

Multiplication in finite fields has been at the heart of much work since the end of the twentieth century [**BPR⁺21**, **Lar19**]. In addition to being of theoretical interest, this subject is of theoretical interest to many applications that need fast arithmetic, such as cryptography. In this paper, the treated problem and the obtained results concern asymptotic theory of finite field multiplication, not algorithms in practice. Different strategies have been studied to build a multiplication algorithm. Among them, interpolation algorithms on algebraic curves, due to D.V. and G.V. Chudnovsky [**CC88**], have been widely studied for their qualities in terms of bilinear complexity [**BPR⁺21**]. Nevertheless, they present a certain number of weaknesses, including their difficulty of construction and use. In this paper, we propose a construction method that allows us to bypass these difficulties, by doing polynomial interpolation, while preserving the benefit of Chudnovsky-type algorithms.

Multiplications in a degree n extension of $\mathbb{F}_q$ require different kinds of operations in $\mathbb{F}_q$. Let $x = \sum_{i=1}^{n} x_i e_i$ and $y = \sum_{i=1}^{n} y_i e_i$ be two elements of $\mathbb{F}_{q^n}$, in a

---

2020 *Mathematics Subject Classification.* Primary 14Q20.
Key words and phrases. Finite fields, bilinear complexity, polynomial interpolation, algebraic function fields.

basis $\{e_1, \ldots, e_n\}$ of $\mathbb{F}_{q^n}$ over $\mathbb{F}_q$. By the usual method, the product of x and y is given by the formula

$$z = xy = \sum_{h=1}^{n} z_h e_h = \sum_{h=1}^{n} \Big(\sum_{i,j=1}^{n} t_{ijh} x_i y_j \Big) e_h, \tag{1}$$

with

$$e_i e_j = \sum_{h=1}^{n} t_{ijh} e_h,$$

where $t_{ijh} \in \mathbb{F}_q$ are constants in $\mathbb{F}_q$. Two different types of multiplications are involved in this product. Scalar multiplications concern the product by a constant in $\mathbb{F}_q$, and bilinear multiplications concern the product of two generic elements, i.e. depend on the two elements being multiplied (i.e. the $x_i y_j$). Of the two, bilinear multiplications are known to be computationally heavier ([**STV92**], see Survey [**BPR$^+$21**]). This explains the motivation to reduce the number of bilinear multiplications in multiplication algorithms and led to the study of the bilinear complexity, that can be defined as follows.

DEFINITION 1. Let $\mathcal{U}$ be an algorithm for multiplication in $\mathbb{F}_{q^n}$ over $\mathbb{F}_q$. Its number of bilinear multiplications is called its bilinear complexity, written $\mu(\mathcal{U})$. The bilinear complexity of the multiplication in $\mathbb{F}_{q^n}$ over $\mathbb{F}_q$, denoted by $\mu_q(n)$, is the quantity:

$$\mu_q(n) = \min_{\mathcal{U}} \mu(\mathcal{U}),$$

where $\mathcal{U}$ runs over all multiplication algorithms in $\mathbb{F}_{q^n}$ over $\mathbb{F}_q$.

1.1. Some known results. From the work of Winograd and De Groote [**DG83**] applied to the multiplication in any finite field $\mathbb{F}_{q^n}$, it is proven that for all n we have $\mu_q(n) \geq 2n - 1$, equality being ensured if and only if $n \leq \frac{1}{2}q + 1$ [**BPR$^+$21**, Theorem 2.2]. Winograd also proved that this lower bound is obtained with interpolation algorithms [**Win79**]. In 1983, Lempel Seroussi and Winograd introduced a recursive construction of polynomial interpolation algorithms, using the Chinese Remainder Theorem, achieving quasi-linear bilinear complexity [**LSW83**]. In 1987, D.V. and G.V. Chudnovsky proposed an interpolation method on algebraic curves [**CC88**], generalizing polynomial interpolation. This method makes it possible to multiply in any extension of degree n of $\mathbb{F}_q$, with a good bilinear complexity, provided that one has an algebraic curve with a sufficient number of rational points. This original algorithm is called the Chudnovsky–Chudnovsky Multiplication Algorithm (CCMA). More generally, a multiplication algorithm using interpolation over algebraic curves is said to be of type Chudnovsky.

For an increasing degree of the extension, the interpolation requires more and more rational points, i.e. rational places of a function field. From the Serre–Weil bound, the number of rational places is bounded for a fixed genus. Hence, the classical strategy is to build these algorithms over function fields of growing genus. Ballet proved that the bilinear complexity is linear in the degree of the extension ([**Bal99**], see [**BPR$^+$21**]) using the original algorithm over an explicit tower of function fields defined by Garcia and Stichtenoth [**GS95**]. However, it is not clear that these algorithms can be constructed in a reasonable time since we have no method to find the place of degree n required to represent $\mathbb{F}_{q^n}$ [**STV92**, Remark 5]. Moreover, there is no generic and deterministic construction for both the divisors and the bases of the Riemann–Roch spaces involved in the algorithms.

The strategy of growing genus was natural since the original algorithm evaluates only at rational places of a function field. But several years later, thanks to the works of Ballet and Rolland [**BR04**], Arnaud [**Arn06**], Cenk and Özbudak [**CÖ10**], and Randriambololona [**Ran12**], the method has been extended to the evaluations at places of higher degrees, and to the use of derivative evaluations. These generalizations led to the introduction of another strategy for constructing the algorithm for asymptotically large extensions. The evaluation at places of higher degrees allows one to fix a function field and to evaluate at places of growing degrees. In [**BBT16**], Ballet, Bonnecaze and Tukumuli built Chudnovsky-type algorithms with interpolation only over elliptic curves, i.e. fixing the genus g of the function field to be equal to 1, and using places of increasing degrees. This work gave a quasi-linear asymptotic bound for the bilinear complexity of these algorithms with respect to the degree of the extension. Moreover, they can be constructed in polynomial time. This latest result is not yet established for the growing genus strategy. Nevertheless, a hybrid strategy between the two approaches, based in particular on some of the results established in the preprint version of this article [**BBP20**], has enabled us to make significant progress in this direction [**BBP23**].

1.2. New results and organization. In this paper, we build Chudnovsky-type algorithms for multiplication in any finite field $\mathbb{F}_{q^n}$, with interpolation only over the projective line, i.e. fixing the genus g to be equal to 0, and using places of increasing degrees. In small degree extensions, the bilinear complexity of the obtained algorithms can achieve the best known upper bound and sometimes even improve it. Compared with the construction over elliptic curves, our work has the advantages of giving an uniform bound for the bilinear complexity of our algorithms and of giving a generic construction of algorithms for multiplication in any finite field. Namely, the implied Riemann–Roch spaces and their associated representations are generic. More precisely, the divisors defining the Riemann–Roch spaces as well as the associated bases are of the same form for all q and n. These are explicitly given and do not have to be computed. Moreover, our setup enables us to interpolate with polynomials. This makes our algorithms closer to well-known algorithms based on polynomial interpolation such as Karatsuba [**Kar63**], Cook [**Bod07**] or especially the construction of Lempel, Seroussi and Winograd [**LSW83**].

This paper begins with an overview of the current generalizations of CCMA. Section 3 focuses on the multiplication in small extensions. We explain how to reach the equality in the Winograd–De Groote bound in the framework of Chudnovsky-type algorithms. Moreover, this construction naturally integrates the trick of Karatsuba algorithm. In Section 4, we give a Recursive Polynomial Generic Construction (RPGC) of algorithms for multiplication in any extension of $\mathbb{F}_q$, and give a natural strategy to build algorithms with a good bilinear complexity, which extends and generalizes the construction of [**LSW83**]. In Section 5, we prove the existence of such algorithms having a quasi-linear uniform bound for their bilinear complexities, with respect to the extension degrees. This bound is obtained using a different approach to that of [**LSW83**], in the sense that it is a geometric approach. We show how one of our arguments can improve their result. In addition, we show that the construction of our algorithms is deterministic, and give a polynomial asymptotic bound for this construction.

2. Chudnovsky and Chudnovsky Multiplication Algorithm

A large description of CCMA and its generalizations is given in [**BPR$^+$21**]. We first recall some basics of function field theory and introduce the notions required for our study. Then, we recall a specialized version of the generalized theorem/algorithm over a function field of arbitrary genus g, which will be useful for the proposed construction.

Let $F/\mathbb{F}_q$ be a function field of genus g over $\mathbb{F}_q$. For $\mathcal{O}$ a valuation ring, the place P is defined to be $P = \mathcal{O} \setminus \mathcal{O}^\times$. We denote by F_P the residue class field at the place P, that is isomorphic to $\mathbb{F}_{q^d}$, d being the degree of the place. A rational place is a place of degree 1. We also denote by $B_d(F/\mathbb{F}_q)$ the number of places of degree d of F over $\mathbb{F}_q$. A divisor $\mathcal{D}$ is a formal sum $\mathcal{D} = \sum_i n_i P_i$, where P_i are places and n_i are relative integers. The support $\operatorname{supp} \mathcal{D}$ of $\mathcal{D}$ is the set of the places P_j for which $n_j \neq 0$, and $\mathcal{D}$ is effective if all the n_i are positive. The degree of $\mathcal{D}$ is defined by $\deg \mathcal{D} = \sum_i n_i$. The Riemann–Roch space associated to the divisor $\mathcal{D}$ is denoted by $\mathcal{L}(\mathcal{D})$. A divisor $\mathcal{D}$ is said to be non-special if $\dim \mathcal{L}(\mathcal{D}) = \deg(\mathcal{D}) + 1 - g$. Details about algebraic function fields can be found in [**Sti08**].

Since Ballet and Rolland [**BR04**], Arnaud [**Arn06**], then Cenk and Özbudak [**CÖ10**] and finally the best current generalization due to Randriambololona [**Ran12**], the algorithm has been extended for instance to the evaluation at places of arbitrary degrees and with multiplicity greater than 1. In the following, we only consider the generalization to the evaluation at places of arbitrary degrees. The statement of the algorithm requires the following definition of the generalized Hadamard product.

Definition 2. Let q be a prime power. The generalized Hadamard product in $\mathbb{F}_{q^{d_1}} \times \cdots \times \mathbb{F}_{q^{d_N}}$, denoted by $\underline{\odot}$, is given for all $(a_1, \ldots, a_N), (b_1, \ldots, b_N) \in \mathbb{F}_{q^{d_1}} \times \cdots \times \mathbb{F}_{q^{d_N}}$ by

$$(a_1, \ldots, a_N) \underline{\odot} (b_1, \ldots, b_N) = (a_1 b_1, \ldots, a_N b_N).$$

Now, we introduce a specialized version of the current generalization of CCMA [**BPR$^+$21**, Theorem 5.3], considering the evaluation at places of arbitrary degrees only. Note that the algorithm presented in this theorem adresses the problem of the multiplication in finite extensions of finite fields, and represents them as residue class fields. Accordingly with the litterature, the theorem is stated multiplying elements of $\mathbb{F}_{q^n}$, even if more precisely they are elements of the residue class field used to represent the finite field.

Theorem 1. *Let*

- *q be a prime power,*
- *n be a positive integer,*
- *$F/\mathbb{F}_q$ be an algebraic function field of genus g,*
- *Q be a degree n place of $F/\mathbb{F}_q$,*
- *$\mathcal{D}$ be a divisor of $F/\mathbb{F}_q$,*
- *$\mathcal{P} = \{P_1, \ldots, P_N\}$ be a set of places of arbitrary degrees of $F/\mathbb{F}_q$ not containing Q.*

We suppose that $\operatorname{supp} \mathcal{D} \cap \{Q, P_1, \ldots, P_N\} = \emptyset$ *and that*

(i) *the evaluation map at* Q

$$\begin{aligned} \operatorname{Ev}_Q : \mathcal{L}(\mathcal{D}) &\to F_Q \\ f &\mapsto f(Q) \end{aligned}$$

is surjective, and

(ii) *the evaluation map at the set* $\mathcal{P}$

$$\begin{aligned} \operatorname{Ev}_{\mathcal{P}} : \mathcal{L}(2\mathcal{D}) &\to \mathbb{F}_{q^{\deg P_1}} \times \cdots \times \mathbb{F}_{q^{\deg P_N}} \\ f &\mapsto \big(f(P_1), \ldots, f(P_N)\big) \end{aligned}$$

is injective.

Then,

(1) *there exists a multiplication algorithm* $\mathcal{U}_{q,n}^{F,\mathcal{P}}(\mathcal{D}, Q)$ *such that for any two elements* α, β *in* $\mathbb{F}_{q^n}$*:*

$$\alpha\beta = E_Q \circ {\operatorname{Ev}_{\mathcal{P}} \mid_{\operatorname{Im} \operatorname{Ev}_{\mathcal{P}}}}^{-1} \left(E_{\mathcal{P}} \circ \operatorname{Ev}_Q^{-1}(\alpha) \underline{\odot} E_{\mathcal{P}} \circ \operatorname{Ev}_Q^{-1}(\beta) \right), \tag{2}$$

where E_Q *denotes the canonical projection from the valuation ring* $\mathcal{O}_Q$ *of the place* Q *to its residue class field* F_Q, $E_{\mathcal{P}}$ *the restriction of* $\operatorname{Ev}_{\mathcal{P}}$ *to* $\mathcal{L}(\mathcal{D})$, ${\operatorname{Ev}_{\mathcal{P}}}_{|\operatorname{Im} \operatorname{Ev}_{\mathcal{P}}}^{-1}$ *the inverse of the restriction of the map* $\operatorname{Ev}_{\mathcal{P}}$ *on its image,* $\underline{\odot}$ *the generalized Hadamard product and* $\circ$ *the standard composition map;*

(2) *the bilinear complexity of* $\mathcal{U}_{q,n}^{F,\mathcal{P}}(\mathcal{D}, Q)$ *is bounded by*

$$\mu(\mathcal{U}_{q,n}^{F,\mathcal{P}}(\mathcal{D}, Q)) \leq \sum_{i=1}^{N} \mu_q(\deg P_i).$$

Existence of the objects satisfying the conditions above is ensured by the following numerical criteria:

(a) *a sufficient condition for the existence of a place* Q *in* $F/\mathbb{F}_q$ *of degree* n *is that* $2g + 1 \leq q^{(n-1)/2}(q^{1/2} - 1)$*, where* g *is the genus of* F*,*

(b) *a sufficient condition for (i) is that the divisor* $D - Q$ *is non-special,*

(c) *a necessary and sufficient condition for (ii) is that the divisor* $2\mathcal{D} - \mathcal{G}$ *is zero-dimensional:*

$$\dim \mathcal{L}(2\mathcal{D} - \mathcal{G}) = 0$$

where $\mathcal{G} = P_1 + \cdots + P_N$.

REMARK 1. The condition that the support of the divisor D and $\{Q, P_1, \ldots, P_n\}$ are disjoint is not necessary because it is always possible to move the support (see [**STV92**, Remark 2.3]). However, this is a simplifying assumption almost always made in the literature.

The following corollary gives sufficient conditions to obtain an algorithm as in Theorem 1.

COROLLARY 1. *Let* q *be a prime power and let* n *be an integer* > 1. *If there exists an algebraic function field* $F/\mathbb{F}_q$ *of genus* g *with a set of places* $\mathcal{P} = \{P_1, \ldots, P_N\}$ *and an effective divisor* $\mathcal{D}$ *of degree* $n + g - 1$ *such that*

1) *there exists a place* Q *of degree* n*,*
2) $\operatorname{supp}(\mathcal{D}) \cap (\mathcal{P} \cup \{Q\}) = \emptyset$*, and* $D - Q$ *is non-special,*
3) $\sum_{i=1}^{N} \deg P_i = 2n + g - 1$ *and* $2\mathcal{D} - \sum P_i$ *is non-special,*

then,

(i) *the evaluation map*

$$\begin{aligned}\mathrm{Ev}_Q : \mathcal{L}(\mathcal{D}) &\to \frac{\mathcal{O}_Q}{Q}\\ f &\mapsto f(Q)\end{aligned}$$

is an isomorphism of vector spaces over $\mathbb{F}_q$,

(ii) *and the evaluation map*

$$\begin{aligned}\mathrm{Ev}_{\mathcal{P}} : \mathcal{L}(2\mathcal{D}) &\to \mathbb{F}_{q^{\deg P_1}} \times \cdots \times \mathbb{F}_{q^{\deg P_N}}\\ f &\mapsto \big(f(P_1), \ldots, f(P_N)\big)\end{aligned}$$

is an isomorphism of vector spaces of dimension $2n + g - 1$ *over* $\mathbb{F}_q$.

Conditions 1), 2) and 3) of Corollary 1 give the conditions a), b) and c) of Theorem 1 respectively. Note that Condition 1) is always satisfied if $2g + 1 \leq q^{\frac{n-1}{2}}(q^{\frac{1}{2}} - 1)$. Moreover, all these new requirements are not necessary to construct the algorithm. Nevertheless, it corresponds to interesting conditions under which we want to build our algorithms. In particular, if one only uses rational places, the following corollary gives a criteria to construct the original CCMA [**Bal99**].

COROLLARY 2. *Let q be a prime power and let n be an integer > 1. If there exists an algebraic function field $F/\mathbb{F}_q$ of genus g satisfying the conditions*

(1) $B_n(F/\mathbb{F}_q) > 0$,

(2) $B_1(F/\mathbb{F}_q) > 2n + 2g - 2$,

then there exists a divisor D of degree $n + g - 1$, a place Q of degree n and a set of rational places $\mathcal{P}$ such that (i) and (ii) of Corollary 1 holds.

In this paper, we specialize these results to the rational function field $F = \mathbb{F}_q(x)$.

3. CCMA and the multiplication in small extensions of $\mathbb{F}_q$

3.1. Polynomial interpolation over rational points. As seen in the introduction, the multiplication in any extension of $\mathbb{F}_q$ of degree $n \leq \frac{1}{2}q + 1$ requires exactly $2n - 1$ bilinear multiplications [**DG83**], and every algorithm reaching this optimal bilinear complexity is of type interpolation [**Win79**]. In this section, we construct Chudnovsky-type algorithms over the projective line using polynomial interpolation, and which have optimal bilinear complexity for q a prime power and $n \leq \frac{1}{2}q + 1$. We begin by making the following setup, i.e. a set of valid input data for the application of Theorem 1, as well as parameters internal to the algorithm, such as the bases of the Riemann-Roch spaces.

PGC : Polynomial Generic Construction

For q a prime power and $n < \frac{1}{2}q + 1$ a positive integer. We set

- Q to be a place of degree n of $\mathbb{F}_q(x)$,
- $\mathcal{D} = (n-1)P_\infty$,
- $\mathcal{P}$ to be a set of rational places distinct from P_∞ of cardinal $\mid \mathcal{P} \mid = 2n - 1$,
- the basis of $\mathcal{L}(\mathcal{D})$ to be $\{1, x, \ldots, x^{n-1}\}$, and
- the basis of $\mathcal{L}(2\mathcal{D})$ to be $\{1, x, \ldots, x^{2n-2}\}$.

In our construction, we set the function field to be $F = \mathbb{F}_q(x)$, and the divisor to be $\mathcal{D} = (n-1)P_\infty$. In order to define an algorithm for multiplication in $\mathbb{F}_{q^n}$ with Theorem 1, the only variables left are the place Q, and the set $\mathcal{P}$. Hence, we denote the algorithm using these parameters by $\mathcal{U}_{q,n}^{\mathcal{P}}(Q)$ to lighten the notations.

PROPOSITION 2. *Let q be a prime power, $n < \frac{1}{2}q+1$ be an integer and $\mathcal{P}$ be a set of rational places distinct from P_∞ of cardinal $\mid \mathcal{P} \mid = 2n-1$. Then, PGC is a valid setup to build a CCMA from Theorem 1, denoted by $\mathcal{U}_{q,n}^{\mathcal{P}}(Q)$, for multiplication in $\mathbb{F}_{q^n}$. This is a polynomial interpolation algorithm that computes $2n-1$ bilinear multiplications in $\mathbb{F}_q$.*

PROOF. First, finding a degree n place Q to construct $\mathbb{F}_{q^n}$ means finding a monic irreducible polynomial $Q(x)$ of degree n over $\mathbb{F}_q$. The residue class field at Q is exactly the quotient $\mathbb{F}_q[x]/(Q(x)) = \mathbb{F}_{q^n}$. Such polynomials exist for all q and n, and the condition 1 of Corollary 2 is verified. Let P_∞ be the place at infinity of $\mathbb{F}_q(x)$, and $\mathcal{D}$ be the divisor defining the Riemann-Roch space. We set $\mathcal{D} = (n-1)P_\infty$. Then, $\{1, x, \ldots, x^{n-1}\}$ is canonically a basis of $\mathcal{L}(\mathcal{D})$, and $\{1, x, \ldots, x^{2n-2}\}$ is a basis of $\mathcal{L}(2\mathcal{D})$, and the interpolation is done with polynomials.

There are $q+1$ rational places in $\mathbb{F}_q(x)$. The condition 2. of Corollary 2 asserts that the algorithm can be built for $n \leq \frac{1}{2}q+1$. But in PGC, the place at infinity P_∞ is used to define the divisor $\mathcal{D}$. Assuming that the support of $\mathcal{D}$ does not contain any of the places in $\mathcal{P}$, it implies that we can evaluate only on q rational places instead of $q+1$, since a function in $\mathcal{L}(\mathcal{D})$ has a pole at P_∞. Thus, we use the more general Corollary 1. The divisor $\mathcal{D} - Q$ is of degree -1 and hence is non-special [**Sti08**, Remark 1.6.11], thus condition 2 is satisfied. Moreover, let $\mathcal{P} = \{P_i\}$, be a set of rational places distinct from P_∞. Condition 3 requires that the cardinality $\mid \mathcal{P} \mid = 2n-1 \leq q$, so $2\mathcal{D} - \sum P_i$ is of degree -1 and thus non-special. Hence, Corollary 1 attests that the algorithm can be built with PGC for $n < \frac{1}{2}q+1$.

Therefore, for all prime powers q, PGC gives CCMA with optimal bilinear complexity using polynomial interpolation for multiplication in an extension of $\mathbb{F}_q$ of degree $n < \frac{1}{2}q+1$. □

REMARK 2. CCMA cannot be constructed with PGC when $n = \frac{1}{2}q+1$.

When q is odd, the equality of Remark 2 never happens, because $\frac{1}{2}q+1$ is not an integer. For an even $q \geq 4$, we can use a place $\mathcal{R}$ of degree $n-1$ to define the divisor, i.e. set $\mathcal{D} = \mathcal{R}$. With this setting, we obtain an algorithm of bilinear complexity $2n-1$, but this is no longer a polynomial interpolation algorithm since it involves rational functions. When $q = 2$ and $n = 2$, the rational function field over $\mathbb{F}_2$ has only 3 rational places, while 4 are required by our construction (3 for the evaluations, and 1 to define $\mathcal{D}$), and we cannot use the previous argument to obtain an algorithm of optimal bilinear complexity.

3.2. The case of $n = \frac{1}{2}q+1$ and polynomial interpolation. We consider the case of Remark 2: the extension of $\mathbb{F}_q$ of degree $n = \frac{1}{2}q+1$. We want to build a Chudnovsky-type algorithm over the rational function field $\mathbb{F}_q(x)$, requiring $\mathcal{D} = (n-1)P_\infty$ to interpolate with polynomials. From the results of Winograd and De Groote, it must be possible to construct such an algorithm with optimal bilinear complexity. We use that the evaluation at a place in the support of the divisor can be defined [**STV92**, Remark 2.3]. In our context, it yields to the following definition of evaluation at P_∞.

DEFINITION 3. Let k be a positive integer and P_∞ be the place at infinity of $\mathbb{F}_q(x)$. Set $\mathcal{L}(\mathcal{D}) = \mathcal{L}(kP_\infty)$, we define the evaluation at P_∞ to be for all $f = \sum_{i=0}^{k-1} f_i x^i \in \mathcal{L}(\mathcal{D})$,

$$f_{\mathcal{D}}(P_\infty) := f_{k-1},$$

that is the leading coefficient of f. We specify the divisor $\mathcal{D}$ in the notation as the evaluation depends on the Riemann–Roch space from which it is defined.

This definition makes sense since

$$f_{\mathcal{D}}(P_\infty)g_{\mathcal{D}}(P_\infty) = (fg)_{2\mathcal{D}}(P_\infty), \tag{3}$$

and using this, the algorithm of Theorem 1 can be specialized to the projective line as follows.

PROPOSITION 3. *Let*

- $\mathbb{F}_q(x)$ *be the rational function field over* $\mathbb{F}_q$,
- n *be a positive integer,*
- Q *be a degree* n *place of* $\mathbb{F}_q(x)$,
- $\mathcal{P} = \{P_\infty, P_0, P_1, \ldots, P_N\}$ *be a set of places of arbitrary degrees of* $\mathbb{F}_q(x)$, *with* P_∞ *the place at infinity and* P_0 *associated to the polynomial* x.

We set $\mathcal{D} = (n-1)P_\infty$. *If*

$$\sum_{P \in \mathcal{P}} \deg P_i = 2n - 1, \tag{4}$$

then

(i) *the evaluation map*

$$\begin{aligned} \mathrm{Ev}_Q : \mathcal{L}(\mathcal{D}) &\to F_Q \\ f &\mapsto f(Q) \end{aligned}$$

is bijective,

(ii) *the evaluation map*

$$\begin{aligned} \tilde{\mathrm{Ev}}_{\mathcal{P}} : \mathcal{L}(2\mathcal{D}) &\to \mathbb{F}_q \times \mathbb{F}_q \times \mathbb{F}_{q^{\deg P_1}} \times \cdots \times \mathbb{F}_{q^{\deg P_N}} \\ f &\mapsto \big(f_{2\mathcal{D}}(P_\infty), f(P_0), f(P_1), \ldots, f(P_N)\big) \end{aligned}$$

is an isomorphism of vector spaces.

Moreover,

(1) *we have a multiplication algorithm* $\mathcal{U}_{q,n}^{\mathcal{P}}(Q)$ *of type polynomial interpolation such that for any two elements* α, β *in* $\mathbb{F}_{q^n}$:

$$\alpha\beta = E_Q \circ \tilde{\mathrm{Ev}}_{\mathcal{P}}^{-1}\left(\tilde{E_{\mathcal{P}}} \circ \mathrm{Ev}_Q^{-1}(\alpha) \underline{\odot} \tilde{E_{\mathcal{P}}} \circ \mathrm{Ev}_Q^{-1}(\beta)\right), \tag{5}$$

where E_Q *denotes the canonical projection from the valuation ring* $\mathcal{O}_Q$ *of the place* Q *in its residue class field* F_Q, $\tilde{\mathrm{Ev}}_{\mathcal{P}}^{-1}$ *the inverse map of* $\tilde{\mathrm{Ev}}_{\mathcal{P}}$, $\underline{\odot}$ *the generalized Hadamard product in* $\mathbb{F}_q \times \mathbb{F}_q \times (\mathbb{F}_{q^{\deg P_1}}) \times \cdots \times (\mathbb{F}_{q^{\deg P_N}})$, $\circ$ *the standard composition map, and*

$$\begin{aligned} \tilde{E_{\mathcal{P}}} : \mathcal{L}(\mathcal{D}) &\to \mathbb{F}_q \times \mathbb{F}_q \times \mathbb{F}_{q^{\deg P_1}} \times \cdots \times \mathbb{F}_{q^{\deg P_N}} \\ f &\mapsto \big(f_{\mathcal{D}}(P_\infty), f(P_0), f(P_1), \ldots, f(P_N)\big), \end{aligned}$$

(2) *the algorithm* $\mathcal{U}_{q,n}^{\mathcal{P}}(Q)$ *defined by* (5) *has bilinear complexity*

$$\mu(\mathcal{U}_{q,n}^{\mathcal{P}}) = \sum_{P \in \mathcal{P}} \mu_q(\deg P).$$

PROOF. Since $\mathcal{D} = (n-1)P_\infty$, $\mathcal{L}(\mathcal{D})$ is isomorphic to $\mathbb{F}_{q^n}$, and we associate elements of $\mathcal{L}(\mathcal{D})$ to elements of $\mathbb{F}_{q^n}$. For $f, g \in \mathcal{L}(\mathcal{D})$, if we compute the Generalized Hadamard product of $\tilde{E_\mathcal{P}}(f)$ and $\tilde{E_\mathcal{P}}(g)$, we get

$$\tilde{E_\mathcal{P}}(f) \odot \tilde{E_\mathcal{P}}(g) = \begin{pmatrix} f_\mathcal{D}(P_\infty) \\ f(P_0) \\ f(P_1) \\ \vdots \\ f(P_N) \end{pmatrix} \underline{\odot} \begin{pmatrix} g_\mathcal{D}(P_\infty) \\ g(P_0) \\ g(P_1) \\ \vdots \\ g(P_N) \end{pmatrix}$$

$$= \begin{pmatrix} f_\mathcal{D}(P_\infty)g_\mathcal{D}(P_\infty) \\ f(P_0)g(P_0) \\ f(P_1)g(P_1) \\ \vdots \\ f(P_N)g(P_N) \end{pmatrix}$$

$$= \begin{pmatrix} (fg)_{2\mathcal{D}}(P_\infty) \\ (fg)(P_0) \\ (fg)(P_1) \\ \vdots \\ (fg)(P_N) \end{pmatrix} = \tilde{\mathrm{Ev}_P}(fg).$$

We have to prove that $\tilde{\mathrm{Ev}_P}$ is bijective. Let $f = a_0 + a_1x + \cdots + a_{2n-2}x^{2n-2}$ be a function in $\mathcal{L}(2\mathcal{D})$ such that $f \in \ker \tilde{\mathrm{Ev}_\mathcal{P}}$. In particular, $f_{2\mathcal{D}}(P_\infty) = 0 = a_{2n-2}$. Then, $f = \sum_{i=0}^{2n-3} a_i x^i \in \mathcal{L}((2n-3)P_\infty)$ and

$$(f(P_0), f(P_1), \ldots, f(P_N)) = (0, 0, \ldots, 0).$$

Hence, $f \in \ker \tilde{\mathrm{Ev}_P} \subseteq \mathcal{L}((2n-3)P_\infty - \sum_{i=0}^N P_i)$. But, the divisor $(2n-3)P_\infty - \sum_{i=0}^N P_i$ is of degree -1 so $f = 0$ and $\ker \tilde{\mathrm{Ev}_\mathcal{P}} = \{0\}$. Thus $\tilde{\mathrm{Ev}_\mathcal{P}}$ is injective, and bijective since between two vector spaces of same dimension. For all $f, g \in \mathcal{L}(\mathcal{D})$, we obtain

$$fg = \tilde{\mathrm{Ev}_P}^{-1}(\tilde{E_\mathcal{P}}(f)\underline{\odot}\tilde{E_\mathcal{P}}(g)),$$

and finally, with $f = \mathrm{Ev}_Q^{-1}(\alpha)$ and $g = \mathrm{Ev}_Q^{-1}(\beta)$ for any $\alpha, \beta \in \mathbb{F}_{q^n}$,

$$\alpha\beta = E_Q \circ \tilde{\mathrm{Ev}_\mathcal{P}}^{-1}\left(\tilde{E_\mathcal{P}} \circ \mathrm{Ev}_Q^{-1}(\alpha)\underline{\odot}\tilde{E_\mathcal{P}} \circ \mathrm{Ev}_Q^{-1}(\beta)\right),$$

where E_Q denotes the canonical projection from the valuation ring $\mathcal{O}_Q$ of the place Q in its residue class field F_Q. Hence, we obtain an algorithm of polynomial interpolation, since functions in the Riemann-Roch spaces are polynomials. Its bilinear multiplications are first given, for $1 \leq i \leq N$, by the $f(P_i)g(P_i)$, that require $\mu_q(\deg P_i)$ bilinear multiplications over $\mathbb{F}_q$ for each $i = 0, 1, \ldots, N$. Then, one more bilinear multiplication is used to compute $f_\mathcal{D}(P_\infty)g_\mathcal{D}(P_\infty)$. These multiplications are independent of the choice of the place Q chosen to represent $\mathbb{F}_{q^n}$. Thus the bilinear complexity of the algorithm $\mathcal{U}_{q,n}^{\mathcal{P}}(Q)$ is the same for all possible Q, and is given by $\mu(\mathcal{U}_{q,n}^{\mathcal{P}}) = \sum_{P \in \mathcal{P}} \mu_q(\deg P)$. □

REMARK 3. In fact, let Q' be another place of degree n. In our construction, Ev_Q and $\mathrm{Ev}_{Q'}$ are both trivial. Then, the only difference between $\mathcal{U}^{\mathcal{P}}_{q,n}(Q)$ and $\mathcal{U}^{\mathcal{P}}_{q,n}(Q')$ is that $E_Q \neq E_{Q'}$. This is why we consider $\mathcal{U}^{\mathcal{P}}_{q,n}$ as an algorithm that can be applied to any place of degree n of $\mathbb{F}_q(x)$, with bilinear complexity $\mu(\mathcal{U}^{\mathcal{P}}_{q,n})$.

Using this result for the construction with rational places, we can include the place at infinity in our setup. We obtain an algorithm of type polynomial interpolation of optimal bilinear complexity when $n = \frac{1}{2}q + 1$.

COROLLARY 3. *Let $q \geq 2$ be an even prime power and $n = \frac{1}{2}q + 1$. Let $\mathcal{P}$ be the set of all rational places of $\mathbb{F}_q(x)$. Given Q a place of degree n, there exists a Chudnovsky-type algorithm over the projective line $\mathcal{U}^{\mathcal{P}}_{q,n}(Q)$ for multiplication in $\mathbb{F}_{q^n}$. This is a polynomial interpolation algorithm that computes $2n-1$ bilinear multiplications in $\mathbb{F}_q$.*

3.3. A particular case: the quadratic extension of $\mathbb{F}_2$. The case of $q = 2$ and $n = 2$ is problematic and interesting. The rational function field $\mathbb{F}_2(x)$ has only three rational places: P_0, the place associated to the polynomial x, P_1, associated to $x - 1$, and P_∞, the place at infinity. The proposed construction requires P_∞ to define the Riemann–Roch space and three other places to evaluate. Proposition 3 gives a Chudnovsky-type algorithm reaching this optimal bilinear complexity, that is exactly Karatsuba's algorithm.

COROLLARY 4. *Let Q be the degree 2 place of $\mathbb{F}_2(x)$ and $\mathcal{P}_2 = \{P_0, P_1, P_\infty\}$, where P_0 and P_1 are the places associated to x and $x-1$ respectively, and P_∞ is the place at infinity. Then, $\mathcal{U}^{\mathcal{P}_2}_{2,2}(Q)$ is a Chudnovsky-type algorithm for multiplication in the quadratic extension of $\mathbb{F}_2$ with bilinear complexity $\mu(\mathcal{U}^{\mathcal{P}_2}_{2,2}) = 3$. Moreover, its bilinear multiplications correspond to those of the Karatsuba Algorithm.*

REMARK 4. Corollary 4 becomes generalized to any prime power q. Let P_0, P_1 and P_∞ be the places of $\mathbb{F}_q(x)$ associated to x, $x - 1$ and at infinity. Let $\mathcal{P}_2 = \{P_0, P_1, P_\infty\}$, and Q be a degree 2 place of $\mathbb{F}_q(x)$. Then, $\mathcal{U}^{\mathcal{P}_2}_{q,2}(Q)$ is a Chudnovsky-type algorithm for multiplication in the quadratic extension of $\mathbb{F}_q$, such that $\mu(\mathcal{U}^{\mathcal{P}_2}_{2,2}) = 3$. The bilinear multiplications of this algorithm are again exactly those of Karatsuba Algorithm i.e. let $f = f_0 + f_1 x$ and $g = g_0 + g_1 x$ be two elements of $\mathcal{L}(P_\infty)$, then

$$\begin{aligned} fg = f(P_0)g(P_0) + \big(f(P_1)g(P_1) - f(P_0)g(P_0) - f_{\mathcal{D}}(P_\infty)g_{\mathcal{D}}(P_\infty)\big)x \\ + f_{\mathcal{D}}(P_\infty)g_{\mathcal{D}}(P_\infty)x^2 \end{aligned}$$

4. Recursive Chudnovsky-type algorithm on $\mathbb{F}_q(x)$

4.1. Recursive Polynomial Generic Construction. Thus far, we have built polynomial interpolation Chudnovsky-type algorithms on the projective line over $\mathbb{F}_q$. They have an optimal bilinear complexity when $n \leq \frac{1}{2}q + 1$, using evaluation at rational places only. However, the evaluations can be done at places of higher degrees of $\mathbb{F}_q(x)$, and we can construct an algorithm for extensions of any degree. In this section, we propose a recursive generic construction of Chudnovsky-type algorithms specialized to the projective line for multiplication in all extensions $\mathbb{F}_{q^n}$, using places of increasing degrees.

Let Q be a place of degree n. According to Proposition 3, we need to construct a set $\mathcal{P}$, containing places such that the sum of their degrees is equal to $2n-1$. We can assume P_0 and P_∞ are always in $\mathcal{P}$. With such a set, Proposition 3 gives an algorithm $\mathcal{U}_{q,n}^{\mathcal{P}}(Q)$ for multiplication in any extension. But at this step, we still do not have any information about how to compute the multiplications of the evaluations at places of an arbitrary degree. Concretely, let $P_i \in \mathcal{P}$ be a place of degree d_i. Then, for $f, g \in \mathcal{L}((n-1)P_\infty)$, the evaluations $f(P_i)$ and $g(P_i)$ are some elements in $\mathbb{F}_{q^{d_i}}$. To compute $(fg)(P_i) = f(P_i)g(P_i)$, we use the Chudnovsky-type algorithm over the projective line $\mathcal{U}_{q,d_i}^{\mathcal{P}_i}(P_i)$. Such an algorithm is called a recursive Chudnovsky-type algorithm over the projective line.

DEFINITION 4. Let q be a prime power and $n > \frac{1}{2}q+1$ be a positive integer. We call a recursive Chudnovsky-type algorithm over the projective line an algorithm, that computes the multiplications of the evaluations at non-rational places with recursively-defined algorithms.

EXAMPLE 1. We consider the multiplication in $\mathbb{F}_{3^6}$. The function field $\mathbb{F}_3(x)$ has 4 rational places, 3 places of degree 2 and 8 places of degree 3. We denote by P_∞ the place at infinity, and by P_0, P_1 and P_2 the three other rational places. The places of degree 2 are P_1^2, P_2^2 and P_3^2, and P^3 is one of the places of degree 3. Moreover, let Q be a place of degree 6. Then, a set of places containing the 4 rational places, 2 places of degree 2 and one of degree 3 is suitable, since we have $4 \times 1 + 2 \times 2 + 1 \times 3 = 11 = 2 \times 6 - 1$. For example, we can take $\mathcal{P} = \{P_\infty, P_0, P_1, P_2, P_1^2, P_2^2, P^3\}$. The products of the evaluations at rational places are multiplications in $\mathbb{F}_q$. The evaluations at places P_i^2 of degree 2 can be multiplied with 3 bilinear multiplications by $\mathcal{U}_{3,2}^{\mathcal{P}_2}$, defined in Remark 4. We also need to construct the algorithm to multiply in the extension of degree 3.

To do this, set $\mathcal{P}_3 = \{P_0, P_1, P_\infty, P_1^2\}$. The sum of the degrees of the places in $\mathcal{P}_3$ is equal to $3 \times 1 + 1 \times 2 = 5 = 2 \times 3 - 1$, and this set is suitable. The algorithm $\mathcal{U}_{3,3}^{\mathcal{P}_3}$ computes 3 bilinear multiplications in $\mathbb{F}_q$, and $\mathcal{U}_{3,2}^{\mathcal{P}_2}(P_1^2)$ computing itself 3 more bilinear multiplications. Finally, its bilinear complexity is $\mu(\mathcal{U}_{3,3}^{\mathcal{P}_3}) = 3\times1+1\times3 = 6$. That is the best-known (and optimal) bound for multiplication in the extensions of degree 3 of $\mathbb{F}_3$. Table 1 illustrates the structure of $\mathcal{U}_{3,3}^{\mathcal{P}_3}$.

Back to the multiplication in extension of degree 6, the algorithm $\mathcal{U}_{3,6}^{\mathcal{P}}(Q)$ computes the multiplications of the evaluations at P_∞, P_0, P_1, P_2, and use $\mathcal{U}_{3,2}^{\mathcal{P}_2}(P_1^2)$, $\mathcal{U}_{3,2}^{\mathcal{P}_2}(P_2^2)$ and $\mathcal{U}_{3,3}^{\mathcal{P}_3}(P^3)$. This gives the bilinear complexity $\mu(\mathcal{U}_{3,6}^{\mathcal{P}}) = 4\times1+2\times3+1\times6 = 16$. Figure 2 gives an illustration of the algorithm.

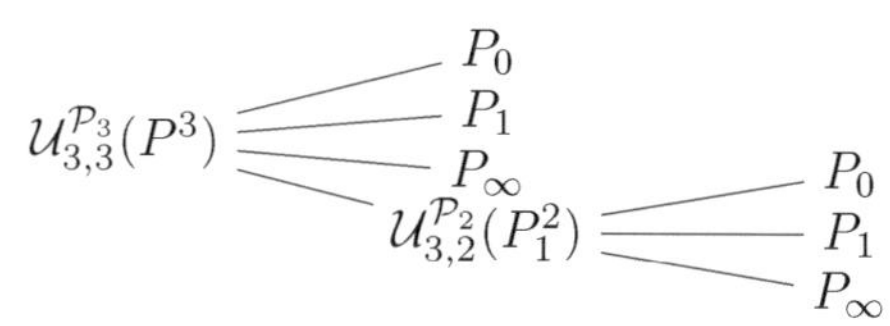

FIGURE 1. Diagram of the construction of $\mathcal{U}_{3,3}^{\mathcal{P}_3}(Q)$.

This setup can be summarized as follows:

RPGC: Recursive Polynomial Generic Construction

For a prime power q and an integer $n \geq 2$, let Q be a place of degree n of $\mathbb{F}_q(x)$. Then, $\mathcal{U}^{\mathcal{P}}_{q,n}(Q)$ is an algorithm for multiplication in $\mathbb{F}_{q^n}$, with the following settings:

- $\mathcal{D} = (n-1)P_\infty$,
- $\mathcal{P} = \{P_1, \ldots, P_N\}$ is a set of places such that $\sum_{i=1}^{N} \deg P_i = 2n-1$,
- the basis of $\mathcal{L}(\mathcal{D})$ is $\{1, x, \ldots, x^{n-1}\}$,
- the basis of $\mathcal{L}(2\mathcal{D})$ is $\{1, x, \ldots, x^{2n-1}\}$, and
- Use recursively RPGC to multiply the evaluations at every non-rational places in $\mathcal{P}$.

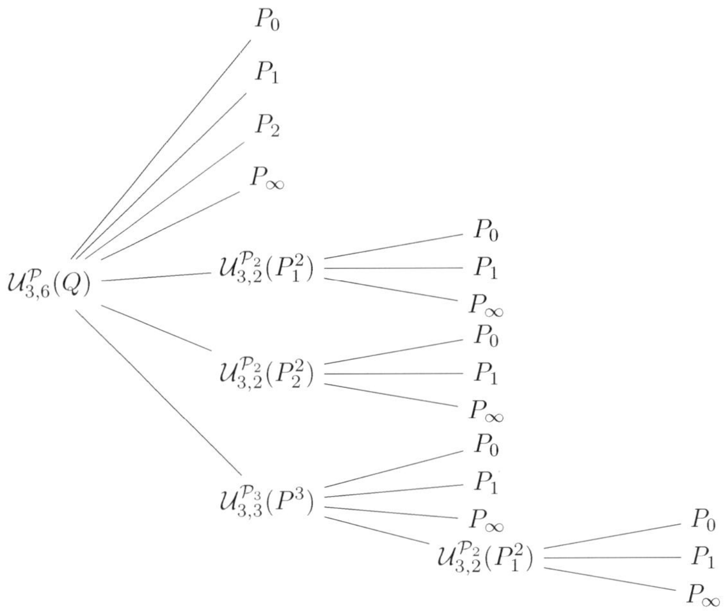

FIGURE 2. Diagram of the construction of $\mathcal{U}^{\mathcal{P}}_{3,6}(Q)$.

PROPOSITION 4. *Let q be a prime power and $n \geq 2$ an integer. RPGC is a valid setup for a recursive Chudnovsky-type algorithm $\mathcal{U}^{\mathcal{P}}_{q,n}(Q)$ over the projective line for multiplication in $\mathbb{F}_{q^n}$, of type polynomial interpolation. Its bilinear complexity is equal to*

$$\mu(\mathcal{U}^{\mathcal{P}}_{q,n}) = \sum_{k}^{d} N_k \mu(\mathcal{U}^{\mathcal{P}_k}_{q,k}).$$

where $d = \max_{P \in \mathcal{P}}\{\deg P\}$ and where N_k is the number of places of degree k in $\mathcal{P}$, and $\mathcal{U}^{\mathcal{P}_k}_{q,k}$ is built using RPGC.

PROOF. Let Q be a degree n place of $\mathbb{F}_q(x)$. We set once again $\mathcal{D} = (n-1)P_\infty$. Let $\mathcal{P}$ be a set of places of $\mathbb{F}_q(x)$, such that the sum of their degrees is equal to

$2n-1$. We can write $\mathcal{P} = \{P_1, \ldots, P_{N_1}, P_1^2, \ldots, P_{N_2}^2, \ldots, P_1^d, \ldots, P_{N_d}^d\}$, where the P_1, …, P_{N_1} are rational places and for all $j = 2, \ldots, d$, P_i^j is a place of degree j of $\mathbb{F}_q(x)$. We suppose that Q is not in $\mathcal{P}$. In particular, the bijectivity condition is given by $\sum kN_k = 2n-1$, where d is running over the degrees of all places in $\mathcal{P}$: the divisor $2\mathcal{D} - \sum_{P\in\mathcal{P}} P$ is of degree -1 and non-special, because of degree greater than $2g-2$, so zero-dimensional. Proposition 3 gives the main algorithm $\mathcal{U}_{q,n}^{\mathcal{P}}(Q)$.

Then, let $\mathcal{P}_2$, …, $\mathcal{P}_d$ be sets of places such that for all $2 \leq k \leq d$, we have $\sum_{P\in\mathcal{P}_k} \deg P = 2k-1$. We construct $\mathcal{U}_{q,k}^{\mathcal{P}_k}$ to multiply the evaluations at the places P_i^k of any degree k. For $f, g \in \mathcal{L}(\mathcal{D})$, each product $f(P_i^k)g(P_i^k)$ is computed using $\mathcal{U}_{q,k}^{\mathcal{P}_k}(P_i^k)$. In particular, the algorithm $\mathcal{U}_{q,1}$ is just a bilinear multiplication in $\mathbb{F}_q$. Finally, the bilinear complexity given by the construction is

$$\mu(\mathcal{U}_{q,n}^{\mathcal{P}}) = \sum_k N_k \mu(\mathcal{U}_{q,k}^{\mathcal{P}_k}). \qquad \square$$

In Section 5 (Proposition 5 and Corollary 5), we show that there exists a set $\mathcal{P}$ such that the maximum degree d of the places in $\mathcal{P}$ is bounded by $\log_q(2n)$.

REMARK 5. An algorithm defined using RPGC is fully defined by $\mathcal{P}$, $\mathcal{P}_2$, …, $\mathcal{P}_d$ and Q.

REMARK 6. Our construction is similar in spirit to the recursive construction using Chinese Remainder Theorem of Lempel, Seroussi and Winograd [**LSW83**]. However, it falls within the more general framework of Chudnovsky-type algorithms, and thus corresponds to a more powerful and evolutionary version. Indeed, the construction described in this paper bridges the gap between the construction of [**LSW83**] and the one of [**BBT16**] using elliptic curves. In Section 1.2 and forthcoming remarks (i.e. Remarks 7 and 10), we explain the advantages obtained with our method.

At this point, we do not know how to construct the set of places $\mathcal{P}$. Finding the set of places that gives the best bilinear complexity is a difficult task when n is large. Hence, we focus on some families of suitable sets to study the bilinear complexity of the recursive Chudnovsky-type algorithms over the projective line.

4.2. Construct $\mathcal{P}$ taking places by increasing degrees. A natural strategy is to construct the set $\mathcal{P}$ by taking every place by growing degrees until the sum of their degrees is equal to $2n-1$. If the sum is bigger than $2n-1$, we remove a place of appropriate degree. Such a set is denoted by $\mathcal{P}^{\deg}$.

DEFINITION 5. We denote by $\mathcal{U}_{q,n}^{\mathcal{P}^{\deg}}$ a recursive Chudnovsky-type algorithm on the projective line for multiplication in $\mathbb{F}_{q^n}$, where $\mathcal{P}^{\deg}$ is built taking places of increasing degrees.

The algorithms of Example 1 were built using this method. Note that this construction is deterministic assuming that the places of $\mathbb{F}_q(x)$ have a given order. We know that the number of rational places over $\mathbb{F}_{q^d}(x)$ is equal to $q^d + 1$. This value is the sum of the kB_k, for k dividing d and $B_k = B_k(\mathbb{F}_q(x))$ ([**Sti08**], (5.40)).

TABLE 1. Bilinear complexity of $\mathcal{U}_{q,n}^{\mathcal{P}^{\deg}}$ in small extensions of $\mathbb{F}_2$, $\mathbb{F}_3$ and $\mathbb{F}_4$.

n	2	3	4	5	6	7	8	9	10	11	12	13	14	15	16	17	18
$\mu(\mathcal{U}_{2,n}^{\mathcal{P}^{\deg}})$	$\underline{3}$	$\underline{6}$	11	15	18	26	29	37	40	48	51	60	65	70	78	81	90
$\mu(\mathcal{U}_{3,n}^{\mathcal{P}^{\deg}})$	$\underline{3}$	$\underline{6}$	$\underline{9}$	12	16	$\underline{19}$	24	28	31	36	40	43	48	52	55	60	64
$\mu(\mathcal{U}_{4,n}^{\mathcal{P}^{\deg}})$	$\underline{3}$	$\underline{5}$	$\underline{8}$	$\underline{11}$	$\underline{14}$	$\underline{17}$	$\underline{20}$	$\underline{23}$	$\underline{27}$	$\underline{30}$	$\underline{33}$	$\underline{37}$	40	43^+	47	50^+	53

Then, the number of places of degree d of $\mathbb{F}_q(x)$ is equal to

$$B_d = \frac{1}{d}\left(q^d + 1 - \sum_{\substack{k|d \\ k\neq d}} kB_k\right). \tag{6}$$

Thus, for a given n we can compute d, the smallest integer such that $\sum_{k=1}^{d} kB_k \geq 2n-1$. Then, the number N_k of places of degree k in $\mathcal{P}^{\deg}$ is given by

$$N_k = \begin{cases} 0 & \text{if } k > d, \\ \lceil \frac{1}{d}(2n - 1 - \sum_{i=1}^{d-1} iB_i) \rceil & \text{if } k = d, \\ B_k - 1 & \text{if } 0 \neq k \equiv -(2n - 1 - \sum_{i=1}^{d-1} iB_i) \pmod{d}, \\ B_k & \text{elsewhere,} \end{cases}$$

where B_k denotes the number of places of degree k of $\mathbb{F}_q(x)$. This allows one to compute iteratively the bilinear complexity since $\mu(\mathcal{U}_{q,n}^{\mathcal{P}^{\deg}}) = \sum_k N_k \mu(\mathcal{U}_{q,k}^{\mathcal{P}^{\deg}})$, where $\mathcal{P}^{\deg}$ is defined in accordance with k or n. Table 1 shows this bilinear complexity for the extensions of degree lower than 18 of $\mathbb{F}_q$, with $q = 2, 3, 4$. We underline the bilinear complexity when it equals the one given by Table 2 of [**BPR$^+$21**], and we denote by $^+$ when we beat this complexity.

REMARK 7. The appendix of [**LSW83**] uses the same construction, but using irreducible polynomials instead of places. The notion of place allows us to use the evaluation at infinity and this explains why the bilinear complexities we obtain are better. This amounts to using Karastuba's trick in our construction, which then integrates the Karatsuba and Cook algorithms with the Lempel-Seroussi-Winograd algorithms.

The previous strategy of construction is not optimal, for instance because the ratio $\frac{\mu(\mathcal{U}_{q,n}^{\mathcal{P}^{\deg}})}{n}$ is not increasing.

EXAMPLE 2. For $q = 2$ and $n = 82$, we need a set of places such that the sum of their degrees is equal to 163. In this case $\mathcal{P}^{\deg}$ contains all places of $\mathbb{F}_2(x)$ of degrees lower than 6, and 8 places of degree 7. The bilinear complexity obtained is hence $\mu(\mathcal{U}_{2,82}^{\mathcal{P}^{\deg}}) = 503$. However, one can check that $\frac{\mu(\mathcal{U}_{2,7}^{\mathcal{P}^{\deg}})}{7} > \frac{\mu(\mathcal{U}_{2,8}^{\mathcal{P}^{\deg}})}{8}$. Hence, it is better to use 7 places of degree 8 than 8 places of degree 7. Let $\mathcal{P}$ be a set containing all the places of degree lower than or equal to 6, and 7 places of degree 8. The algorithm obtained using this set (with the same sub-algorithms) has bilinear complexity $\mu(\mathcal{U}_{2,82}^{\mathcal{P}}) = 499$.

REMARK 8. Furthermore, the use of evaluations with multiplicity can also provide algorithms over the projective line with a better bilinear complexity. However, this is not involved in the asymptotic construction proposed in the next section. The authors believe that this generalization can possibly refine the results, but does not change them profoundly. It was therefore chosen not to include evaluations with multiplicity in this paper.

5. Asymptotical study for algorithms obtained using RPGC

5.1. An explicit construction for the asymptotical study. Hitherto, the bilinear complexity of algorithms obtained using RPGC can be computed for a fixed n, but we do not have an estimate relative to n yet. First, note that given a Chudnovsky-type algorithm $\mathcal{U}_{q,n}^{\mathcal{P}_n}$ over the projective line for multiplication in $\mathbb{F}_{q^n}$, we can define an algorithm of the same bilinear complexity for an extension of degree lower than n.

LEMMA 1. *Let $\mathcal{U}_{q,n}^{\mathcal{P}_n}$ be a Chudnovsky-type algorithm over the projective line for multiplication in $\mathbb{F}_{q^n}$. Then, for all $m < n$, there exists a Chudnovsky-type algorithm over the projective line $\mathcal{U}_{q,m}^{\mathcal{P}_m}$ for multiplication in $\mathbb{F}_{q^m}$ such that*

$$\mu(\mathcal{U}_{q,m}^{\mathcal{P}_m}) \leq \mu(\mathcal{U}_{q,n}^{\mathcal{P}_n}).$$

PROOF. Let $\mathcal{U}_{q,n}^{\mathcal{P}_n}$ be a Chudnovsky-type algorithm on the projective line for multiplication in $\mathbb{F}_{q^n}$. We have $\sum_{P\in\mathcal{P}_n} \deg P = 2n-1 > 2m-1$. Set $\mathcal{P}_m = \mathcal{P}_n$. Without loss of generality, we can assume that the place at infinity is not in $\mathcal{P}_m$ (in fact, if $P_\infty \in \mathcal{P}_n$, we can consider $\mathcal{P}_m = \mathcal{P}_n \setminus \{P_\infty\}$, and $\sum_{P\in\mathcal{P}_m} \deg P = 2n-2 > 2m-1$). Then, we build the algorithm $\mathcal{U}_{q,m}^{\mathcal{P}_m}$ from Theorem 1, with Q a place of degree m of $\mathbb{F}_q(x)$ and $\mathcal{D} = (m-1)P_\infty$. The map E_Q gives an isomorphism between $\mathcal{L}(\mathcal{D})$ and $\mathbb{F}_{q^m}$. Since $\sum_{P\in\mathcal{P}_m} \deg P > 2m-1$, the evaluation map $\mathrm{Ev}_{\mathcal{P}_m} : \mathcal{L}(2\mathcal{D}) \longrightarrow \mathbb{F}_q^N$ is injective. The conditions of Theorem 1 are satisfied, and we obtain an algorithm $\mathcal{U}_{q,m}^{\mathcal{P}_m}$. Moreover, if we compute the multiplications of the evaluations at the places in $\mathcal{P}_m$ as in $\mathcal{U}_{q,n}^{\mathcal{P}_n}$, we obtain that $\mu(\mathcal{U}_{q,m}^{\mathcal{P}_m}) \leq \mu(\mathcal{U}_{q,n}^{\mathcal{P}_n})$. □

Let us consider a convenient construction for the asymptotical study. For any integer l, recall that the places of $\mathbb{F}_q(x)$ of degree dividing l correspond to the q^l+1 rational places of $\mathbb{F}_{q^l}(x)$, and that $B_k = B_k(\mathbb{F}_q(x))$ denotes the number of places of degree k of $\mathbb{F}_q(x)$. Let d be the smallest integer d such that $q^d \geq 2n$, i.e. such that $d-1 \leq \log_q(2n) \leq d$. Then, the sum $\sum_{k|d} kB_k = q^d+1$ is greater than $2n-1$. We construct the set $\mathcal{P}$ in two steps. First, we include only places of degrees dividing d by increasing degrees, while the sum of their degrees is lower than $2n-1$. In anticipation of a future calculation, we do not include one of the rational places. Then, the second step is to adjust the set to obtain a sum exactly equals to $2n-1$. More precisely, consider

$$S = \Big(\sum_{\substack{k\,|\,d\\ k\neq d}} kB_k\Big) - 1. \tag{7}$$

The quantity S is the sum of the degrees of the places over all places of degrees strictly dividing d, minus a rational place that we do not count in the calculation. Now, consider $\delta = 2n-1-S \pmod d$. If $\delta = 0$, then $\sum_{P\in\mathcal{P}} \deg P = 2n-1$ after the first step, and the algorithm is done. Elsewhere, we have to increase this sum by δ. If $\delta \nmid d$, then we add a place of degree δ in $\mathcal{P}^{\mathrm{div}}$. If $\delta \mid d$, the places of degree δ are already in $\mathcal{P}$. For ℓ the largest divisor of d non equal to d, we adjust $\mathcal{P}$ by adding a place of degree $\ell+\delta$ and by removing a place of degree ℓ. The correctness of this construction will be proven in Proposition 5. The set obtained is denoted by $\mathcal{P}^{\text{div}}$, and its construction is formalized in the following algorithm.

Algorithm 1 Construction of $\mathcal{P}^{\text{div}}$

INPUT: q, $n > \frac{1}{2}q + 1$.
OUTPUT: $\mathcal{P}^{\text{div}}$.

(i) Let d be the smallest integer greater than $\log_q(2n)$, set $S = (\sum_{\substack{k|d \\ k \neq d}} kB_k) - 1$, and let ℓ be the greatest proper divisor of d. Initialize $\mathcal{P}^{\text{div}}$ including N_k places of degree k, for all k dividing d, with N_k as follows:

$$N_k = \begin{cases} \lfloor \frac{1}{d}(2n-1-S) \rfloor & \text{if } k = d, \\ q & \text{if } k = 1, \\ B_k & \text{if } k \mid d \text{ and } k \neq 1, d, \\ 0 & \text{elsewhere;} \end{cases}$$

(ii) Consider $\delta = 2n - 1 - S \pmod{d}$. Then,
- If $\delta = 0$, the algorithm is done;
- If $\delta \nmid d$, add to $\mathcal{P}^{\text{div}}$ a place of degree δ;
- If $0 \neq \delta \mid d$ add to $\mathcal{P}^{\text{div}}$ a place of degree $\ell + \delta$ and remove a place of degree ℓ.

DEFINITION 6. We denote by $\mathcal{U}_{q,n}^{\mathcal{P}^{\text{div}}}$ a recursive Chudnovsky-type algorithm on the projective line for multiplication in $\mathbb{F}_{q^n}$, with $\mathcal{P}^{\text{div}}$ given by Algorithm 1.

REMARK 9. For the asymptotical study, we can construct Chudnovsky-type algorithms using RPGC with $\mathcal{P}^{\text{div}}$, and use the algorithms from the previous section in the recursion. For instance, we can use the bounds of Table 1.

PROPOSITION 5. *Let q be a prime power, n be an integer greater than $\frac{1}{2}q + 1$ and Q a place of degree n of $\mathbb{F}_q(x)$. The recursive Chudnovsky-type algorithm over the projective line $\mathcal{U}_{q,n}^{\mathcal{P}^{\text{div}}}(Q)$ has bilinear complexity*

$$\mu(\mathcal{U}_{q,n}^{\mathcal{P}^{\text{div}}}) \leq \sum_{k|d} \frac{q^k}{k} \mu(\mathcal{U}_{q,k}^{\mathcal{P}_k}),$$

where $d = \lceil \log_q(2n) \rceil$, and $\mathcal{U}_{q,k}^{\mathcal{P}_k}$ is an algorithm for multiplication in $\mathbb{F}_{q^k}$.

PROOF. We have to check that $\mathcal{P}^{\text{div}}$ is well defined. After the first step, $\mathcal{P}^{\text{div}}$ contains only places of degree dividing d. Consider the quantities δ, S and ℓ from Algorithm 1. If $\delta = 0$ the algorithm is done. If $\delta \nmid d$, we can add a place of degree δ in $\mathcal{P}^{\text{div}}$. If $0 \neq \delta \mid d$, we have to argue differently since the places of degree δ are already in $\mathcal{P}^{\text{div}}$. We can always add a place of degree $\ell + \delta$ since either it does not divide d or it is equal to d. In fact, ℓ is less than or equal to $d/2$ and $\delta \leq d/2$ as well. Hence, $\ell < \ell + \delta \leq d$. Suppose that $\ell + \delta \neq d$. Since ℓ is the greatest non-trivial divisor of d, $\ell + \delta$ does not divide d, and we can add a place of degree $\ell + \delta$ in $\mathcal{P}^{\text{div}}$. If $\delta = \ell = d/2$, then $\delta + \ell = d$, and we have to verify that the number N_d of places of degree d used after the first step is strictly lower than B_d, so that one can add a place of degree d. Let us show that $N_d < B_d$. After step (i), the sum of the degrees of the places in $\mathcal{P}^{\text{div}}$ is equal to $\sum_{k|d} dN_d = S + d\lfloor \frac{1}{d}(2n-1-S) \rfloor \leq 2n - 1$.

Suppose that $N_d = B_d$, i.e. all places of degree d are included after step (i). The condition $d \geq \log_q(2n)$ attests that $\sum_{k|d} kB_k = q^d + 1$ is greater than $2n+1$. After step (i), the sum of the degrees of the places is hence greater than $2n$, since we do not count a rational place. But this sum has to be lower than or equal to $2n-1$, which gives a contradiction. Finally, the number N_d of places of degree d after the first step is

$$N_d = \lfloor \frac{1}{d}(2n-1-S) \rfloor < B_d, \tag{8}$$

and we can add a place of degree d. Now, Algorithm 1 is complete, and we denote by N_k^{div} the number of places of degree k in $\mathcal{P}^{\mathrm{div}}$ after (i) and (ii). It remains to verify that the sum of the kN_k^{div} is equal to $2n-1$. Thanks to the euclidean division of $2n-1-S$ by d, we have

$$\sum_{P \in \mathcal{P}^{\mathrm{div}}} \deg P = \sum kN_k^{\mathrm{div}} = S + \delta + d\lfloor \frac{1}{d}(2n-1-S) \rfloor = 2n-1.$$

Hence, $\mathcal{P}^{\mathrm{div}}$ is well defined, and we obtain a recursive Chudnovsky-type algorithm over the projective line. Its bilinear complexity verifies

$$\mu(\mathcal{U}_{q,n}^{\mathcal{P}^{\mathrm{div}}}) = \sum_k N_k^{\mathrm{div}} \mu(\mathcal{U}_{q,k}^{\mathcal{P}_k}) = \sum_{k|d} N_k^{\mathrm{div}} \mu(\mathcal{U}_{q,k}^{\mathcal{P}_k}) + \Delta,$$

where

$$\Delta = \begin{cases} \mu(\mathcal{U}_{q,\delta}^{\mathcal{P}_\delta}) & \text{if } \delta \nmid d, \\ \mu(\mathcal{U}_{q,\ell+\delta}^{\mathcal{P}_{\ell+\delta}}) & \text{if } 0 \neq \delta \mid d \text{ and } \ell+\delta \neq d, \\ 0 & \text{if } \delta = 0 \text{ or } \ell + \delta = d. \end{cases}$$

Recall that $\ell \leq \ell + \delta \leq d$. Thanks to Lemma 1, we define $\mathcal{U}_{q,\delta}^{\mathcal{P}_\delta}$ (resp. $\mathcal{U}_{q,\ell+\delta}^{\mathcal{P}_{\ell+\delta}}$) using $\mathcal{P}_\delta = \mathcal{P}_d$ and $\mathcal{P}_{\ell+\delta} = \mathcal{P}_d$. Hence, $\mu(\mathcal{U}_{q,\delta}^{\mathcal{P}_d}) \leq \mu(\mathcal{U}_{q,d}^{\mathcal{P}_d})$, and $\mu(\mathcal{U}_{q,\ell+\delta}^{\mathcal{P}_d}) \leq \mu(\mathcal{U}_{q,d}^{\mathcal{P}_d})$. If $\Delta \neq 0$, the inequality (8) attests that $N_d^{\mathrm{div}} < B_d$. Since $\Delta \leq \mu(\mathcal{U}_{q,d}^{\mathcal{P}_d})$, we have that $N_d^{\mathrm{div}} \mu(\mathcal{U}_{q,d}^{\mathcal{P}_d}) + \Delta \leq B_d \mu(\mathcal{U}_{q,d}^{\mathcal{P}_d})$. For all k, we have $N_k^{\mathrm{div}} \leq B_k$. Moreover, for all $k \geq 2$ we have $B_k \leq \frac{q^k}{k}$. Recalling that $N_1^{\mathrm{div}} \leq q$, we obtain

$$\mu(\mathcal{U}_{q,n}^{\mathcal{P}^{\mathrm{div}}}) \leq \sum_{k|d} \frac{q^k}{k} \mu(\mathcal{U}_{q,k}^{\mathcal{P}_k}). \qquad \square$$

COROLLARY 5. *Proposition* 5 *shows that there exists an algorithm in Proposition* 4 *with* $d = \lfloor \log_q(2n) \rfloor$.

5.2. Bound for the bilinear complexity of algorithms obtained using RPGC. Our bound for the bilinear complexity requires to introduce the iterated logarithm.

DEFINITION 7. For all integer n, the iterated logarithm of n, denoted by $\log^*(n)$, is defined by the following recursive function:

$$\log^*(n) = \begin{cases} 0 & \text{if } n \leq 1 \\ 1 + \log^*(\log(n)) & \text{elsewhere.} \end{cases}$$

This value corresponds to the number of times the logarithm is iteratively applied from n in order to obtain a result lower than or equal to 1.

To get a uniform bound, we have to specify the basis of the logarithm. Here, we will use q and finally $\sqrt{q}$ as basis. When the basis of the logarithm is a real number greater than $e^{\frac{1}{e}}$, this function is well defined. But the iterated logarithm can be defined with any base a, for a real a strictly greater than 1. Actually, we will deal with the case of the base $\sqrt{2}$, which is between 1 and $e^{\frac{1}{e}} \simeq 1.44467\ldots$. For $1 < a < e^{\frac{1}{e}}$, there exists $x_0 > 1$ such that $x_0 = \sup\{x \mid x = \log_a(x)\}$. Hence, the values obtained by applying successively log_a to $x \geq x_0$ converge to $x_0 > 1$, and the stopping step of the function defined above cannot be reached. We can define $\log_a^*$ for any a between 1 and $e^{\frac{1}{e}}$ by changing the stopping step to $\log_a^*(n) = 0$ if $n \leq \lfloor x_0 \rfloor + 1$. For $a = \sqrt{2}$, this number is $4 = \log_{\sqrt{2}}(4)$, and the iterated logarithm is given by

$$\log^*_{\sqrt{2}}(n) = \begin{cases} 0 & \text{if } n \leq 5 \\ 1 + \log^*_{\sqrt{2}}(\log_{\sqrt{2}}(n)) & \text{elsewhere.} \end{cases} \tag{9}$$

These are very slow-growing functions [**BBT16**, Table II]. The study of the bound also requires the following lemma:

LEMMA 2. *For all prime powers $q \geq 2$ and all integers $d \geq 1$, we have $dq^{\frac{d}{2}+1} - q^d \leq q^{d+1}$.*

PROOF. Let q be a prime power and $d \geq 1$ an integer. The inequality above is equivalent to $d \leq q^{\frac{d-2}{2}}(q+1)$, and then to $\log_q(\frac{d}{q+1}) \leq \frac{d-2}{2}$. Hence we have to show that $d - 2 - 2\log_q(d) + 2\log_q(q+1) \geq 0$. We can write this inequality $d \ln q - 2\ln q - 2\ln d + 2\ln(q+1) \geq 0$. For all $d \in [1\,;+\infty[$, we define

$$g_q(d) = d\ln q - 2\ln q - 2\ln d + 2\ln(q+1).$$

This function is derivable over $[1\,;+\infty[$ and its derivative is $g_q'(d) = \ln q - \frac{2}{d}$. Hence g_q has a minimum in $\frac{2}{\ln q}$. For $q \geq 8$, $\frac{2}{\ln q} \leq 1$, so g_q is growing over $[1\,;+\infty[$ and its minimum is given by

$$g_q(1) = \ln q - 2\ln q + 2\ln(q+1) \geq 0.$$

For $q \leq 7$, g_q reaches its minimum in $\frac{2}{\ln q} > 1$ and it can be checked that these minima are positive. Then, the inequality of the lemma is holding for all prime power q and all $d \geq 1$. □

We can finally state the following result:

THEOREM 6. *Let q be a prime power and $n \geq 2$ a positive integer. Then, there exists a recursive Chudnovsky-type algorithm $\mathcal{U}_{q,n}^{\mathcal{P}}$ over the projective line for multiplication in $\mathbb{F}_{q^n}$ over $\mathbb{F}_q$ with a uniform upper bound for its bilinear complexity*

$$\mu(\mathcal{U}_{q,n}^{\mathcal{P}}) \leq Cn\left(\frac{4q^2}{q-1}\right)^{\log^*_{\sqrt{q}}(2n)},$$

where $C = 1$ for $q \geq 3$ and $C = 3$ for $q = 2$.

PROOF. When $n \leq \frac{1}{2}q + 1$, the construction of Section 3 gives a Chudnovsky-type algorithm of optimal bilinear complexity over the project line, and the bound is verified. For $n > \frac{1}{2}q + 1$, We construct the set $\mathcal{P} = \mathcal{P}^{\text{div}}$ with Algorithm 1 and use RPGC. Then, Lemma 1 and Proposition 5 give

$$
\begin{aligned}
\mu(\mathcal{U}_{q,n}^{\mathcal{P}^{\mathrm{div}}}) &\leq \sum_{k|d_1} \frac{q^k}{k} \mu(\mathcal{U}_{q,k}^{\mathcal{P}^{\mathrm{div}}}) \\
&\leq \left(\frac{q^{d_1}}{d_1} + \sum_{k=1}^{d_1/2} \frac{q^k}{k} \right) \mu(\mathcal{U}_{q,d_1}^{\mathcal{P}^{\mathrm{div}}}) \\
&\leq \left(\frac{q^{d_1}}{d_1} + \frac{q^{\frac{d_1}{2}+1} - 1}{q-1} \right) \mu(\mathcal{U}_{q,d_1}^{\mathcal{P}^{\mathrm{div}}}) \\
&= \left(\frac{q^{d_1+1} - q^{d_1} + d_1 q^{\frac{d_1}{2}+1} - d_1}{d_1(q-1)} \right) \mu(\mathcal{U}_{q,d_1}^{\mathcal{P}^{\mathrm{div}}}).
\end{aligned}
$$

Thanks to Lemma 2, for all $q \geq 2$ and all $d_1 \geq 1$, we have $d_1 q^{\frac{d_1}{2}+1} - q^{d_1} \leq q^{d_1+1}$. Hence,

$$
\begin{aligned}
\mu(\mathcal{U}_{q,n}^{\mathcal{P}^{\mathrm{div}}}) &\leq \left(\frac{2q^{d_1+1}}{d_1(q-1)} \right) \mu(\mathcal{U}_{q,d_1}^{\mathcal{P}^{\mathrm{div}}}) \\
&\leq \left(\frac{4q^2 n}{d_1(q-1)} \right) \mu(\mathcal{U}_{q,d_1}^{\mathcal{P}^{\mathrm{div}}}),
\end{aligned}
$$

because d_1 is such that $d_1 - 1 < \log_q(2n) \leq d_1$.

We can similarly give a bound for $\mu(\mathcal{U}_{q,d_1})$, using $d_2 - 1 < \log_q(2d_1) \leq d_2$, and $\mathcal{P}^{\mathrm{div}}$ for the extension of degree d_1. It follows that

$$
\mu(\mathcal{U}_{q,d_1}^{\mathcal{P}^{\mathrm{div}}}) \leq \left(\frac{4q^2 d_1}{d_2(q-1)} \right) \mu(\mathcal{U}_{q,d_2}^{\mathcal{P}^{\mathrm{div}}}).
$$

and hence,

$$
\mu(\mathcal{U}_{q,n}^{\mathcal{P}^{\mathrm{div}}}) \leq \frac{n}{d_2} \left(\frac{4q^2}{q-1} \right)^2 \mu(\mathcal{U}_{q,d_2}^{\mathcal{P}^{\mathrm{div}}}).
$$

Doing i times this process, we get

$$
\mu(\mathcal{U}_{q,n}^{\mathcal{P}^{\mathrm{div}}}) \leq \frac{n}{d_i} \left(\frac{4q^2}{q-1} \right)^i \mu(\mathcal{U}_{q,d_i}^{\mathcal{P}^{\mathrm{div}}}),
$$

with $d_j - 1 < \log_q(2d_{j-1}) \leq d_j$, for all j between 2 and i. We now have to find i such that the recursion stops. We want $d_i < 1$, but $d_i > \log_q(2d_{i-1})$, then it suffices that $\log_q(2d_{i-1}) < 1$. Actually, for $2 \leq j \leq i$, we have $\log_q(2d_{j-1}) < d_j$. Hence, we are looking for i such that

$$
\underbrace{\log_q(2\log_q(\ldots(2\log_q(2n))\ldots))}_{i \text{ terms}} < 1.
$$

Notice that for all $a \in \mathbb{R}_+^*$ and $q > 1$, $\log_{\sqrt{q}}(a) = 2\log_q(a)$. Then,

$$
\underbrace{\log_q(2\log_q(\ldots(2\log_q(2n))\ldots))}_{i \text{ terms}} = \frac{1}{2} \underbrace{\log_{\sqrt{q}}(\log_{\sqrt{q}}(\ldots(\log_{\sqrt{q}}(2n))\ldots))}_{i \text{ terms}}.
$$

First, we consider $q \geq 3$. By Definition 7 of $\log^*$, $i = \log^*_{\sqrt{q}}(2n)$ is convenient. We finally get the bound

$$\mu(\mathcal{U}_{q,n}^{\mathcal{P}^{\mathrm{div}}}) \leq n \left(\frac{4q^2}{q-1}\right)^{\log^*_{\sqrt{q}}(2n)}, \text{ for } q \geq 3.$$

For $q = 2$, we process similarly using the iterated logarithm with base $\sqrt{2}$ as in (9), so we stop the recursion at $n \leq 5$. We can use the algorithms given in Section 4 for these small extensions. From Table 1, recall that $\mu(\mathcal{U}_{2,5}^{\mathcal{P}^{\mathrm{deg}}}) = 15$. For $q = 2$ and $n \geq 5$ we obtain the bound

$$\mu(\mathcal{U}_{2,n}^{\mathcal{P}^{\mathrm{div}}}) \leq n \frac{\mu(\mathcal{U}_{2,5}^{\mathcal{P}^{\mathrm{deg}}})}{5} \left(\frac{4q^2}{q-1}\right)^{\log^*_{\sqrt{2}}(2n)} = 3n \left(\frac{4q^2}{q-1}\right)^{\log^*_{\sqrt{2}}(2n)}.$$

Moreover, we can check in Table 1 the bilinear complexity for the algorithms of multiplication in extensions of degree lower than 5. Tables 1 gives $\mu(\mathcal{U}_{2,2}^{\mathcal{P}^{\mathrm{deg}}}) = 3$, $\mu(\mathcal{U}_{2,3}^{\mathcal{P}^{\mathrm{deg}}}) = 6$ and $\mu(\mathcal{U}_{2,4}^{\mathcal{P}^{\mathrm{deg}}}) = 11$. These values verify the latter bound. □

The given bound is not obtained by taking into account all the places in the construction, nor by using derivative evaluations. Consequently, this bound is not tight, but it does have the advantage of being explicit in n. Moreover, the construction of the $\mathcal{P}^{\mathrm{div}}$ is more restrictive than $\mathcal{P}^{\mathrm{deg}}$, and we can think that $\mu(\mathcal{U}_{q,n}^{\mathcal{P}^{\mathrm{deg}}}) \leq \mu(\mathcal{U}_{q,n}^{\mathcal{P}^{\mathrm{div}}})$, but this result is neither proven nor has a counter-example yet.

REMARK 10. In [**LSW83**, Theorem 2], Lempel et al. proved that their algorithm has a bilinear complexity, that we denote here by $\mu_{LSW}(n)$, satisfying

$$\mu_{LSW}(n) \leq f_q(n)n,$$

where $f_q(n)$ is a very slow growing function defined recursively by

$$f_q(n) = 2f_q(\lceil \log_q(2(q-1)n) \rceil), \; n \geq 4, \; q \geq 2,$$

and for $n \leq 4$ $f_q(n)$ is defined as follows:

$$f_q(n) = \begin{cases} 1, & n = 1, \; q \geq 2, \\ \frac{3}{2}, & n = 2, \; q \geq 2, \\ \frac{5}{3}, & n = 3, \; q \geq 4, \\ 2, & n = 3, \; 2 \leq q \leq 3. \end{cases}$$

This bound is obtain with a different approach, and also holds in the case of our construction. Moreover, it can be shown using Corollary 5 that f_q can actually be recursively defined by $f_q(n) = 2f_q(\lceil \log_q(2n) \rceil))$. Following [**LSW83**], denote by $s(n)$ the least positive integer such that $\sum_{k=1}^{s(n)} kI_q(k) \geq 2n-1$. In [**LSW83**, Proof of Theorem 2], the recursive definition of f_q is obtained by bounding $s(n)$ by $\lceil \log_q(2(q-1)n) \rceil$). However, by Corollary 5, there exist algorithms where the maximum degree of the places used is $d = \lfloor \log_q(2n) \rfloor$, and $s(n)$ can be bounded by d.

PROPOSITION 7. *Asymptotically with respect to n, the recursive Chudnovsky-type algorithm $\mathcal{U}_{q,n}^{\mathcal{P}^{\mathrm{div}}}$ for multiplication in $\mathbb{F}_{q^n}$ of Theorem 6 has bilinear complexity*

$$\mu(\mathcal{U}_{q,n}^{\mathcal{P}^{\mathrm{div}}}) \in \mathcal{O}(n(2q)^{\log^*_q(n)}).$$

PROOF. In the proof of Theorem 6, we consider that asymptotically the number of places of degree d of $\mathbb{F}_q(x)$ is in $\mathcal{O}(\frac{q^d}{d})$. Then, we have that $\mu(\mathcal{U}_{q,n}^{\mathcal{P}^{\text{div}}}) \in \mathcal{O}\left(\frac{q^{d_1}}{d_1}\mu(\mathcal{U}_{q,d_1}^{\mathcal{P}^{\text{div}}})\right)$. Remaining calculations are similar, and we obtain $\mu(\mathcal{U}_{q,n}^{\mathcal{P}^{\text{div}}}) \in \mathcal{O}(n(2q)^{\log_q^*(n)})$. Details can be found in the proof of the elliptic case [**BBT16**]. □

This asymptotic bound is the same as for the construction of [**BBT16**], i.e. with algorithms constructed over elliptic curves. On some examples, when the extension is large enough and with a good choice of curve, the algorithm over an elliptic curve might have a lower bilinear complexity, because a function field of genus 1 can have more places of a fixed degree than the projective line. However, it is not clear that such a generic construction can be obtained by using elliptic curves. Moreover, there is no uniform bound for the bilinear complexity of the construction with elliptic curves yet.

The generalized Karatsuba algorithm, which is a Divide and Conquer construction based on the multiplication of two polynomials of degree 1 (i.e. $\mathcal{U}_{q,2}^{\mathcal{P}_2}$ here), requires $\mathcal{O}(n^{\log_2 3})$ multiplications, all of them bilinear. This is much more than the bilinear complexity obtained with our construction.

REMARK 11. Note that [**HHL17**, Section 9.5] gives a similar bound on the bilinear complexity, but that is asymptotical. It would be relevant to compare our algorithms more thoroughly with the recent results by Harvey, van der Hoeven and Lecerf [**HHL17, HvdH22, HVDH19**]. But this work of comparison is sufficiently important to require a further work of its own.

5.3. Complexity of the construction of $\mathcal{U}_{q,n}^{\mathcal{P}^{\text{div}}}$. We can now describe the complexity of the construction of our algorithms. More precisely, we focus on the family of algorithms $\mathcal{U}_{q,n}^{\mathcal{P}^{\text{div}}}$ obtained using the sets of places $\mathcal{P}^{\text{div}}$ (Algorithm 1) to estimate the asymptotic behavior of the algebraic complexity of the construction. Our construction has the advantage of using generic parameters. In particular, the divisor and the bases of the Riemann-Roch spaces are explicitly given. The construction of our algorithms mainly consists in the construction of an irreducible polynomial of degree n over $\mathbb{F}_q$ as well as the matrices corresponding to the evaluation maps. This complexity is given by the number of elementary operations in $\mathbb{F}_q$. We use the standard Landau notation $\mathcal{O}$. Our result requires the following lemma.

LEMMA 3. *Let q be a prime power and n, d be positive integers. Let P be a place of degree d of $\mathbb{F}_q(x)$. The $d \times n$ matrix of the evaluations of x^i at P, for $i = 0$, $\dots$, $n-1$, can be computed with $\mathcal{O}(dn)$ operations in $\mathbb{F}_q$.*

PROOF. Let $P(x) = \sum_{j=0}^{d} a_j x^j$ be the monic irreducible polynomial associated to the place P. For any positive integer i, the evaluation of x^i at P is by definition its class in the residue class field (Cf. [**Sti08**]), i.e. in $\frac{\mathbb{F}_q[x]}{(P(x))} = \mathbb{F}_{q^d}$. Let M be the $d \times n$ matrix over $\mathbb{F}_q$ of the evaluations of x^i at P, for $i = 0, \dots, n-1$. More precisely, the i–th column of M is the evaluation of x^{i-1} at P in the basis $\{1, \alpha, \dots, \alpha^{d-1}\}$ of $\mathbb{F}_{q^d}$, with α a root of $P(x)$. We denote by $M_{k,l}$ the coefficient in the k–th line and in the l–th column of M. The d first columns are given by the identity matrix. Then, the $d+1$–th column is given by the opposite of the coefficients of $P(x)$, i.e. $M_{k,d+1} = -a_{k-1}$, for $k = 1, \dots, d$, since $\alpha^d = -\sum_{j=0}^{d-1} a_j \alpha^j$. Then, every column can be computed from the previous ones and $P(x)$ with $2d$ operations. Indeed, Let

l be strictly greater than $d+1$. Then,

$$\begin{aligned}\alpha^l &= \alpha \times \alpha^{l-1} \\ &= \alpha \times \sum_{k=1}^{d} M_{k,l-1}\alpha^{k-1} \\ &= -M_{d,l-1}a_{k-1} + \sum_{k=2}^{d}(M_{k-1,l-1} - M_{d,l-1}a_{k-1})\alpha^{k-1}.\end{aligned}$$

Hence $M_{1,l} = -M_{d,l-1}a_{k-1}$, and for $k \geq 2$ the coefficient $M_{k,l}$ is given by $M_{k,l} = M_{k-1,l-1} - a_{k-1}M_{d,l-1}$. Thus, it takes $2d$ operations to compute each of the $n-d-1$ columns. Finally, the matrix M can be computed in time $\mathcal{O}(dn)$. □

As first stated by Shparlinski, Tsfasman and Vlăduţ [**STV92**], the most expensive part of constructing Chudnovsky-type algorithms is to find a degree n place in the function field. In our case, this means constructing a monic irreducible polynomial of degree n over $\mathbb{F}_q$, and there exist algorithms to do it in polynomial time.

THEOREM 8. *The recursive Chudnovsky-type algorithm over the projective line* $\mathcal{U}_{q,n}^{\mathcal{P}^{\text{div}}}$ *of Theorem 6 built using RPGC is constructible deterministically and in time* $\mathcal{O}(n^4)$.

PROOF. To obtain the degree n place of $\mathbb{F}_q(x)$, we compute a monic irreducible polynomial of degree n over $\mathbb{F}_q$ thanks to Shoup [**Sho97**] in time $\mathcal{O}(n^4)$. We use as a divisor $\mathcal{D} = (n-1)P_\infty$, thus we set $\{1, x, \ldots, x^{n-1}\}$ as the basis of $\mathcal{L}(\mathcal{D})$, and $\{1, x, \ldots, x^{2n-2}\}$ as basis of $\mathcal{L}(2\mathcal{D})$.

From Theorem 6, constructing the places of degrees dividing d, for $\log_q(2n) \leq d \leq \log_q(2n) + 1$, is enough. Asymptotically, the base of the logarithm makes no difference and we can consider that $\mathcal{O}(\frac{q^d}{d})$ places of degree d are required, with $d \in \mathcal{O}(\log(2n))$. Hence, we have to construct the rational places of $\mathbb{F}_{q^d}(x)$ and group them in places of $\mathbb{F}_q(x)$ by applying them the iterated Frobenius. From Von zur Gathen and Gerhard book [**GG13**], Algorithm 14.26 computes the iterated Frobenius, which for a given $\alpha \in \mathbb{F}_{q^d}$ outputs $\alpha, \alpha^q, \ldots, \alpha^{q^k}$, for k dividing d. Its complexity is in time $\mathcal{O}(M(d)^2\log(d)\log(k))$, where $M(d) \in \mathcal{O}(d^\omega)$, with ω the best exponent for multiplication of two matrices of size $d \times d$. Currently, $\omega \simeq 2.37286\ldots$ [**AW21**]. With this algorithm, the $\mathcal{O}(\frac{q^d}{d})$ places of degree d can be constructed with $\mathcal{O}\big(\frac{q^d}{d}M(d)^2\log(d)\log(d)\big) = \mathcal{O}\big(2n(\log(2n))^{2\omega-1}\log(\log 2n)^2\big)$ operations.

Then, we construct the matrices of $\tilde{E_P}$, $\tilde{\text{Ev}_P}^{-1}$, E_Q and Ev_Q^{-1} of $\mathcal{U}_{q,n}^{\mathcal{P}^{\text{div}}}$. The matrix Ev_Q^{-1} is the identity. The construction of E_Q is given by the evaluation at Q of the basis of $\mathcal{L}(2\mathcal{D})$ and can be computed in $\mathcal{O}((2n)^2)$ operations thanks to Lemma 3. The matrix $\tilde{\text{Ev}_P}$ is obtained with $\mathcal{O}(\frac{q^d}{d})$ matrices $d \times 2n-1$ as in Lemma 3. Hence, it can be computed in $\mathcal{O}(\frac{q^d}{d}2nd) = \mathcal{O}(4n^2)$ operations. All other required matrices are sub-matrices or inverses of $\tilde{\text{Ev}_P}$. The inverses of the matrices can be performed in $\mathcal{O}\left(M(2n)\right)$, since a matrix inversion has asymptotically the same cost as a matrix multiplication.

Finally, the algorithm $\mathcal{U}_{q,n}^{\mathcal{P}^{\text{div}}}$ involves recursively defined algorithms to compute multiplications in intermediate extensions, and we need to construct $\mathcal{U}_{q,d}^{\mathcal{P}^{\text{div}}}(Q_d)$ for

Q_d traversing the set of places used for the interpolation. Asymptotically, recall that we consider $\mathcal{O}(\frac{q^d}{d})$ places of degree $d \in \mathcal{O}(\log(2n))$. For a given degree d place Q_d, the construction $\mathcal{U}_{q,d}^{\mathcal{P}^{\text{div}}}(Q_d)$ is negligible, since $d \in \mathcal{O}(\log(2n))$. Nevertheless, the matrices E_{Q_d} have to be computed for $\mathcal{O}(\frac{q^d}{d})$ places of degree d. Each of these can be obtained in $\mathcal{O}(d \log d)$ operations by Lemma 3, and the construction of all these matrices costs $\mathcal{O}(q^d \log d) = \mathcal{O}(2n \log\log 2n)$ operations in the base field.

Putting this all together, we obtain a complexity of construction in $\mathcal{O}(n^4)$ operations. □

This bound is for a deterministic construction. In practice, we can construct the monic irreducible polynomial of degree n using the Las Vegas-type algorithm by Couveignes and Lercier [**CL13**], with running time $n^{1+\epsilon(n)} \times (\log q)^{5+\epsilon(q)}$, where $\epsilon(x)$ are functions in $\mathcal{O}(1)$, or in expected time $\tilde{\mathcal{O}}(n^2 \log q)$ with Ben-Or's algorithm [**GG13**, Theorem 14.42].

THEOREM 9. *Using a probabilistic algorithm to compute the monic irreducible polynomial of degree n, the recursive Chudnovsky-type algorithm over the projective line $\mathcal{U}_{q,n}^{\mathcal{P}^{\text{div}}}$ of Theorem* 6 *obtained using RPGC is constructible in expected time* $\mathcal{O}(M(2n))$.

PROOF. In the previous proof, we replace Shoup's algorithm by the one of Couveignes and Lercier [**CL13**] or Ben-Or [**GG13**, Theorem 14.42] to find the monic irreducible polynomial. Then, the heaviest part of the complexity consists in computing the inverse of the matrices. □

COROLLARY 6. *If a place Q of degree n is given, the recursive Chudnovsky-type algorithm over the projective line $\mathcal{U}_{q,n}^{\mathcal{P}^{\text{div}}}$ of Theorem* 6 *obtained using RPGC is constructible in expected time* $\mathcal{O}(M(2n))$.

REMARK 12. The construction over elliptic curves [**BBT16**] is also polynomial but more expensive. For instance, the construction of the degree n place requires to compute the iterated Frobenius of a nontrivial rational point of the curve over $\mathbb{F}_{q^n}$. With the algorithm from [**GG13**], as in the proof of Theorem 8, it takes $\mathcal{O}(n^{2\omega} \log(n)^2)$ operations to construct this degree n place. This is already above the complexity of the construction of our algorithms. Moreover, there is no precise estimation of this complexity yet.

Acknowledgment

The authors would like to thank the referee for his many valuable comments, which helped to significantly improve the article.

References

[Arn06] Nicolas Arnaud, *Évaluations dérivées, multiplication dans les corps finis et codes correcteurs*, Ph.D. thesis, Université de la Méditerranée, Institut de Mathématiques de Luminy, 2006.

[AW21] Josh Alman and Virginia Vassilevska Williams, *A refined laser method and faster matrix multiplication*, Proceedings of the 2021 ACM-SIAM Symposium on Discrete Algorithms (SODA), [Society for Industrial and Applied Mathematics (SIAM)], Philadelphia, PA, 2021, pp. 522–539, DOI 10.1137/1.9781611976465.32. MR4262465

[Bal99] Stéphane Ballet, *Curves with many points and multiplication complexity in any extension of* $\mathbf{F}_q$, Finite Fields Appl. **5** (1999), no. 4, 364–377, DOI 10.1006/ffta.1999.0255. MR1711833

[BBP20] Stéphane Ballet, Alexis Bonnecaze, and Bastien Pacifico, *Multiplication in finite fields with Chudnovsky-type algorithms on the projective line*, Preprint `arXiv:2007.16082`, August 2020, 28 pages.

[BBP23] Stéphane Ballet, Alexis Bonnecaze, and Bastien Pacifico, *Polynomial constructions of Chudnovsky-type algorithms for multiplication in finite fields with linear bilinear complexity*, Arithmetic of finite fields, Lecture Notes in Comput. Sci., vol. 13638, Springer, Cham, [2023] ©2023, pp. 35–52, DOI 10.1007/978-3-031-22944-2_3. MR4540578

[BBT16] Stéphane Ballet, Alexis Bonnecaze, and Mila Tukumuli, *On the construction of elliptic Chudnovsky-type algorithms for multiplication in large extensions of finite fields*, J. Algebra Appl. **15** (2016), no. 1, 1650005, 26, DOI 10.1142/S0219498816500055. MR3393934

[Bod07] Marco Bodrato, *Towards optimal Toom-Cook multiplication for univariate and multivariate polynomials in characteristic 2 and 0*, Arithmetic of finite fields, Lecture Notes in Comput. Sci., vol. 4547, Springer, Berlin, 2007, pp. 116–133, DOI 10.1007/978-3-540-73074-3_10. MR2373888

[BPR$^+$21] S. Ballet, J. Pieltant, M. Rambaud, H. Randriambololona, R. Rolland, and J. Chaumine, *On the tensor rank of multiplication in finite extensions of finite fields and related issues in algebraic geometry* (Russian, with Russian summary), Uspekhi Mat. Nauk **76** (2021), no. 1(457), 31–94, DOI 10.4213/rm9928; English transl., Russian Math. Surveys **76** (2021), no. 1, 29–89. MR4223937

[BR04] S. Ballet and R. Rolland, *Multiplication algorithm in a finite field and tensor rank of the multiplication*, J. Algebra **272** (2004), no. 1, 173–185, DOI 10.1016/j.jalgebra.2003.09.031. MR2029030

[CC88] D. V. Chudnovsky and G. V. Chudnovsky, *Algebraic complexities and algebraic curves over finite fields*, J. Complexity **4** (1988), no. 4, 285–316, DOI 10.1016/0885-064X(88)90012-X. MR974928

[CL13] Jean-Marc Couveignes and Reynald Lercier, *Fast construction of irreducible polynomials over finite fields*, Israel J. Math. **194** (2013), no. 1, 77–105, DOI 10.1007/s11856-012-0070-8. MR3047063

[CÖ10] Murat Cenk and Ferruh Özbudak, *On multiplication in finite fields*, J. Complexity **26** (2010), no. 2, 172–186, DOI 10.1016/j.jco.2009.11.002. MR2607731

[DG83] Hans F. de Groote, *Characterization of division algebras of minimal rank and the structure of their algorithm varieties*, SIAM J. Comput. **12** (1983), no. 1, 101–117, DOI 10.1137/0212007. MR687805

[GG13] Joachim von zur Gathen and Jürgen Gerhard, *Modern computer algebra*, 3rd ed., Cambridge University Press, Cambridge, 2013, DOI 10.1017/CBO9781139856065. MR3087522

[GS95] Arnaldo García and Henning Stichtenoth, *A tower of Artin-Schreier extensions of function fields attaining the Drinfel′d-Vlăduţ bound*, Invent. Math. **121** (1995), no. 1, 211–222, DOI 10.1007/BF01884295. MR1345289

[HHL17] David Harvey, Joris van der Hoeven, and Grégoire Lecerf, *Faster polynomial multiplication over finite fields*, J. ACM **63** (2017), no. 6, Art. 52, 23, DOI 10.1145/3005344. MR3614860

[HVDH19] David Harvey and Joris van der Hoeven, *Faster polynomial multiplication over finite fields using cyclotomic coefficient rings*, J. Complexity **54** (2019), 101404, 18, DOI 10.1016/j.jco.2019.03.004. MR3983215

[HvdH22] David Harvey and Joris van der Hoeven, *Polynomial multiplication over finite fields in time* $O(n \log n)$, J. ACM **69** (2022), no. 2, Art. 12, 40, DOI 10.1145/3505584. MR4433320

[Kar63] Anatolii Karatsuba, *Multiplication of multidigit number on automata*, Soviet Physics Doklady **7** (1963), 595–596.

[Lar19] Robin Larrieu, *Fast finite fields arithmetic*, Ph.D. thesis, École polytechnique, Université Paris-Saclay, December 2019.

[LSW83] A. Lempel, G. Seroussi, and S. Winograd, *On the complexity of multiplication in finite fields*, Theoret. Comput. Sci. **22** (1983), no. 3, 285–296, DOI 10.1016/0304-3975(83)90108-1. MR693061

[Ran12] Hugues Randriambololona, *Bilinear complexity of algebras and the Chudnovsky-Chudnovsky interpolation method*, J. Complexity **28** (2012), no. 4, 489–517, DOI 10.1016/j.jco.2012.02.005. MR2925903

[Sho97] Victor Shoup, *New algorithms for finding irreducible polynomials over finite fields*, Math. Comp. **54** (1990), no. 189, 435–447, DOI 10.2307/2008704. MR993933

[Sti08] Henning Stichtenoth, *Algebraic function fields and codes*, 2nd ed., Graduate Texts in Mathematics, vol. 254, Springer-Verlag, Berlin, 2009. MR2464941

[STV92] Igor E. Shparlinski, Michael A. Tsfasman, and Serge G. Vladut, *Curves with many points and multiplication in finite fields*, Coding theory and algebraic geometry (Luminy, 1991), Lecture Notes in Math., vol. 1518, Springer, Berlin, 1992, pp. 145–169, DOI 10.1007/BFb0087999. MR1186422

[Win79] S. Winograd, *On multiplication in algebraic extension fields*, Theoret. Comput. Sci. **8** (1979), no. 3, 359–377, DOI 10.1016/0304-3975(79)90017-3. MR532476

Aix Marseille Univ, CNRS, I2M, Marseille, France
Email address: `Stephane.Ballet@univ-amu.fr`

Aix Marseille Univ, CNRS, I2M, Marseille, France
Email address: `Alexis.Bonnecaze@univ-amu.fr`

Aix Marseille Univ, CNRS, I2M, Marseille, France
Email address: `Bastien.Pacifico@univ-amu.fr`

Contemporary Mathematics
Volume **832**, 2026
https://doi.org/10.1090/conm/832/16606

Explicit Riemann-Roch spaces in the Hilbert class field

Jean-Marc Couveignes and Jean Gasnier

ABSTRACT. Let $\mathbf{K}$ be a finite field, X and Y two curves over $\mathbf{K}$, and $Y \to X$ an unramified abelian cover with Galois group G. Let D be a divisor on X and E its pullback on Y. Under mild conditions the linear space associated with E is a free $\mathbf{K}[G]$-module. We study the algorithmic aspects and applications of these modules.

CONTENTS

1. Introduction

Given a curve Y over a field $\mathbf{K}$, and two divisors E and Q on Y, with Q effective and disjoint from E, the evaluation map $e : H^0(Y, \mathcal{O}_Y(E)) \to H^0(Y, \mathcal{O}_Y/\mathcal{O}_Y(-Q))$ is a natural $\mathbf{K}$-linear datum of some importance for various algorithmic problems such as efficient computing in the Picard group of Y (see [**28, 29**]), constructing good error correcting codes [**16, 18, 49**], or bounding the bilinear complexity of multiplication in finite fields [**3, 4, 9, 38, 45, 46**]. Assume that G is a finite group of automorphisms of $Y/\mathbf{K}$, and the divisors E and Q are G-equivariant (they are equal to their pullback by any element of G). The evaluation map e is then a $\mathbf{K}[G]$-linear map between two $\mathbf{K}[G]$-modules. In some cases these modules can be shown to be both free. Their rank as $\mathbf{K}[G]$-modules is then smaller than their dimension as $\mathbf{K}$-vector spaces, by a factor $\mathfrak{o}$, the order of G. This is helpful when G is commutative, because multiplication in $\mathbf{K}[G]$ is achieved in quasi-linear time using a discrete Fourier transform, and the advantage of lowering dimension is stronger than the

2020 *Mathematics Subject Classification.* Primary 11G45, 14H30, 14G50, 65T50.

disadvantage of dealing with a larger ring of scalars. We will focus on free $\mathbf{K}[G]$-modules arising from commutative groups acting freely on a curve. This special case has a rich mathematical background and produces interesting constructions. For example Theorem 3 states the existence of excellent algebraic geometry codes that can be encoded in quasi-linear time and decoded in quasi-quadratic time in their length.

In Section 2 we review elementary properties of $\mathbf{K}[G]$-modules when $\mathbf{K}$ is a commutative field and G a finite group. We recall in Section 3 how unramified fibers of Galois covers of curves produce free $\mathbf{K}[G]$-modules and we introduce natural bases for these modules. We study the abelian unramified case in Section 4. Theorem 1 states that in this case, the Riemann-Roch space associated to a G-equivariant divisor of large enough degree is a free $\mathbf{K}[G]$-module. Evaluating at another G-equivariant divisor then produces a $\mathbf{K}[G]$-linear map between two free $\mathbf{K}[G]$-modules. This makes it possible to treat evaluation and interpolation as $\mathbf{K}[G]$-linear problems. We introduce the matrices associated to these problems. Section 5 is devoted to the definition and computation of Padé approximants in this context. The complexity of arithmetic operations in $\mathbf{K}[G]$ is bounded in Section 6 using various classical discrete Fourier transforms. Theorem 2 states that the complexity of multiplication in $\mathbf{K}[G]$ is quasi-linear when G is commutative. In Section 7 we use effective class field theory and the algorithmics of curves and jacobian varieties to compute the evaluation and interpolation matrices introduced in Section 4. Section 8 is concerned with two applications of interpolation with $\mathbf{K}[G]$-modules: multiplication in finite fields and geometric codes. The asymptotic properties of the codes constructed in this way are studied in Section 9. We thank the anonymous referee for their comments and suggestions. The calculation in Section 7.2 has been implemented using SageMath (Version 9.4) the Sage Mathematics Software System [**48**].

2. Duality for $\mathbf{K}[G]$-modules

In this section $\mathbf{K}$ is a commutative field and G is a finite group. We state elementary properties of $\mathbf{K}[G]$-modules and their duals. In Section 2.1 we describe the natural correspondence between G-invariant $\mathbf{K}$-bilinear forms and $\mathbf{K}[G]$-bilinear forms. We see in Section 2.2 that the orthogonal of a $\mathbf{K}[G]$-submodule for either form is the same. Sections 2.3 is concerned with the canonical bilinear form relating a $\mathbf{K}[G]$-module and its dual. The ring $\mathbf{K}[G]$ has the Frobenius property [**13**, Chapter IX]. We recall in Section 2.4 a convenient consequence of it.

2.1. Invariant bilinear forms. Let M be a right $\mathbf{K}[G]$-module. Let N be a left $\mathbf{K}[G]$-module. Let

$$< .,. >: M \times N \to \mathbf{K}$$

be a $\mathbf{K}$-bilinear form. We assume that this form is invariant by the action of G in the sense that

$$< m.\sigma, n > = < m, \sigma.n >$$

for every m in M, n in N, and σ in G. We define a map

$$(1) \qquad (.,.) \quad : \quad N \times M \longrightarrow \mathbf{K}[G]$$
$$n, m \longmapsto (n, m) = \textstyle\sum_{\sigma \in G} < m.\sigma^{-1}, n > \sigma$$

PROPOSITION 1. *The map* $(.,.)$ *in Equation* (1) *is* $\mathbf{K}[G]$*-bilinear.*

PROOF. Indeed for any τ in G, m in M, and n in N

$$\begin{aligned}(\tau.n, m) &= \sum_{\sigma \in G} < m.\sigma^{-1}, \tau.n > \sigma \\ &= \sum_{\sigma \in G} < m.\sigma^{-1}\tau^{-1}, \tau.n > \tau\sigma \\ &= \sum_{\sigma \in G} < m.\sigma^{-1}, n > \tau\sigma \\ &= \tau \sum_{\sigma \in G} < m.\sigma^{-1}, n > \sigma \\ &= \tau(n, m).\end{aligned}$$

And

$$\begin{aligned}(n, m.\tau) &= \sum_{\sigma \in G} < m.\tau\sigma^{-1}, n > \sigma \\ &= \sum_{\sigma \in G} < m.\tau\tau^{-1}\sigma^{-1}, n > \sigma\tau \\ &= \sum_{\sigma \in G} < m.\sigma^{-1}, n > \sigma\tau \\ &= (n, m)\tau.\end{aligned}$$

□

2.2. Orthogonality. In the situation of Section 2.1 we consider a right submodule U of the $\mathbf{K}[G]$-module M. Call

$$U^{\perp} = \{n \in N \mid < U, n >= 0\}$$

the orthogonal to U in N for the $< .,. >$ form. This is a $\mathbf{K}$-vector space. Since U is stable by the action of G, its orthogonal $U^{\perp}$ is a left $\mathbf{K}[G]$-module. And $U^{\perp}$ is the orthogonal to U for the $(.,.)$ form:

$$U^{\perp} = \{n \in N \mid (n, U) = 0\}.$$

We consider similarly a left $\mathbf{K}[G]$-submodule V of N and let

$$V^{\circ} = \{m \in M \mid < m, V >= 0\}$$

be the orthogonal to V in M for the $< .,. >$ form. This is a right $\mathbf{K}[G]$-module. And V° is the orthogonal to V for the $(.,.)$ form:

$$V^{\circ} = \{m \in M \mid (V, m) = 0\}.$$

We have $U \subset (U^{\perp})^{\circ}$ and $V \subset (V^{\circ})^{\perp}$. These inclusions are equalities when M and N are finite dimensional and $< .,. >$ is perfect.

2.3. The dual of a $\mathbf{K}[G]$-module. Let N be a left $\mathbf{K}[G]$-module. We can see N as a $\mathbf{K}$-vector space and let $\hat{N}$ be its dual. This is a right $\mathbf{K}[G]$-module. For every φ in $\hat{N}$ and σ in G we set $\varphi.\sigma = \varphi \circ \sigma$. We consider the canonical $\mathbf{K}$-bilinear form defined by

$$< \varphi, n >= \varphi(n)$$

for every n in N and φ in $\hat{N}$. For every σ in G we have

$$< \varphi.\sigma, n >= \varphi(\sigma.n) =< \varphi, \sigma.n >$$

so $< .,. >$ is invariant by G. Following Section 2.1 we define a $\mathbf{K}[G]$-bilinear form

$$(.,.) : N \times \hat{N} \to \mathbf{K}[G]$$

by

$$(n, \varphi) = \sum_{\sigma \in G} \varphi(\sigma^{-1}.n)\sigma. \tag{2}$$

We define a map from $\hat{N}$ to the dual $\check{N}$ of N as a $\mathbf{K}[G]$-module, by sending φ to the map

$$\varphi^G : n \mapsto (n, \varphi). \tag{3}$$

We prove that this map is a bijection. First $\varphi \mapsto \varphi^G$ is trivially seen to be an injection. As for surjectivity, let $\psi : N \to \mathbf{K}[G]$ be a $\mathbf{K}[G]$-linear map. Writing

$$\psi(n) = \sum_{\sigma \in G} \psi_\sigma(n)\sigma$$

we define a $\mathbf{K}$-linear coordinate form ψ_σ on N for every σ in G. We deduce from the $\mathbf{K}[G]$-linearity of ψ that $\psi_\sigma(n) = \psi_1(\sigma^{-1}.n)$ where $1 \in G$ is the identity element. So $\psi(n) = (n, \psi_1)$ for every n in N. So $\psi = (\psi_1)^G$.

2.4. Free submodules of a $\mathbf{K}[G]$-module. The ring $\mathbf{K}[G]$ may not be semisimple. Still free $\mathbf{K}[G]$-submodules of finite rank are direct summands.

PROPOSITION 2. *Let G be finite group, $\mathbf{K}$ a commutative field, and N a left $\mathbf{K}[G]$-module. Let V a submodule of N. If V is free of finite rank then it is a direct summand: there exists a submodule W of N such that $N = V \oplus W$. Such a W is called a complementary submodule to V.*

PROOF. Let r be the rank of V. Let $v_1, v_2, \ldots, v_r$ be a basis of V. Let $\varphi_1, \varphi_2, \ldots, \varphi_r$ be the dual basis. For every i such that $1 \leqslant i \leqslant n$, the coordinate form $\varphi_{i,1}$ associated to the identity element 1 in G belongs to $\hat{V}$. Let ψ_i be a $\mathbf{K}$-linear form on N whose restriction to V is $\varphi_{i,1}$. Let $\psi_i^G \in \check{N}$ be the associated $\mathbf{K}[G]$-linear form according to Equations (2) and (3). The restriction of ψ_i^G to V is $\varphi_{i,1}^G$ and this is φ_i. The map

$$\begin{array}{cccc} \psi & : & N & \longrightarrow V \\ & & n & \longmapsto \sum_{1 \leqslant i \leqslant r} \psi_i^G(n).v_i \end{array}$$

is a $\mathbf{K}[G]$-linear projection onto V. Its kernel is a complementary $\mathbf{K}[G]$-submodule to V. □

Proposition 2 is a consequence of the Frobenius property which is known to be satisfied by $\mathbf{K}[G]$. See [**13**, Chapter IX]. The proof above provides an algorithm to compute a complementary module.

3. Curves with a group action

Let $\mathbf{K}$ be a commutative field. Let p be the characteristic of $\mathbf{K}$. Let X and Y be two smooth, projective, absolutely integral curves over $\mathbf{K}$. Let g_X be the genus of X and let g_Y be the genus of Y. Let $\tau : Y \to X$ be a Galois cover with Galois group G. Let $\mathfrak{o}$ be the order of G. There is a natural left action of G on $\mathbf{K}(Y)$ defined by

$$\sigma.f = f \circ \sigma^{-1} \quad \text{for} \quad f \in \mathbf{K}(Y) \quad \text{and} \quad \sigma \in G. \tag{4}$$

There is a natural right action of G on meromorphic differentials defined by

$$\omega.\sigma = \sigma^{\star}\omega \quad \text{for} \quad \omega \in \Omega_{\mathbf{K}(Y)/\mathbf{K}} \quad \text{and} \quad \sigma \in G. \tag{5}$$

These are $\mathbf{K}(X)$-linear actions. And the two actions are compatible in the sense that

$$(\omega.\sigma)(\sigma^{-1}.f) = (\omega f).\sigma \tag{6}$$

We study some free $\mathbf{K}[G]$-modules that arise naturally in this context.

3.1. The residue ring of a non-ramified fiber. Let P be a prime divisor (a place) on X. Let t_P be a uniformizing parameter at P. Let

$$a = \deg(P).$$

This is the degree over $\mathbf{K}$ of the residue field

$$\mathbf{K}_P = H^0(P, \mathcal{O}_P) = H^0(X, \mathcal{O}_X/\mathcal{O}_X(-P)).$$

We assume that τ is not ramified above P and let Q_1 be a place above P. Let G_1 be the decomposition group of Q_1. This is the stabilizer of Q_1 in G. Places above P are parametrized by left cosets in G/G_1. We write the fiber above P

$$Q = \sum_{\sigma \in G/G_1} Q_\sigma \quad \text{with} \quad Q_\sigma = \sigma(Q_1).$$

Let

$$\mathfrak{b} = [G : G_1]$$

be the number of places above P and let

$$c = \mathfrak{o}/\mathfrak{b} = |G_1|$$

be the residual degree, that is the degree of

$$\mathbf{K}_\sigma = H^0(Q_\sigma, \mathcal{O}_{Q_\sigma})$$

over $\mathbf{K}_P$ for all $\sigma \in G/G_1$. Let

$$\mathbf{R}_Q = H^0(Q, \mathcal{O}_Q) = H^0(Y, \mathcal{O}_Y/\mathcal{O}_Y(-Q))$$

be the residue ring at Q. We have

$$\mathbf{R}_Q = \bigoplus_{\sigma \in G/G_1} \mathbf{K}_\sigma.$$

The action of G on $\mathbf{R}_Q$ makes it a free left $\mathbf{K}[G]$-module of rank a. Indeed it is a free $\mathbf{K}_P[G]$-module of rank 1. A basis for it consists of any normal element θ in $\mathbf{K}_1/\mathbf{K}_P$.

If m is a positive integer, Taylor expansion provides an isomorphism of modules over $\mathbf{K}_P[G]$

$$H^0(Y, \mathcal{O}_Y/\mathcal{O}_Y(-mQ)) \simeq \mathbf{R}_Q[t_P]/t_P^m$$

between the residue ring at mQ and the ring of truncated series in t_P. So the former is a free left $\mathbf{K}_P[G]$-module of rank m. A basis for it is made of the θt_P^k for $0 \leqslant k < m$.

3.2. The residue ring of a non-ramified G-equivariant divisor. We take P an effective divisor on X. We assume that τ does not ramify above P and let Q be the pullback of P by τ. We write

$$P = \sum_{1 \leqslant i \leqslant I} m_i P_i.$$

Let t_i be a uniformizing parameter at P_i. Let a_i be the degree of the place P_i. Let b_i be the number of places of Y above P_i. Let $c_i = \mathfrak{o}/b_i$. For every $1 \leqslant i \leqslant I$ we choose a place $Q_{i,1}$ above P_i and let $G_{i,1}$ be the decomposition group at $Q_{i,1}$. Let Q_i be the pullback of P_i by τ and write

$$Q_i = \sum_{\sigma \in G/G_{i,1}} Q_{i,\sigma} \quad \text{with} \quad Q_{i,\sigma} = \sigma(Q_{i,1}) \tag{7}$$

its decomposition as a sum of b_i places. Let $\mathbf{K}_{i,\sigma}$ be the residue field at $Q_{i,\sigma}$. We denote by $\mathbf{A}$ the residue algebra $H^0(Q, \mathcal{O}_Q)$. Taylor expansion induces an isomorphism of $\mathbf{K}$-algebras

$$\mathbf{A} = H^0(Q, \mathcal{O}_Q) = H^0(Y, \mathcal{O}_Y/\mathcal{O}_Y(-Q)) \simeq \bigoplus_{i=1}^{I} \bigoplus_{\sigma \in G/G_{i,1}} \mathbf{K}_{i,\sigma}[t_i]/t_i^{m_i}$$

which is compatible with the left actions of G as defined by Equations (4) and (7). In the special case when all the places P_i have degree one, a basis for $H^0(Q, \mathcal{O}_Q)$ as a $\mathbf{K}[G]$-module is made of the $\theta_i t_i^{k_i}$ for $1 \leqslant i \leqslant I$ and $0 \leqslant k_i < m_i$ where θ_i is a normal element in the extension $\mathbf{K}_{i,1}/\mathbf{K}$. The proposition below follows from the discussion in this section and the previous one.

PROPOSITION 3. *Assume the hypotheses at the beginning of Section 3. Let P be an effective divisor on X. Assume that τ is not ramified above P and let Q be the pullback of P by τ. The residue ring $H^0(Q, \mathcal{O}_Q)$ is a free $\mathbf{K}[G]$-module of rank the degree of P.*

3.3. Duality. We need a dual of $\mathbf{A}$ as a $\mathbf{K}$-vector space. We set

$$\hat{\mathbf{A}} = H^0(Y, \Omega_{Y/\mathbf{K}}(-Q)/\Omega_{Y/\mathbf{K}}) \simeq \bigoplus_{i=1}^{I} \bigoplus_{\sigma \in G/G_{i,1}} (\mathbf{K}_{i,\sigma}[t_i]/t_i^{m_i}) \frac{dt_i}{t_i^{m_i}}.$$

For $f \in \mathbf{A}$ and $\omega \in \hat{\mathbf{A}}$ we write $< \omega, f >$ for the sum of the residues of ωf at all the geometric points of Q. This is a $\mathbf{K}$-bilinear form. We deduce from Equation (6) that this form is invariant by the action of G

$$< \omega.\sigma, f > = < \omega, \sigma.f >$$

We define a $\mathbf{K}[G]$-bilinear form using the construction in Section 2.1

$$(f, \omega) = \sum_{\sigma \in G} < \omega.\sigma^{-1}, f > \sigma \in \mathbf{K}[G]. \tag{8}$$

These two bilinear forms turn $\hat{\mathbf{A}}$ into the dual of $\mathbf{A}$ as a $\mathbf{K}$-vector space (resp. as a $\mathbf{K}[G]$-module). In the special case when all the places P_i have degree one, the dual basis to the basis introduced before Proposition 3 is made of the $\mu_i t_i^{-k_i} dt_i/t_i$ for $1 \leqslant i \leqslant I$ and $0 \leqslant k_i < m_i$ where μ_i is the dual to the normal element θ_i in the extension $\mathbf{K}_{i,1}/\mathbf{K}$.

4. Free commutative actions

We study the situation at the beginning of Section 3 in the special case when the Galois cover $\tau : Y \to X$ is abelian and unramified. We prove that large enough equivariant Riemann-Roch spaces are free $\mathbf{K}[G]$-modules. To this end we prove in Section 4.2 that evaluation at some fibers induces an isomorphism with one of the $\mathbf{K}[G]$-modules studied in Section 3.2. We need a criterion for an equivariant divisors on Y to be non-special. We recall such a criterion in Section 4.1. We introduce in Section 4.3 the evaluation, interpolation and checking matrices whose existence follows from the freeness of the considered modules.

4.1. Special invariant divisors. The pullback by τ of a degree $g_X - 1$ divisor on X is a degree $g_Y - 1$ divisor on Y according to the Riemann-Hurwitz formula. We need a criterion for the latter divisor to be special. We will say that a divisor class is *effective* if it contains an effective divisor. When the degree of the class is the genus of the curve minus one, being effective is equivalent to being special.

PROPOSITION 4. *Assume the hypotheses at the beginning of Section 3 with τ abelian and unramified and $\mathbf{K}$ algebraically closed. Write $\mathfrak{o} = \mathfrak{o}_p \times \mathfrak{o}_{p'}$ where $\mathfrak{o}_p$ is the largest power of p dividing $\mathfrak{o}$. Let c be a divisor class of degree $g_X - 1$ on X and let $\tau^\star(c)$ be its pullback on Y. If the class $\tau^\star(c)$ is effective then c is the sum of an effective class of degree $g_X - 1$ and a class of degree 0 annihilated by $\tau^\star$ and by $\mathfrak{o}_{p'}$.*

PROOF. From [**11**, §14]. Let D be a divisor in c and let E be the pullback of D by τ. We assume that $\tau^\star(c)$ is effective. The space $H^0(Y, \mathcal{O}_Y(E))$ is non-zero and is acted on by G. Recall that a finite set of commuting endomorphisms of a finite dimensional vector space over an algebraically closed field has a common eigenvector. Let f be such an eigenvector for the action of G. The divisor of f is $J - E$ where J is effective and stable under the action of G. So there exists an effective divisor I on X such that J is the pullback of I by τ. And the class of $I - D$ is annihilated by $\tau^\star$. It is also annihilated by $\mathfrak{o}_{p'}$ because $f^{\mathfrak{o}_{p'}}$ is invariant by G. □

4.2. Riemann-Roch spaces. Let E be a divisor on Y defined over $\mathbf{K}$ and invariant by G. The Riemann-Roch space $H^0(Y, \mathcal{O}_Y(E))$ is a $\mathbf{K}[G]$-module. This module is free provided the degree of E is large enough.

THEOREM 1. *Assume the hypotheses at the beginning of Section 3 with τ abelian and unramified. Let D be a divisor on X with degree $\geqslant 2g_X - 1$. Let E be the pullback of D by τ. The $\mathbf{K}$-vector space $H^0(Y, \mathcal{O}_Y(E))$ is a free $\mathbf{K}[G]$-module of rank* $\deg(D) - g_X + 1$.

PROOF. The statement is empty if $g_X = 0$. We assume that $g_X \geqslant 1$. Because of the Noether-Deuring theorem [**8**, §2, Section 5], we can assume that $\mathbf{K}$ is algebraically closed. Let $k = \deg(D) - g_X + 1$. We note that $k \geqslant g_X$. By dimension count, there exist k points

$$P_1, P_2, \ldots, P_k \quad \text{on} \quad X$$

such that the class of $D - P_1 - P_2 - \cdots - P_k$ is not the sum of an effective class of degree $g_X - 1$ and a class annihilated by

$$\tau^\star : \mathrm{Pic}(X) \to \mathrm{Pic}(Y).$$

Indeed every divisor class of degree $g_X - 1$ contains a divisor of the form $D - P_1 - P_2 - \cdots - P_k$ because k is greater than or equal to the dimension g_X of Pic^{g_X-1}. On the other hand the set of effective classes of degree $g_X - 1$ has dimension $g_X - 1$. And the kernel of τ^* is finite. So the set of bad classes has codimension 1 in Pic^{g_X-1}.

Let P be the divisor sum of all P_i and let Q be its pullback by τ. According to Proposition 4 the class of $E - Q$ is ineffective. Thus the evaluation map

$$H^0(Y, \mathcal{O}_Y(E)) \to H^0(Y, \mathcal{O}_Y(E)/\mathcal{O}_Y(E-Q)) \simeq H^0(Q, \mathcal{O}_Q)$$

is an isomorphism of $\mathbf{K}[G]$-modules. Proposition 3 then implies that $H^0(Q, \mathcal{O}_Q)$ is a free $\mathbf{K}[G]$-module of rank k. □

When the degree of D is smaller than $2g_X$–1 it is not granted that $H^0(Y, \mathcal{O}_Y(E))$ is free. We mention two useful partial results.

Proposition 5. *Assume the hypotheses at the beginning of Section 3 with τ abelian and unramified. Assume that p does not divide $\mathfrak{o}$. Let D be a divisor on X with degree $\geqslant g_X$. Let E be the pullback of D by τ. Then $H^0(Y, \mathcal{O}_Y(E))$ contains a free $\mathbf{K}[G]$-module of rank $\deg(D) - g_X + 1$.*

Proof. The ring $\mathbf{K}[G]$ is semi-simple. Let $\mathcal{L}(E) = H^0(Y, \mathcal{O}_Y(E))$. Let m be the smallest among the multiplicities in $\mathcal{L}(E)$ of irreducible representations of G. This is the smallest among the multiplicities of multiplicative characters of G in $\mathcal{L}(E) \otimes \bar{\mathbf{K}}$ where $\bar{\mathbf{K}}$ is an algebraic closure of $\mathbf{K}$. It is clear that $\mathcal{L}(E)$ contains m copies of the regular representation of G. On the other hand let $\chi : G \to \bar{\mathbf{K}}$ be a multiplicative character. By the normal basis theorem there exists an eigenfunction r in $\bar{\mathbf{K}}(Y)$ associated with χ. The divisor of r is the pullback by τ of a divisor R on X. Let $\mathcal{L}(E)_\chi$ be the eigenspace in $\mathcal{L}(E)$ associated with χ. The map $f \mapsto f/r$ is a bijection between $\mathcal{L}(E)_\chi$ and $H^0(X, \mathcal{O}_X(D+R))$. The dimension of the latter is at least $\deg(D) - g_X + 1$. Thus $m \geqslant \deg(D) - g_X + 1$. □

We can say something also when G is a p-group and $\mathbf{K}$ a finite field.

Proposition 6. *Assume the hypotheses at the beginning of Section 3 with τ abelian and unramified. Assume that $\mathbf{K}$ is a finite field with at least four elements. Assume that $\mathfrak{o}$ is a power of p. Assume that $g_X \geqslant 2$. Let $d \geqslant g_X$ be an integer. Let $r = d - g_X + 1$. Assume that there exists an effective divisor on X with degree r and defined over $\mathbf{K}$. Then there exists a divisor D on X such that D is defined over $\mathbf{K}$, D has degree d, and $H^0(Y, \mathcal{O}_Y(E))$ is a free $\mathbf{K}[G]$-module of rank $r = d - g_X + 1$ where E is the pullback of D by τ.*

Proof. Set $r = d - g_X + 1$. Let P be an effective divisor on X with degree r and defined over $\mathbf{K}$. According to [**2**, Theorem 11] by Ballet and Le Brigand, there exists a degree $g_X - 1$ non-special divisor I defined over $\mathbf{K}$. Set $D = I + P$. Let E, J, and Q be the pullbacks of D, I, and P by τ. The class of the divisor J is ineffective according to Proposition 4. So the evaluation map $H^0(Y, \mathcal{O}_Y(E)) \to H^0(Q, \mathcal{O}_Q)$ is a bijection. And the latter is a free $\mathbf{K}[G]$-module according to Proposition 3. □

Theorem 1 translates into a similar statement for differentials.

Proposition 7. *Assume the hypotheses at the beginning of Section 3 with τ abelian and unramified. Let D be a divisor on X with $\deg D < 0$. Let E be the pullback of D by τ. The $\mathbf{K}$-vector space $H^0(Y, \Omega_{Y/\mathbf{K}}(E))$ is a free $\mathbf{K}[G]$-module of rank $g_X - 1 - \deg(D)$.*

PROOF. The statement is trivial if $g_X = 0$. We assume that $g_X \geqslant 1$. Let ω_0 be a non-zero holomorphic differential on X. The pullback of ω_0 on Y by τ is denoted by ω_0 also. The map $\omega \mapsto \omega/\omega_0$ is an isomorphism of $\mathbf{K}$-vector spaces between $H^0(Y, \Omega_{Y/\mathbf{K}}(E))$ and $H^0(Y, \mathcal{O}_Y((\omega_0)-E))$. According to Equation (6) this isomorphism is compatible with the actions of G on either sides given by Equations (4) and (5). Since the degree of $(\omega_0) - D$ is at least $2g_X - 1$ we can apply Theorem 1 to prove that $H^0(Y, \mathcal{O}_Y((\omega_0) - E))$ is free and deduce that $H^0(Y, \Omega_{Y/\mathbf{K}}(E))$ is free as well. □

4.3. The orthogonal submodule. In the situation of the beginning of Section 3 and assuming that τ is abelian and unramified we let D and P be divisors on X with P effective. We assume that D and P are disjoint. We assume that

$$2g_X - 1 \leqslant \deg(D) \leqslant \deg(P) - 1. \tag{9}$$

Let E be the pullback of D by τ and let Q be the pullback of P. We write

$$\mathcal{L}(E) = H^0(Y, \mathcal{O}_Y(E)) \quad \text{and} \quad \Omega(-Q+E) = H^0(Y, \Omega_{Y/\mathbf{K}}(-Q+E)).$$

Theorem 1, Proposition 7, and Equation (9) imply that these two $\mathbf{K}[G]$-modules are free. And the evaluation maps

$$\mathcal{L}(E) \longrightarrow \mathbf{A} \quad \text{and} \quad \Omega(-Q+E) \longrightarrow \hat{\mathbf{A}} \quad \text{are injective.}$$

So $\mathcal{L}(E)$ can be seen as a free submodule of $\mathbf{A}$ and $\Omega(-Q+E)$ as a free submodule of $\hat{\mathbf{A}}$. For dimension reasons and due to the residue theorem, these two $\mathbf{K}[G]$-modules are orthogonal to each other for the form introduced in Equation (8). Proposition 2 implies that $\mathcal{L}(E)$ has a complementary submodule in $\mathbf{A}$ that is isomorphic to the dual of $\Omega(-Q+E)$ and is thus a free submodule. Similarly $\Omega(-Q+E)$ has a free complementary submodule in $\hat{\mathbf{A}}$ that is isomorphic to the dual of $\mathcal{L}(E)$.

In the special case when all the places P_i have degree one, we have introduced a natural basis for $\mathbf{A}$ before Proposition 3 and its dual basis $\hat{\mathbf{A}}$ in Section 3.3, using Taylor expansions at the places above the P_i.

We choose $\mathbf{K}[G]$-bases for $\mathcal{L}(E)$ and $\Omega(-Q+E)$. We denote by $\mathcal{E}_E$ the $\deg(P) \times (\deg(D) - g_X + 1)$ matrix with coefficients in $\mathbf{K}[G]$ of the evaluation map $\mathcal{L}(E) \to \mathbf{A}$ in the chosen bases. We denote by $\mathcal{C}_E$ the $\deg(P) \times (\deg(P) - \deg(D) + g_X - 1)$ matrix of the map $\Omega(-Q+E) \to \hat{\mathbf{A}}$ in the chosen bases. The matrix $\mathcal{C}_E$ checks that a vector in $\mathbf{A}$ belongs to $\mathcal{L}(E)$. Its left kernel is the image of $\mathcal{E}_E$. So

$$\mathcal{C}_E^t \times \mathcal{E}_E = 0,$$

a zero $(\deg(P) - \deg(D) + g_X - 1) \times (\deg(D) - g_X + 1)$ matrix with entries in $\mathbf{K}[G]$.

We choose a $\mathbf{K}[G]$-linear projection $\mathbf{A} \to \mathcal{L}(E)$ and denote by $\mathcal{I}_E$ the matrix of this projection. This is a $(\deg(D) - g_X + 1) \times \deg(P)$ matrix with coefficients in $\mathbf{K}[G]$. This is an interpolation matrix since it recovers a function in $\mathcal{L}(E)$ from its evaluation at Q. Equivalently

$$\mathcal{I}_E \times \mathcal{E}_E = 1$$

the $(\deg(D) - g_X + 1) \times (\deg(D) - g_X + 1)$ identity matrix with coefficients in $\mathbf{K}[G]$. We note that applying either of the matrices $\mathcal{E}_E$, $\mathcal{C}_E$, $\mathcal{I}_E$ requires at most a constant times $\deg(P)^2$ operations in $\mathbf{K}[G]$.

5. Padé approximants

In the situation of the beginning of Section 3 and assuming that τ is abelian and unramified we let D_0, D_1 and P be divisors on X with P effective. We assume that D_0 and D_1 are disjoint from P. Let E_0, E_1, and Q be the pullbacks of D_0, D_1, and P by τ. We assume that

$$2g_X - 1 \leqslant \deg(D_1) \leqslant \deg(P) - 1, \tag{10}$$

$$g_X \leqslant \deg(D_0) \leqslant \deg(P) - 1. \tag{11}$$

Equation (10) implies that the $\mathbf{K}[G]$-modules $\mathcal{L}(E_1)$ and $\Omega(-Q + E_1)$ are free and the evaluation maps into $\mathbf{A}$ and $\hat{\mathbf{A}}$ are injective. We assume that $\mathcal{L}(E_0)$ contains a free $\mathbf{K}[G]$-module of rank $\deg(D_0) - g_X + 1$ and denote by $\mathcal{L}(E_0)_{\mathbf{fr}}$ such a submodule.

Given r in $\mathbf{A}$, $a_0 \neq 0$ in $\mathcal{L}(E_0)$ and a_1 in $\mathcal{L}(E_1)$ such that

$$a_0 r - a_1 = 0 \in \mathbf{A},$$

we say that (a_0, a_1) is a **Padé approximant** of r and we say that a_0 is a **denominator** for r. Denominators for r are non-zero a_0 in $\mathcal{L}(E_0) \subset \mathbf{A}$ such that

$$a_0 r \in \mathcal{L}(E_1).$$

Equivalently

$$(a_0 r, \omega) = 0 \quad \text{for every} \quad \omega \in \Omega(-Q + E_1). \tag{12}$$

Denominators are thus non-zero solutions of a $\mathbf{K}$-linear system of equations. We note that this is not a $\mathbf{K}[G]$-linear system in general because multiplication by r is not $\mathbf{K}[G]$-linear. In Section 5.1 we show that one can be a bit more explicit in some cases. We consider the problem of computing Padé approximants in Section 5.2.

5.1. The split case.

Assume that $P = P_1 + \cdots + P_n$ is a sum of n pairwise distinct rational points over $\mathbf{K}$. Assume that the fiber of τ above each P_i decomposes as a sum of $\mathfrak{o}$ rational points over $\mathbf{K}$. We choose a point $Q_{i,1}$ above each P_i and set

$$Q_{i,\sigma} = \sigma(Q_{i,1}) \quad \text{for every} \quad \sigma \in G.$$

For every $1 \leqslant i \leqslant n$ let α_i be the function in $\mathbf{A}$ that takes value 1 at $Q_{i,1}$ and zero everywhere else. We thus form a basis

$$\mathcal{A}_G = (\alpha_i)_{1 \leqslant i \leqslant n}$$

of $\mathbf{A}$ over $\mathbf{K}[G]$. We note $\hat{\mathcal{A}}_G$ its dual basis. For every $1 \leqslant i \leqslant n$ and $\sigma \in G$ let

$$\alpha_{i,\sigma} = \sigma.\alpha_i = \alpha_i \circ \sigma^{-1}$$

be the function in $\mathbf{A}$ that takes value 1 at $Q_{i,\sigma}$ and zero everywhere else. We thus form a basis

$$\mathcal{A}_{\mathbf{K}} = (\alpha_{i,\sigma})_{1 \leqslant i \leqslant n,\, \sigma \in G}$$

of $\mathbf{A}$ over $\mathbf{K}$. The coordinates of r in the $\mathbf{K}[G]$-basis $\mathcal{A}_G$ are

$$r_G = (\sum_{\sigma \in G} r(Q_{i,\sigma})\sigma)_{1 \leqslant i \leqslant n}$$

and the coordinates of $r \in \mathbf{A}$ in the $\mathbf{K}$-basis $\mathcal{A}_{\mathbf{K}}$ are

$$r_{\mathbf{K}} = (r(Q_{i,\sigma}))_{1 \leqslant i \leqslant n,\, \sigma \in G}.$$

Multiplication by r is a $\mathbf{K}$-linear map from $\mathbf{A}$ to $\mathbf{A}$. Let

$$\mathcal{R}_{\mathbf{K}} \in \mathcal{M}_{\mathfrak{o}.n,\mathfrak{o}.n}(\mathbf{K})$$

be the $\mathfrak{o}.n \times \mathfrak{o}.n$ diagonal matrix of this map in the basis $\mathcal{A}_{\mathbf{K}}$.

We choose a $\mathbf{K}[G]$-basis $\mathcal{Z}_G$ for $\mathcal{L}(E_0)_{\mathbf{fr}}$ and denote by $\mathcal{E}_G^0$ the $\deg(P) \times (\deg(D_0) - g_X + 1)$ matrix of the $\mathbf{K}[G]$-linear injective map

$$\mathcal{L}(E_0)_{\mathbf{fr}} \to \mathbf{A} \tag{13}$$

in the bases $\mathcal{Z}_G$ and $\mathcal{A}_G$. We denote by $\mathcal{Z}_{\mathbf{K}}$ the $\mathbf{K}$-basis of $\mathcal{L}(E_0)_{\mathbf{fr}}$ obtained by letting G act on $\mathcal{Z}_G$. Let $\mathcal{E}_{\mathbf{K}}^0$ be the matrix of the map (13) in the bases $\mathcal{Z}_{\mathbf{K}}$ and $\mathcal{A}_{\mathbf{K}}$. The matrix $\mathcal{E}_{\mathbf{K}}^0$ is obtained from $\mathcal{E}_G^0$ by replacing each $\mathbf{K}[G]$ entry by the corresponding $\mathfrak{o} \times \mathfrak{o}$ circulant-like matrix with entries in $\mathbf{K}$.

Let $\hat{\mathcal{A}}_G$ be the basis of the $\mathbf{K}[G]$-module $\hat{\mathbf{A}}$, dual to $\mathcal{A}_G$. We choose a $\mathbf{K}[G]$-basis $\mathcal{U}_G$ for $\Omega(-Q + E_1)$ and denote by $\mathcal{C}_G^1$ the matrix of the injective map

$$\Omega(-Q + E_1) \to \hat{\mathbf{A}} \tag{14}$$

in the bases $\mathcal{U}_G$ and $\hat{\mathcal{A}}_G$. This is a $\deg(P) \times (\deg(P) - \deg(D_1) + g_X - 1)$ matrix with entries in $\mathbf{K}[G]$. Let $\mathcal{U}_{\mathbf{K}}$ be the $\mathbf{K}$-basis of $\Omega(-Q + E_1)$ obtained by letting G act on $\mathcal{U}_G$. Let $\hat{\mathcal{A}}_{\mathbf{K}}$ be the basis of the $\mathbf{K}$-vector space $\hat{\mathbf{A}}$, dual to $\mathcal{A}_{\mathbf{K}}$. The matrix of the map (14) in the bases $\mathcal{U}_{\mathbf{K}}$ and $\hat{\mathcal{A}}_{\mathbf{K}}$ is called $\mathcal{C}_{\mathbf{K}}^1$.

Let a_0 in $\mathcal{L}(E_0)_{\mathbf{fr}}$ and let x_G be the coordinates of a_0 in the $\mathbf{K}[G]$-basis $\mathcal{Z}_G$. This is a column of height $\deg(D_0) - g_X + 1$. We let $x_{\mathbf{K}}$ be the coordinates of a_0 in the $\mathbf{K}$-basis $\mathcal{Z}_{\mathbf{K}}$. This is a column of height $\mathfrak{o}.(\deg(D_0) - g_X + 1)$ obtained from x_G by replacing each entry by its $\mathfrak{o}$ coefficients in the canonical basis of $\mathbf{K}[G]$. We deduce from Equation (12) that a_0 is a denominator for r if and only if $x_{\mathbf{K}}$ is in the kernel of the matrix

$$\mathcal{D}_r = (\mathcal{C}_{\mathbf{K}}^1)^t \times \mathcal{R}_{\mathbf{K}} \times \mathcal{E}_{\mathbf{K}}^0 \in \mathcal{M}_{\mathfrak{o}.(\deg P - deg D_1 + g_X - 1) \times \mathfrak{o}.(\deg D_0 - g_X + 1)}(\mathbf{K}).$$

PROPOSITION 8. *Assume that we are in the context of the beginning of Section 5. In particular assume Equations (10) and (11), assume that P is a sum of n pairwise distinct $\mathbf{K}$-rational points, and that the n corresponding fibers of τ split over $\mathbf{K}$. Assume that we are given the matrices $\mathcal{E}_{\mathbf{K}}^0$ and $\mathcal{C}_{\mathbf{K}}^1$. On input an $r = (r(Q_{i,\sigma}))_{1 \leqslant i \leqslant n,\, \sigma \in G}$ in $\mathbf{A}$ and some a_0 in $\mathcal{L}(E_0)_{\mathbf{fr}}$, given by its coordinates $x_{\mathbf{K}}$ in the basis $\mathcal{Z}_{\mathbf{K}}$, one can check if $a_0 r \in \mathcal{L}(E_1)$ at the expense of $\mathcal{Q}.n^2$ operations in $\mathbf{K}[G]$ (addition, multiplication) and $\mathcal{Q}.\mathfrak{o}.n$ operations in $\mathbf{K}$ (addition, multiplication) where $\mathcal{Q}$ is some absolute constant.*

PROOF. We first multiply $x_{\mathbf{K}}$ by $\mathcal{E}_{\mathbf{K}}^0$ or rather x_G by $\mathcal{E}_G^0$. This requires less than $2\deg(P) \times (\deg(D_0) - g_X + 1)$ operations in $\mathbf{K}[G]$. We then multiply the result by $\mathcal{R}_{\mathbf{K}}$. This requires less than $\mathfrak{o}.\deg(P)$ operations in $\mathbf{K}$ because $\mathcal{R}_{\mathbf{K}}$ is diagonal. We finally multiply the result by $(\mathcal{C}_{\mathbf{K}}^1)^t$ or rather $(\mathcal{C}_G^1)^t$. This requires less than $2\deg(P) \times (\deg(P) - \deg(D_1) + g_X - 1)$ operations in $\mathbf{K}[G]$. □

5.2. Computing Padé approximants. According to Proposition 8 one can efficiently check a denominator. As a consequence, one can find a random denominator, assuming that there is some in $\mathcal{L}(E_0)_{\mathbf{fr}}$, using an iterative method as in [**25, 53**]. Recall that an $\ell \times n$ **black box** matrix A with coefficients in a field $\mathbf{K}$ is an oracle that on input an $n \times 1$ vector x returns Ax.

PROPOSITION 9 (Wiedemann, Kaltofen, Saunders). *There exists a probabilistic (Las Vegas) algorithm that takes as input an* $\ell \times n$ *black box matrix* A *and an* $\ell \times 1$ *vector* b *with entries in a finite field* $\mathbf{K}$ *and returns a uniformly distributed random solution* x *to the system* $Ax = b$*, if there is some, with probability of success* $\geqslant 1/2$ *at the expense of* $\mathcal{Q}.m.\log m$ *calls to the black box for* A *and* $\mathcal{Q}.m^2.(\log(m))^2$ *operations in* $\mathbf{K}$ *(addition, multiplication, inversion, picking a random element) where* $\mathcal{Q}$ *is some absolute constant and* $m = \max(\ell, n)$.

Using Proposition 9 and bounding the cost of a call to the black box with the help of Proposition 8 we deduce

PROPOSITION 10. *Under the hypotheses of Proposition 8 and on input a vector* $r = (r(Q_{i,j})_{i,j})$ *in* $\mathbf{A}$ *one can find a uniformly distributed random denominator for* r*, if there is some in* $\mathcal{L}(E_0)_{\mathbf{fr}}$*, with probability of success* $\geqslant 1/2$*, at the expense of* $\mathcal{Q}.\mathfrak{o}.n^3.\log(\mathfrak{o}.n)$ *operations in* $\mathbf{K}[G]$ *(addition, multiplication) and* $\mathcal{Q}.(\mathfrak{o}.n.\log(\mathfrak{o}.n))^2$ *operations in* $\mathbf{K}$ *(addition, multiplication, inversion, picking a random element) where* $\mathcal{Q}$ *is some absolute constant.*

Once we have found a denominator a_0 for r we set $a_1 = ra_0$ and recover the coordinates of a_1 applying the interpolation matrix associated to E_1.

6. Computing in the group algebra

Given a finite commutative group G and a finite field $\mathbf{K}$ we will need efficient algorithms to multiply in $\mathbf{K}[G]$. This is classically achieved using a discrete Fourier transform when G is cyclic and $\mathbf{K}$ contains enough roots of unity. The complexity analysis requires some care in general. This is the purpose of this section. We recall in Section 6.1 the definition of the Fourier transform in the setting of commutative finite groups. The most classical case of cyclic groups is studied in Section 6.2 from an algorithmic point of view. The general case follows by induction as explained in Section 6.3. The complexity of the resulting multiplication algorithm in $\mathbf{K}[G]$ is bounded in Section 6.4.

6.1. Fourier transforms. Let G be a finite commutative group. Let $\mathfrak{o}$ be the order of G. Let e be its exponent. Let $\mathbf{K}$ be a commutative field containing a primitive e-th root of unity. In particular e and $\mathfrak{o}$ are non-zero in $\mathbf{K}$. Let $\hat{G}$ be the dual of G defined as the group of characters $\chi : G \to \mathbf{K}^*$. We define a map from the group algebra of G to the algebra of functions on G

$$\top \quad : \quad \begin{array}{ccc} \mathbf{K}[G] & \longrightarrow & \mathrm{Hom}_{\mathbf{set}}(G, \mathbf{K}) \\ \sum_{\sigma \in G} a_\sigma \sigma & \longmapsto & \sigma \mapsto a_\sigma \end{array}$$

This is an isomorphism of $\mathbf{K}$-vector spaces. We let $\perp : \mathrm{Hom}_{\mathbf{set}}(G, \mathbf{K}) \to \mathbf{K}[G]$ be the reciprocal map. We dually define

$$\hat{\top} \quad : \quad \begin{array}{ccc} \mathbf{K}[\hat{G}] & \longrightarrow & \mathrm{Hom}_{\mathbf{set}}(\hat{G}, \mathbf{K}) \\ \sum_{\chi \in \hat{G}} a_\chi \chi & \longmapsto & \chi \mapsto a_\chi \end{array}$$

and its reciprocal map $\hat{\perp}$. We let

$$\iota_G : \mathbf{K}[G] \to \mathbf{K}[G]$$

be the $\mathbf{K}$-linear involution that maps σ onto σ^{-1}. We define the Fourier transform

$$\begin{array}{rcrcl} \mathrm{FT}_G & : & \mathbf{K}[G] & \longrightarrow & \mathrm{Hom}_{\mathbf{set}}(\hat{G}, \mathbf{K}) \\ & & \sum_{\sigma \in G} a_\sigma \sigma & \longmapsto & \chi \mapsto \sum_\sigma a_\sigma \chi(\sigma) \end{array}$$

The Fourier transform evaluates an element in the group algebra at every character. The Fourier transform of the dual group

$$\begin{array}{rcrcl} \mathrm{FT}_{\hat{G}} & : & \mathbf{K}[\hat{G}] & \longrightarrow & \mathrm{Hom}_{\mathbf{set}}(G, \mathbf{K}) \\ & & \sum_{\chi \in \hat{G}} a_\chi \chi & \longmapsto & \sigma \mapsto \sum_\chi a_\chi \chi(\sigma) \end{array}$$

provides an inverse for FT_G in the sense that

$$\perp \circ \mathrm{FT}_{\hat{G}} \circ \hat{\perp} \circ \mathrm{FT}_G = \mathfrak{o}.\iota_G$$

is the $\mathbf{K}$-linear invertible map that sends σ to $\mathfrak{o}.\sigma^{-1}$.

Let M be a finite dimensional $\mathbf{K}$-vector space. We set

$$M[G] = M \otimes_{\mathbf{K}} \mathbf{K}[G]$$

and note that

$$\mathrm{Hom}_{\mathbf{set}}(\hat{G}, M) = M \otimes_{\mathbf{K}} \mathrm{Hom}_{\mathbf{set}}(\hat{G}, \mathbf{K}).$$

We define a Fourier transform on M

$$\begin{array}{rcrcl} \mathrm{FT}_M & : & M[G] & \longrightarrow & \mathrm{Hom}_{\mathbf{set}}(\hat{G}, M) \\ & & \sum_{\sigma \in G} m_\sigma \otimes \sigma & \longmapsto & \chi \mapsto \sum_\sigma \chi(\sigma) m_\sigma \end{array}$$

It turns a free $\mathbf{K}[G]$-module into a free $\mathrm{Hom}_{\mathbf{set}}(\hat{G}, \mathbf{K})$-module.

6.2. Univariate Fourier transforms. We assume in this section that the group G is cyclic of order $\mathfrak{o}$. We choose a primitive $\mathfrak{o}$-th root of unity ω in $\mathbf{K}$. We choose a generator in $\hat{G}$ and deduce the following identifications

$$\mathrm{Hom}_{\mathbf{set}}(\hat{G}, \mathbf{K}) = \mathbf{K}^{\mathfrak{o}} \quad \text{and} \quad \mathbf{K}[G] = \mathbf{K}[x]/(x^{\mathfrak{o}} - 1).$$

Let M be a finite dimensional $\mathbf{K}$-vector space. Setting

$$M[x] = M \otimes_{\mathbf{K}} \mathbf{K}[x] \quad \text{and} \quad M[G] = M \otimes_{\mathbf{K}} \mathbf{K}[x]/(x^{\mathfrak{o}} - 1).$$

the Fourier transform is

$$\begin{array}{rcrcl} \mathrm{FT}_M & : & M[G] & \longrightarrow & M^{\mathfrak{o}} \\ & & m & \longmapsto & (m(1), m(\omega), m(\omega^2), \dots, m(\omega^{\mathfrak{o}-1})) \end{array}$$

Given m in $M[G] = M \otimes_{\mathbf{K}} \mathbf{K}[x]/(x^{\mathfrak{o}} - 1)$ the computation of $\mathrm{FT}_M(m)$ reduces to the multiplication of a polynomial of degree $2\mathfrak{o} - 2$ in $\mathbf{K}[x]$ and a vector of degree $\mathfrak{o} - 1$ in $M[x]$ using formulae by Rabiner, Schafer, Rader, and Bluestein [**6, 37**].

PROPOSITION 11. *Let $\mathbf{K}$ be a commutative field. Let M be a finite dimensional $\mathbf{K}$-vector space. Let $\mathfrak{o} \geqslant 2$ be an integer. Assume that $\mathbf{K}$ contains a primitive $\mathfrak{o}$-th root of unity ω and a primitive root of unity of order a power of two that is bigger than $3\mathfrak{o} - 3$. Let*

$$m = m_0 \otimes 1 + m_1 \otimes x + \cdots + m_{\mathfrak{o}-1} \otimes x^{\mathfrak{o}-1} \bmod x^{\mathfrak{o}} - 1 \in M \otimes_{\mathbf{K}} \mathbf{K}[x]/(x^{\mathfrak{o}} - 1).$$

One can compute $\mathrm{FT}_M(m)$ *at the expense of* $\mathcal{Q}.\mathfrak{o}.\log\mathfrak{o}$ *additions, multiplications and inversions in* $\mathbf{K}$*, additions and scalar multiplications in* M*, where* $\mathcal{Q}$ *is an absolute constant.*

PROOF. We adapt the notation from [**7**, I.5.4, Proposition 5.10]. For every $0 \leqslant i \leqslant 2\mathfrak{o}-2$ let

$$t_i = i(i-1)/2 \quad \text{and} \quad \beta_i = \omega^{t_i}.$$

We note that

$$t_{i+1} = t_i + i \quad \text{and} \quad \beta_{i+1} = \beta_i \omega^i.$$

So one can compute the β_i for $0 \leqslant i \leqslant 2\mathfrak{o}-2$ at the expense of $4\mathfrak{o}$ operations in $\mathbf{K}$. We then compute the inverse of every β_i. For every $0 \leqslant i \leqslant \mathfrak{o}-1$ let

$$n_i = \beta_i^{-1} m_i.$$

These can be computed at the expense of $\mathfrak{o}$ scalar multiplications in M. Let

$$n(x) = n_{\mathfrak{o}-1} + n_{\mathfrak{o}-2} \otimes x + \cdots + n_0 \otimes x^{\mathfrak{o}-1} \in M[x]$$

and let

$$b(x) = \beta_0 + \beta_1 x + \cdots + \beta_{2\mathfrak{o}-2} x^{2\mathfrak{o}-2} \in \mathbf{K}[x].$$

Let

$$r(x) = b(x).n(x) = \sum_{0 \leqslant i \leqslant 3\mathfrak{o}-3} r_i \otimes x^i \in M[x].$$

From the identity

$$t_{i+j} = t_i + t_j + ij$$

we deduce

$$\omega^{ij}\beta_i\beta_j = \beta_{i+j} \quad \text{for} \quad 0 \leqslant i, j \leqslant \mathfrak{o}-1$$

and

$$\sum_{j=0}^{\mathfrak{o}-1} \omega^{ij} m_j = \beta_i^{-1} \sum_{j=0}^{\mathfrak{o}-1} \beta_{i+j} n_j.$$

We deduce that $\mathrm{FT}_M(m) = (\beta_0^{-1} r_{\mathfrak{o}-1}, \beta_1^{-1} r_{\mathfrak{o}}, \beta_2^{-1} r_{\mathfrak{o}+1}, \ldots, \beta_{\mathfrak{o}-1}^{-1} r_{2\mathfrak{o}-2})$. Since $\mathbf{K}$ contains a primitive root of unity of order a power of two that is bigger than $3\mathfrak{o}-3$, the coefficients in the product $r(x) = b(x).n(x)$ can be computed at the expense of $\mathcal{Q}.\mathfrak{o}.\log\mathfrak{o}$ operations in $\mathbf{K}$, additions in M and products of a vector in M by a scalar in $\mathbf{K}$. See [**7**, I.2.4, Algorithme 2.3]. □

6.3. Multivariate Fourier transforms. Let $(\mathfrak{o}_i)_{1 \leqslant i \leqslant I}$ be integers such that $2 \leqslant \mathfrak{o}_1 | \mathfrak{o}_2 | \ldots | \mathfrak{o}_I$. Let $C_i = \mathbf{Z}/\mathfrak{o}_i\mathbf{Z}$ and $G = \Pi_{1 \leqslant i \leqslant I} C_i$. For $1 \leqslant i \leqslant I$ set

$$A_i = \mathbf{K}[C_i] \quad \text{and} \quad B_i = \mathrm{Hom}_{\mathbf{set}}(\hat{C}_i, \mathbf{K}).$$

For $0 \leqslant i \leqslant I$ set

$$M_i = \bigotimes_{j \leqslant i} B_j \otimes \bigotimes_{j > i} A_j.$$

So $M_0 = \mathbf{K}[G]$ and $M_I = \mathrm{Hom}_{\mathbf{set}}(\hat{G}, \mathbf{K})$. For $0 \leqslant i \leqslant I-1$ write

$$M_i = \bigotimes_{j \leqslant i} B_j \otimes \mathbf{K}[C_{i+1}] \otimes \bigotimes_{j > i+1} A_j$$

as a $\mathbf{K}[C_{i+1}]$-module and let

$$F_i : M_i \to M_{i+1}$$

be the corresponding Fourier transform as defined in Section 6.2. We check that

$$\mathrm{FT}_G = F_{I-1} \circ F_{I-2} \circ \cdots \circ F_0.$$

Using Proposition 11 we deduce

PROPOSITION 12. *Let $(\mathfrak{o}_i)_{1\leqslant i\leqslant I}$ be integers such that $2 \leqslant \mathfrak{o}_1|\mathfrak{o}_2|\dots|\mathfrak{o}_I$. Let $G = \prod_{1\leqslant i\leqslant I}(\mathbf{Z}/\mathfrak{o}_i\mathbf{Z})$. Let $\mathfrak{o}$ be the order of G. Let $e = \mathfrak{o}_I$ be the exponent of G. Let $\mathbf{K}$ be a commutative field containing a primitive root of unity of order e and a primitive root of unity of order a power of two that is bigger than $3e-3$. Given an element $a = \sum_{\sigma\in G} a_\sigma \sigma$ in $\mathbf{K}[G]$ one can compute $\mathrm{FT}_G(a)$ in $\mathrm{Hom}_{\mathbf{set}}(\hat{G}, \mathbf{K})$ at the expense of $\mathcal{Q}.\mathfrak{o}.\log \mathfrak{o}$ additions, multiplications and inversions in $\mathbf{K}$. Here $\mathcal{Q}$ is some absolute constant.*

6.4. Fast multiplication in $\mathbf{K}[G]$. Let G, $\mathfrak{o}$, e be as in Section 6.3. Let $\mathbf{K}$ be a commutative field. In this section we study the algorithmic complexity of computing the product of two given elements

$$a = \sum_{\sigma\in G} a_\sigma \sigma \quad \text{and} \quad b = \sum_{\sigma \in G} b_\sigma \sigma \quad \text{in} \quad \mathbf{K}[G]. \tag{15}$$

It will depend on the field $\mathbf{K}$. We first treat the case when $\mathbf{K}$ has enough roots of unity.

PROPOSITION 13. *In the context of the beginning of Section 6.4 assume that $\mathbf{K}$ contains a primitive root of unity of order e and a primitive root of unity of order a power of two that is bigger than $3e-3$. One can compute the product $ab \in \mathbf{K}[G]$ at the expense of $\mathcal{Q}.\mathfrak{o}.\log \mathfrak{o}$ operations in $\mathbf{K}$ where $\mathcal{Q}$ is some absolute constant.*

PROOF. We compute $A = \mathrm{FT}_G(a)$ and $B = \mathrm{FT}_G(b)$ as in Section 6.3. We then compute $C = AB$ in $\mathrm{Hom}_{\mathbf{set}}(\hat{G}, \mathbf{K}^*)$ at the expense of $\mathfrak{o}$ multiplications in $\mathbf{K}$. We then deduce $c = ab$ applying FT_G^{-1} to C. The cost of this computation is bounded using Proposition 12. □

We now consider the case when $\mathbf{K}$ is $\mathbf{Z}/p\mathbf{Z}$ where p is a prime integer. We miss roots of unity in $\mathbf{K}$ in general. So we transport the problem into another ring using non-algebraic maps. We let t be the smallest power of 2 that is bigger than $3e-3$. Let p' be the smallest prime integer congruent to 1 modulo $\mathfrak{o}.(p-1)^2.t$. We set $\mathbf{K}' = \mathbf{Z}/p'\mathbf{Z}$ and note that $\mathbf{K}'$ contains a primitive root of unity of order e and a primitive root of order a power of two bigger than $3e-3$. Also

$$p' > \mathfrak{o}.(p-1)^2.$$

By a result of Heath-Brown, the exponent in Linnik's theorem for primes in arithmetic progressions can be taken to be 11/2. See [**20**] and the recent improvement [**54**]. We deduce that there exists an absolute constant $\mathcal{Q}$ such that

$$p' \leqslant \mathcal{Q}(\mathfrak{o}.p)^{11}.$$

For c a congruence class in $\mathbf{K} = \mathbf{Z}/p\mathbf{Z}$ we denote by $\ell(c)$ the lift of c, that is the unique integer in the intersection of c with the interval $[0, p[$. We write

$$\uparrow(c) = \ell(c) \bmod p'. \tag{16}$$

We thus define maps $\ell : \mathbf{K} \to \mathbf{Z}$ and $\uparrow : \mathbf{K} \to \mathbf{K}'$. We similarly define the lifting map $\ell' : \mathbf{K}' \to \mathbf{Z}$ and $\downarrow : \mathbf{K}' \to \mathbf{K}$ by

$$\downarrow(c) = \ell'(c) \bmod p \quad \text{for} \quad c \in \mathbf{K}'. \tag{17}$$

These four maps can be extended to the corresponding group algebras by coefficientwise application. Given a and b as in Equation (15) we define

$$A = \ell(a) = \sum_{\sigma \in G} \ell(a_\sigma)\sigma \quad \text{and} \quad B = \ell(b) = \sum_{\sigma \in G} \ell(b_\sigma)\sigma \quad \text{in } \mathbf{Z}[G] \quad \text{and} \quad C = AB.$$

The coefficients in C belong to the interval $[0, \mathfrak{o}.(p-1)^2]$. So

$$C = \ell'((A \mod p') \times (B \mod p')) \quad \text{and} \quad ab = \downarrow(\uparrow(a)\uparrow(b)).$$

Using Proposition 13 we deduce

PROPOSITION 14. *There exists an absolute constant $\mathcal{Q}$ such that the following is true. Let G, $\mathfrak{o}$, e be as in Section 6.3. Let $\mathbf{K} = \mathbf{Z}/p\mathbf{Z}$ be a prime field. There exists a prime integer $p' \leqslant \mathcal{Q}(\mathfrak{o}.p)^{11}$ and a straight-line program of length smaller than $\mathcal{Q}.\mathfrak{o}.\log \mathfrak{o}$ that computes the product $c = \sum_g c_g[g]$ of two elements $a = \sum_g a_g[g]$ and $b = \sum_g b_g[g]$ in $\mathbf{K}[G]$ given by their coefficients $(a_g)_g$ and $(b_g)_g$. The operations in this straight-line program are additions and multiplications in $\mathbf{Z}/p'\mathbf{Z}$ and evaluations of the maps $\uparrow$ and $\downarrow$ defined in Equations* (16) *and* (17).

Now let $\mathbf{L}$ be a field extension of degree d of $\mathbf{K} = \mathbf{Z}/p\mathbf{Z}$. We assume that elements in $\mathbf{L}$ are represented by their coordinates in some $\mathbf{K}$-basis of $\mathbf{L}$. The bilinear part of one multiplication in $\mathbf{L}[G]$ reduces to $\mu_p(d)$ multiplications in $\mathbf{K}[G]$ where $\mu_p(d)$ is the $\mathbf{K}$-bilinear complexity of multiplication in $\mathbf{L}$. Work by Chudnovsky [**10**], Shparlinski, Tsfasmann, Vladut [**46**], Shokrollahi [**45**], Ballet and Rolland [**3**, **4**], Chaumine [**9**], Randriambololona [**38**] and others imply that $\mu_p(d)$ is bounded by an absolute constant times d. We deduce the following theorem.

THEOREM 2. *There exists an absolute constant $\mathcal{Q}$ such that the following is true. Let G be a finite commutative group of order $\mathfrak{o}$ and exponent e. Let $\mathbf{K} = \mathbf{Z}/p\mathbf{Z}$ and $\mathbf{L}$ a field extension of degree d of $\mathbf{K}$. There exists a prime integer $p' \leqslant \mathcal{Q}(\mathfrak{o}.p)^{11}$ and a straight-line program of length $\leqslant \mathcal{Q}(d.\mathfrak{o}.\log \mathfrak{o} + d^2.\mathfrak{o})$ that computes the product $c = \sum_g c_g[g]$ of two elements $a = \sum_g a_g[g]$ and $b = \sum_g b_g[g]$ in $\mathbf{L}[G]$ given by their coefficients $(a_g)_g$ and $(b_g)_g$. The operations in this straight-line program are additions and multiplications in $\mathbf{Z}/p\mathbf{Z}$ and in $\mathbf{Z}/p'\mathbf{Z}$ and evaluations of the maps $\uparrow$ and $\downarrow$ defined in Equations* (16) *and* (17).

REMARK 1. The $d^2.\mathfrak{o}$ summand in the complexity comes from the linear part in the Chudnovsky algorithm for multiplication in finite field extensions.

7. Constructing functions in the Hilbert class field

We have defined in Section 4 matrices $\mathcal{E}$, $\mathcal{C}$ and $\mathcal{I}$ for the evaluation and interpolation of global sections of a G-equivariant invertible sheaf on a curve Y acted on freely by a commutative group G. We have seen in Sections 4, 5, and 6 how to efficiently compute with these matrices. In this section we consider the problem of computing these matrices.

We recall in Section 7.1 the necessary background from class field theory of function fields over a finite field. We illustrate the constructive aspects of class fields on a small example in section 7.2. An important feature of this method is that we only work with divisors and functions on X, the quotient of Y by G. This is of some importance since in the applications presented in Sections 8 and 9 the genus of Y is much larger (e.g. exponentially) than the genus of X.

7.1. Class field theory and the jacobian variety. Let X be a projective, smooth, absolutely integral curve over a finite field $\mathbf{K}$ of characteristic p. Let $\bar{\mathbf{K}}$ be an algebraic closure of $\mathbf{K}$. We need an abelian unramified cover $\tau : Y \to X$ over $\mathbf{K}$, with Y absolutely integral. We will require that Y has a $\mathbf{K}$-rational point Q_1. This implies that τ is completely split above $P_1 = \tau(Q_1)$.

According to class field theory [**39, 43**] there is a maximal abelian unramified cover of X over $\mathbf{K}$ that splits totally above P_1. We briefly recall its geometric construction. Let J_X be the jacobian variety of X and let

$$j_X : X \to J_X$$

be the Jacobi map with origin P_1. Let

$$F_{\mathbf{K}} : J_X \to J_X$$

be the Frobenius endomorphism of degree $|\mathbf{K}|$, the cardinality of $\mathbf{K}$. The endomorphism

$$\wp = F_{\mathbf{K}} - 1 : J_X \to J_X$$

is an unramified Galois cover between $\mathbf{K}$-varieties with Galois group $J_X(\mathbf{K})$. We denote by

$$\tau_{\max} : Y_{\max} \to X$$

the pullback of $\wp$ along j_X. This is the maximal abelian unramified cover of X that splits totally above P_1. Any such cover $\tau : Y \to X$ is thus a quotient of $\tau_{\max}$ by some subgroup H of $J_X(\mathbf{K})$. We set $G = J_X(\mathbf{K})/H$ and notice that G is at the same time the fiber of τ above P_1 and its Galois group, acting by translations in J_X/H.

$$\begin{array}{ccccc}
J_X(\mathbf{K}) & \hookrightarrow & Y_{\max} & \hookrightarrow & J_X \\
\downarrow & & \downarrow & & \downarrow{\scriptstyle H} \\
G = J_X(\mathbf{K})/H & \hookrightarrow & Y & \hookrightarrow & J_X/H \\
\downarrow & & \downarrow{\scriptstyle \tau} & & \downarrow{\scriptstyle G} \\
0 = P_1 & \hookrightarrow & X & \hookrightarrow & J_X
\end{array}
\qquad (\wp : J_X \to J_X)$$

Let P be a $\mathbf{K}$-rational point on X and let $Q_{\max}$ be any point on $Y_{\max}(\bar{\mathbf{K}})$ such that

$$\tau_{\max}(Q_{\max}) = \wp(Q_{\max}) = P.$$

We have $F_{\mathbf{K}}(Q_{\max}) = Q_{\max} + P$. So the Artin map and the Jacobi map coincide, and the decomposition group of any place on Y above P is the subgroup of G generated by P itself. In particular the fiber of τ above P splits over $\mathbf{K}$ if and only if P is sent into H by the Jacobi map. Equivalently the class of $P - P_1$ belongs to H.

7.2. An example. In this section $\mathbf{K}$ is the field with three elements and X is the plane projective curve with homogeneous equation

$$Y^2Z^3 = X(X - Z)(X^3 + X^2Z + 2Z^3).$$

This is a smooth absolutely integral curve of genus 2. The characteristic polynomial of the Frobenius of $X/\mathbf{K}$ is

$$\chi_{\mathbf{K}}(t) = t^4 + t^3 + 2t^2 + 3t + 9. \tag{18}$$

The characteristic polynomial of the Frobenius of a curve over a finite field (given by a reasonable model) can be computed in time polynomial in $p.g.n$ where p is the characteristic of the field, n its degree over the prime field, and g the genus of the curve, using p-adic methods introduced by Kato-Lubkin [**26**], Satoh [**40**], Mestre [**33**], Kedlaya [**27**], Lauder and Wan [**31**] and widely extended since then.

When the genus of the curve is fixed, the characteristic polynomial of the Frobenius can be computed in time polynomial in the logarithm of the cardinality of $\mathbf{K}$, using ℓ-adic methods introduced by Schoof [**41**] and generalized by Pila [**35**].

We deduce from Equation (18) that the jacobian variety J_X of X has

$$\chi_{\mathbf{K}}(1) = 16$$

rational points. There are 5 places of degree 1 on X. We let P_1 be the unique place at $(0, 1, 0)$ and let

$$P_2 = (0, 0, 1),\ P_3 = (1, 0, 1),\ P_4 = (2, 2, 1),\ P_5 = (2, 1, 1).$$

The Picard group $J_X(\mathbf{K})$ is the direct sum of a subgroup of order 8 generated by the class of $P_4 - P_1$ and a subgroup of order 2 generated by $P_2 - P_1$. The class of $4(P_4 - P_1)$ is the class of $P_3 - P_1$. The classes of $P_2 - P_1$ and $P_3 - P_1$ generate a subgroup H of $\mathrm{Pic}^0(X)$ isomorphic to $(\mathbf{Z}/2\mathbf{Z})^2$. The quotient group

$$G = J_X(\mathbf{K})/H = \mathrm{Pic}^0(X)/H$$

is cyclic of order 4 generated by $P_4 - P_1$. So the subcover $\tau : Y \to X$ of $Y_{\max}$ associated with H is cyclic of order 4. And the fibers above P_1, P_2, and P_3 in this cover all split over $\mathbf{K}$. We will work with this cover.

According to Kummer theory, there is a duality (as group schemes) between the prime to p part of $\mathrm{Pic}^0(X)$ and the étale part of the kernel of $F_{\mathbf{K}} - p$. Associated to the quotient $G = \mathrm{Pic}^0(X)/H$ there must be a subgroup scheme isomorphic to μ_4 inside the latter kernel.

We let ζ be a primitive fourth root of unity in $\bar{\mathbf{K}}$ and denote by $\mathbf{L}$ the degree two extension of $\mathbf{K}$ generated by ζ. In order to find the group of order 4 we are interested in, we use algorithms to compute the kernels of $F_{\mathbf{K}} - 1$ and $F_{\mathbf{K}} - p$ described in [**14**, Chapter 13]. The idea is to pick random elements in $J_X(\mathbf{L})$ and project them onto the relevant characteristic subspaces for the action of $F_{\mathbf{K}}$, using our knowledge of the characteristic polynomial $\chi_{\mathbf{K}}$. We set

$$P_6 = (2\zeta, 2) \quad \text{and} \quad \Gamma = 2(P_6 - P_4)$$

and find that the class γ of Γ is of order 4 and satisfies

$$F_{\mathbf{K}}(\gamma) = 3\gamma.$$

Thus γ generates the group we were looking for. There is a unique function R in $\mathbf{L}(X)$ with divisor 4Γ and taking value 1 at P_1. The cover $\tau : Y \to X$ we are interested in is obtained by adding a 4-th root r of R to $\mathbf{L}(X)$. To be quite precise this construction produces the base change to $\mathbf{L}$ of the cover we are interested in. This will be fine for our purpose. So we let

$$r = R^{1/4}$$

be the 4-th root of R taking value 1 at Q_1. Equivalently we define Q_1 to be the point over P_1 where r takes the value 1. With the notation of Section 4.3 we take

$$D = 2P_5 \quad \text{and} \quad P = P_1 + P_2 + P_3.$$

We let E be the pullback of D by τ and Q the pullback of P. We expect

$$\mathcal{L}(E) = H^0(Y, \mathcal{O}_Y(E))$$

to be a free $\mathbf{K}[G]$-module of rank

$$\deg(D) - g_X + 1 = 1.$$

This will be confirmed by our computations. Because the fibers above P_1, P_2 and P_3 all split over $\mathbf{K}$, the evaluation map $\mathcal{L}(E) \to \mathbf{A}$ is described by a 3×1 matrix with coefficients in $\mathbf{K}[G]$.

For every $2 \leqslant i \leqslant 3$ we choose a 4-th root of $R(P_i)$ in $\mathbf{L}$. This amounts to choosing a point $Q_{i,1}$ in the fiber of τ above P_i. We let σ be the unique element in G that sends r to $\zeta.r$ so

$$G \ni \sigma : r \mapsto \zeta.r.$$

The $\mathbf{K}$-vector space $\mathcal{L}(E)$ decomposes over $\mathbf{L}$ as a sum of four eigenspaces associated to the four eigenvalues 1, ζ, $\zeta^2 = -1$, $\zeta^3 = -\zeta$ of σ. Let $0 \leqslant j \leqslant 3$ and let f be an eigenfunction in $\mathcal{L}(E)$ associated with the eigenvalue ζ^j. Then the quotient f/r^j is invariant by G and its divisor satisfies

$$(f/r^j) \geqslant -E - j.(r) = -E - j.\tau^*(\Gamma).$$

So f/r^j can be seen as a function on X with divisor bigger than or equal to $-D-j\Gamma$. The eigenspace $\mathcal{L}(E)_j$ associated to ζ^j is thus obtained as the image of the map

$$\begin{array}{rcl} H^0(X, \mathcal{O}_X(D + j\Gamma)) & \longrightarrow & \mathcal{L}(E)_j \\ F & \longmapsto & f = Fr^j \end{array}$$

Evaluating f at $Q_{i,1}$ for $1 \leqslant i \leqslant 3$ then reduces to evaluating $F = f/r^j$ at P_i and multiplying the result by the chosen 4-th root of $R(P_i)$, raised to the power j.

This remark enables us to compute a $\mathbf{K}$-basis of $\mathcal{L}(E)$ consisting of eigenfunctions of σ and to evaluate the functions in this basis at the $(Q_{i,1})_{1 \leqslant i \leqslant 3}$ without ever writing equations for Y. We only need to compute the Riemann-Roch spaces associated to the divisors $D + j\Gamma$ on X for $0 \leqslant j \leqslant 3$. The Riemann-Roch space of a divisor $D = D_+ - D_-$ on a curve X is computed in time polynomial in the genus of X and the degrees of the positive and negative parts D_+ and D_- of D, using Brill-Noether algorithm and its many variants. See [**21, 22, 52**] and the most efficient general algorithm due to Makdisi [**28, 29**]. In case the exponent of G is large, we may have to compute linear spaces like $H^0(X, \mathcal{O}_X(D + j\Gamma))$ for large j. In that case, one should use the method introduced by Menezes, Okamoto, and Vanstone [**32**] in the context of pairing computation, in order to replace j by its logarithm in the complexity.

Passing from the values of the eigenfunctions to the evaluation matrix $\mathcal{E}$ reduces to applying an inverse Fourier transform. We find

$$\mathcal{E} = \begin{pmatrix} 1 \\ e_{1,2} \\ e_{1,3} \end{pmatrix} \quad \text{with } e_{1,1} = 1,\ e_{1,2} = 1 + 2\sigma + 2\sigma^2 + 2\sigma^3,\ e_{1,3} = 2 + 2\sigma + 2\sigma^2 + \sigma^3.$$

Having a unit for $e_{1,1}$ is quite convenient. In general one says that $\mathcal{E}$ is systematic when the top square submatrix is the identity. This is possible when the first points $Q_{i,1}$ form a basis for the dual of $\mathcal{L}(E)$. This situation is generic in some sense but not granted. From a systematic matrix $\mathcal{E}$ it is trivial to deduce the associated checking and interpolation matrices

$$\mathcal{C} = \begin{pmatrix} e_{1,2} & e_{1,3} \\ -1 & 0 \\ 0 & -1 \end{pmatrix} \quad \text{and } \mathcal{I} = \begin{pmatrix} 1 & 0 & 0 \end{pmatrix}.$$

REMARK 2. We may wonder how general is the method presented above. The approach via Kummer theory applies as long as the order $\mathfrak{o}$ of G is prime to p. In case the order of G is a power of p, one may try to use Hasse-Witt theory instead, following the rather effective presentation in Serre [**42**]. When $\mathfrak{o}$ is neither prime to p nor a power of p we do not know any better method than the general purpose algorithm in [**21**].

8. Interpolation on algebraic curves

In this section we recall two classical applications of interpolation on algebraic curves over finite fields and illustrate the benefit of $\mathbf{K}[G]$-module structures in this context. Section 8.1 is concerned with the multiplication tensor in finite fields. In Sections 8.2 and 8.3 we see that geometric codes associated to G-equivariant divisors can be encoded in quasi-linear time and decoded in quasi-quadratic time if G is commutative, acts freely, and the code is long enough.

8.1. The complexity of multiplication in finite fields. The idea of using Lagrange interpolation over an algebraic curve to multiply two elements in a finite field is due to Chudnovsky [**10**] and has been developed by Shparlinski, Tsfasmann and Vladut [**46**], Ballet and Rolland [**3**], Chaumine [**9**], Randriambololona [**38**] and others.

Let $\mathbf{K}$ be a finite field and let $\mathfrak{o} \geqslant 2$ be an integer. Let Y be a smooth, projective, absolutely integral curve over $\mathbf{K}$ and B a place of degree $\mathfrak{o}$ on Y. Let $\mathbf{L} = H^0(B, \mathcal{O}_B)$ be the residue field at B. We choose a divisor E disjoint from B and assume that the evaluation map

$$e_B : H^0(Y, \mathcal{O}_Y(E)) \to \mathbf{L}$$

is surjective so that elements in $\mathbf{L}$ can be represented by functions in $H^0(Y, \mathcal{O}_Y(E))$. The latter functions will be characterized by their values at a collection $(Q_i)_{1 \leqslant i \leqslant N}$ of $\mathbf{K}$-rational points on Y. We denote by

$$e_Q : H^0(Y, \mathcal{O}_Y(2E)) \to \mathbf{K}^N$$

the evaluation map at these points which we assume to be injective. The multiplication of two elements $e_B(f_1)$ and $e_B(f_2)$ in $\mathbf{L}$ can be achieved by evaluating f_1 and f_2 at the Q_i, then multiplying each $f_1(Q_i)$ by the corresponding $f_2(Q_i)$, then finding the unique function f_3 in $H^0(Y, \mathcal{O}_Y(2E))$ taking value $f_1(Q_i)f_2(Q_i)$ at Q_i, then computing $e_B(f_3)$. The number of bilinear multiplications in $\mathbf{K}$ in the whole process is equal to N.

This method uses curves over $\mathbf{K}$ with arbitrarily large genus having a number of $\mathbf{K}$-points bigger than some positive constant times their genus. It bounds the bilinear complexity of multiplication in $\mathbf{L}/\mathbf{K}$ by an absolute constant times the degree $\mathfrak{o}$ of $\mathbf{L}$ over $\mathbf{K}$, but it says little abound the linear part of the algorithm, that is evaluation of the maps e_B and e_Q and their right (resp. left) inverses.

Now assume that the group of $\mathbf{K}$-automorphisms of Y contains a cyclic subgroup G of order $\mathfrak{o}$ acting freely on Y. Let $\tau : Y \to X$ be the quotient by G map. Assume that B is the fiber of τ above some rational point a on X. Assume that

E (resp. Q) is the pullback by τ of a divisor D (resp. P) on X. Under mild conditions, all the linear spaces above become free $\mathbf{K}[G]$-modules and the evaluation maps are G-equivariant. A computational consequence is that the linear part in the Chudnovsky algorithm becomes quasi-linear in the degree $\mathfrak{o}$ of the extension $\mathbf{L}/\mathbf{K}$. This remark has been exploited in [**11, 12**] to bound the complexity of multiplication of two elements in a finite field given by their coordinates in a normal basis. The decompositions of the multiplication tensor that are proven to exist in [**11**] can be actually computed using the techniques presented in Section 7.

8.2. Geometric codes. The construction of error correcting codes by evaluating functions on algebraic curves of higher genus is due to Goppa [**16, 17**]. Let Y be a smooth, projective, absolutely integral curve over a finite field $\mathbf{K}$ of characteristic p. Let d be the degree of $\mathbf{K}$ over the prime field $\mathbf{Z}/p\mathbf{Z}$. Let g_Y be the genus of Y. Let $Q_1, \ldots, Q_N$ be pairwise distinct $\mathbf{K}$-rational points on Y. Let t_i be a uniformizing parameter at Q_i. Let E be a divisor that is disjoint from $Q = Q_1 + \cdots + Q_N$. Assume that

$$2g_Y - 1 \leqslant \deg(E) \leqslant \deg(Q) - 1. \tag{19}$$

Let

$$\mathbf{A} = H^0(Q, \mathcal{O}_Q) = H^0(Y, \mathcal{O}_Y/\mathcal{O}_Y(-Q)) \simeq \mathbf{K}^N$$

be the residue algebra at Q. Let

$$\hat{\mathbf{A}} = H^0(Y, \Omega_{Y/\mathbf{K}}(-Q)/\Omega_{Y/\mathbf{K}}) \simeq \bigoplus_{i=1}^{N} \mathbf{K}\frac{dt_i}{t_i} \simeq \mathbf{K}^N$$

be the dual of $\mathbf{A}$. Evaluation at the Q_i defines an injective linear map

$$\mathcal{L}(E) = H^0(Y, \mathcal{O}_Y(E)) \to \mathbf{A}.$$

We similarly define an injective linear map

$$\Omega(-Q+E) = H^0(Y, \Omega_{Y/\mathbf{K}}(-Q+E)) \to \hat{\mathbf{A}}.$$

The two vector subspaces $\mathcal{L}(E)$ and $\Omega(-Q+E)$ are orthogonal to each other for the canonical duality pairing. They can be considered as linear codes over $\mathbf{K}$ and denoted by $C_{\mathcal{L}}$ and C_{Ω} respectively. The code $C_{\mathcal{L}}$ has length N, dimension

$$K = \deg(E) - g_Y + 1$$

and minimum distance greater than or equal to $N - \deg(E)$. Given a basis of $\mathcal{L}(E)$ one defines the generating matrix $\mathcal{E}_E$ of the code $C_{\mathcal{L}}$ to be the $N \times K$-matrix of the injection $\mathcal{L}(E) \to \mathbf{A} = \mathbf{K}^N$. One similarly defines the parity-check matrix $\mathcal{C}_E$ to be the $N \times (N-K)$-matrix of $\Omega(-Q+E) \to \hat{\mathbf{A}}$. We finally denote by $\mathcal{I}_E$ the $K \times N$-matrix of some projection of $\mathbf{A}$ onto $C_{\mathcal{L}}$. A message of length K is encoded by multiplying the corresponding column on the left by $\mathcal{E}_E$. The received word is checked by multiplying it on the left by the transpose of $\mathcal{C}_E$. And the initial message is recovered from a correct codeword applying the interpolation matrix $\mathcal{I}_E$. In full generality, coding, testing and interpolating respectively require $2NK$, $2N(N-K)$ and $2KN$ operations in $\mathbf{K}$.

Assume now that the group of $\mathbf{K}$-automorphisms of Y contains a finite commutative subgroup G of order $\mathfrak{o}$ acting freely on Y. Let $\tau : Y \to X$ be the quotient by G map. Assume that $\mathfrak{o}$ divides N and let

$$n = N/\mathfrak{o}.$$

Assume that Q is the pullback by τ of a divisor

$$P = P_1 + \cdots + P_n$$

on X. Assume that E is the pullback of some divisor D on X. We are thus in the situation of Section 4. The code $C_{\mathcal{L}}$ is a free $\mathbf{K}[G]$-submodule of $\mathbf{A}$ of rank

$$k = K/\mathfrak{o}$$

and $\mathcal{C}_{\Omega}$ is its orthogonal module for the $\mathbf{K}[G]$-bilinear form defined in Section 3.3.

The matrices $\mathcal{E}_E$, $\mathcal{C}_E$, and $\mathcal{I}_E$ can be seen as matrices with coefficients in $\mathbf{K}[G]$ of respective sizes $n \times k$, $n \times (n-k)$, and $k \times n$. Coding now requires $2nk$ operations in $\mathbf{K}[G]$ rather than $2NK$ operations in $\mathbf{K}$. According to Theorem 2, each such operation requires less than $\mathcal{Q}.d^2.\mathfrak{o}.\log \mathfrak{o}$ operations in $\mathbf{Z}/p\mathbf{Z}$ and $\mathbf{Z}/p'\mathbf{Z}$ where $p' \leqslant \mathcal{Q}.(\mathfrak{o}.p)^{11}$ for some absolute constant $\mathcal{Q}$. The total cost of coding is thus bounded by an absolute constant times

$$\frac{NK}{\mathfrak{o}^2}.d^2.\mathfrak{o}.\log(\mathfrak{o}).(\log p + \log \mathfrak{o})^2 = N.d^2.\log(\mathfrak{o}).k.(\log p + \log \mathfrak{o})^2$$

elementary operations.

Remark 3. Assuming that $\log \mathfrak{o}$ is bigger than k times a positive constant, the cost of coding is quasi-linear in the length N of the code. The same holds for parity-checking and interpolating. Indeed the action of a large commutative group G provides a significant computational advantage. We shall see in Section 9 that geometric class field theory produces examples of free commutative group actions meeting this condition.

8.3. Basic decoding. Assume that we are in the situation of the beginning of Section 8.2, and that we have received a message r in $\mathbf{A} = \mathbf{K}^N$. Let c be the closest codeword to r in $C_{\mathcal{L}}$ for the Hamming distance in $\mathbf{K}^N$. Write

$$r = c + \epsilon$$

and ϵ the error vector. Let f be the unique function in $\mathcal{L}(E)$ such that $f = c \bmod Q$. The support of the error vector ϵ is the effective divisor $\mathrm{Supp}(\epsilon)$ consisting of all points Q_i where ϵ is not-zero. The degree of $\mathrm{Supp}(\epsilon)$ is the number of errors in r.

The principle of the basic decoding algorithm [**24**, **47**] is: if a_0 is a small degree function vanishing at every point in the support $\mathrm{Supp}(\epsilon)$ then $a_0 r = a_0 c \mod Q$ is the residue modulo Q of an algebraic function $a_0 f$ of not too large degree. This function can be recovered from its values at Q if N is large enough. More concretely we let E_0 be some auxiliary divisor on Y with degree at least g_Y and set

$$E_1 = E + E_0.$$

Let $\mathcal{P}$ be the subspace of $\mathcal{L}(E_0)$ consisting of all a_0 such that there exists a_1 in $\mathcal{L}(E_1)$ with $a_0 r = a_1 \bmod Q$. Non-zero elements in $\mathcal{P}$ are denominators for r in the sense of Section 5. We just saw that every function in $\mathcal{L}(E_0)$ vanishing at every point in the support of ϵ belongs to $\mathcal{P}$.

Conversely if a_0 is in $\mathcal{P}$ then $a_0 r$ belongs to $\mathcal{L}(E_1)$ modulo Q. But $a_0 c$ belongs to $\mathcal{L}(E_1)$ modulo Q also because a_0 is in $\mathcal{L}(E_0)$ modulo Q and c is in $\mathcal{L}(E)$ modulo Q. So $a_0(r - c) = a_0 \epsilon$ belongs to $\mathcal{L}(E_1)$ modulo Q. There is a function in $\mathcal{L}(E_1)$ that is $a_0 \epsilon$ modulo Q. This function has $N - \deg(\mathrm{Supp}(\epsilon))$ zeros and degree at most $\deg(E_1) = \deg(E) + \deg(E_0)$. If we assume that

$$\deg(\mathrm{Supp}(\epsilon)) \leqslant N - 1 - \deg(E) - \deg(E_0) \tag{20}$$

then the latter function must be zero. So a_0 vanishes at $\mathrm{Supp}(\epsilon)$. Assuming Equation (20) we thus have $\mathcal{P} = \mathcal{L}(E_0 - \mathrm{Supp}(\epsilon))$. Assuming further that

$$\deg(\mathrm{Supp}(\epsilon)) \leqslant \deg(E_0) - g \tag{21}$$

this space is non-zero. Computing it is a matter of linear algebra and requires a constant times N^3 operations in $\mathbf{K}$. Given any non-zero element a_0 in $\mathcal{P}$ we denote by A_0 the divisor consisting of all Q_i where a_0 vanishes. The degree of A_0 is bounded by $\deg E_0$. The error ϵ is an element in $\mathbf{A}$ with support contained in A_0 and such that $r - \epsilon$ belongs to $C_{\mathcal{L}}$. Finding ϵ is a linear problem in $\leqslant \deg E_0$ unknows and $N - \deg(E) + g_Y - 1$ equations. The solution is unique because the difference of two solutions is in $C_{\mathcal{L}}$ and has at least $N - \deg(E_0)$ zeros. And this is strictly greater than $\deg(E)$ by Equation (20).

Combining Equations (20) and (21) we see that the basic decoding algorithm corrects up to d_{basic} errors where

$$d_{\mathrm{basic}} = \frac{N - \deg(E) - 1 - g_Y}{2}. \tag{22}$$

Assume now that the group of $\mathbf{K}$-automorphisms of Y contains a finite commutative subgroup G of order $\mathfrak{o}$ acting freely on Y. Let $\tau : Y \to X$ be the quotient by G map. Assume that $\mathfrak{o}$ divides N and let $n = N/\mathfrak{o}$. Assume that Q is the pullback by τ of a divisor

$$P = P_1 + \cdots + P_n$$

on X. Assume that E is the pullback of some divisor D on X. Assume that E_0 is the pullback of some divisor D_0 on X. Assume that $\mathcal{L}(E_0)$ contains a free module of rank $\deg(D_0) - g_X + 1$ over $\mathbf{K}[G]$. According to Proposition 5, such an E_0 exists if the order $\mathfrak{o}$ of G is prime to p. According to Proposition 6, such an E_0 exists if the order $\mathfrak{o}$ of G is a power of p, and the cardinality q of $\mathbf{K}$ is at least 4, and the genus of X is at least 2. Another sufficient condition if that $\deg(D_0) \geqslant 2g_X - 1$. According to Proposition 10 we can find a denominator a_0 at the expense of $\mathcal{Q}.(\mathfrak{o}.n.\log(\mathfrak{o}.n))^2$ operations in $\mathbf{K}$ and $\mathcal{Q}.\mathfrak{o}.n^3\log(\mathfrak{o}.n)$ operations in $\mathbf{K}[G]$. According to Theorem 2, each operation in $\mathbf{K}[G]$ requires less than

$$\mathcal{Q}.d^2.\mathfrak{o}.\log(\mathfrak{o}).(\log p + \log \mathfrak{o})^2$$

elementary operations. The total cost of finding a denominator is thus bounded by an absolute constant times

$$N^2.n.d^2.\log^4(\mathfrak{o}.n.p)$$

elementary operation.

REMARK 4. Assuming that $\log \mathfrak{o}$ is bigger than n times a positive constant, the cost of finding a denominator is quasi-quadratic in the length N of the code. Once found a denominator, the error can be found at the same cost.

9. Good geometric codes with quasi-linear encoding

In this section we specialize the constructions presented in Sections 8.2 and 8.3 using curves with many points and their Hilbert class fields. We quickly review in Section 9.1 some standard useful results and observations which we apply in Section 9.2 to the construction of families of good geometric codes having quasi-linear encoding and a quasi-quadratic decoder. Recall that a family of codes over a fixed alphabet is said to be good when the length tends to infinity while both the rate and the minimum distance have a strictly positive lim inf.

9.1. Controlling the class group and the Artin map. We keep the notation from Section 7.1. In particular P_1 is a $\mathbf{K}$-rational point on X and

$$j_X : X \to J_X$$

is the Jacobi map with origin P_1. For the applications we have in mind we need some control on the $\mathbf{K}$-rational points on X, on the group $\mathrm{Pic}^0(X)$ and most importantly on the image of $X(\mathbf{K})$ in $\mathrm{Pic}^0(X)$ by the Jacobi map. A typical advantageous situation would be:

(1) X has enough $\mathbf{K}$-rational points, that is a fixed positive constant times its genus g_X,
(2) a fixed positive proportion of these points are mapped by j_X into a subgroup H,
(3) H is not too large i.e. the quotient $\log |H| / \log |\mathrm{Pic}^0(X)|$ is smaller than a fixed constant smaller than 1.

A range of geometric techniques relevant to that problem is presented in Serre's course [**44**] with the related motivation of constructing curves with many points. One says that (a family of) curves over a fixed finite field of cardinality q have many points when the ratio of the number of rational points by the genus tends to $\sqrt{q}-1$. Modular curves $X_0(N)$ have many points over finite fields with p^2 elements, corresponding to supersingular moduli, as was noticed by Ihara [**23**] and by Tzfasman, Vladut, and Zink [**50**]. These authors also found families of Shimura curves having many points over fields with cardinality a square. Garcia and Stichtenoth [**15**] constructed for every square q an infinite tower of algebraic curves over $\mathbf{F}_q$ such that the quotient of the number of $\mathbf{F}_q$-points by the genus converges to $\sqrt{q}-1$, and the quotient of the genera of two consecutive curves converges to q.

As for conditions (2) and (3) above, it is noted in [**44**, 5.12.4] that the images by j_X of $P_2, \dots, P_n$ generate a subgroup H with at most $n-1$ invariant factors. If the class group $J_X(\mathbf{K})$ has $I \geqslant n-1$ invariant factors then the size of the quotient G is bigger than or equal to the product of the $I-(n-1)$ smallest invariant factors of $J_X(\mathbf{K})$.

Another favourable situation exploited in [**19, 34, 36, 51**] is when $\mathbf{K}$ has a strict subfield $\mathbf{k}$ and X is defined over $\mathbf{k}$ and P_1 is $\mathbf{k}$-rational. Then the Jacobi map sends the points in $X(\mathbf{k})$ into the subgroup $J_X(\mathbf{k})$ of $J_X(\mathbf{K})$. We will use this remark in the next section.

9.2. A construction. Let $\mathbf{k}$ be a finite field with characteristic p. Let q be the cardinality of $\mathbf{k}$. We assume that q is a square. We consider a family of curves $(X_k)_{k \geqslant 1}$ over $\mathbf{k}$ having many points over $\mathbf{k}$. For example we may take X_k to be the k-th curve in the Garcia-Stichtenoth tower associated with q. We denote by g_X the genus of X_k. We omit the index k in the sequel because there is no risk of confusion. We denote by n the number of $\mathbf{k}$-rational points on X. We denote these points by $P_1, \dots, P_n$ and let P be the effective divisor sum of all these points. We let $\mathbf{K}$ be a non-trivial extension of $\mathbf{k}$. We will assume that the degree of $\mathbf{K}$ over $\mathbf{k}$ is 2 because higher values seem to bring nothing but disadvantages. We denote by T the quotient

$$T = J_X(\mathbf{K})/J_X(\mathbf{k}).$$

We denote by T_p the p-Sylow subgroup of T. We denote by $T_{p'}$ the complement subgroup of T_p in T. Let G be the bigger among T_p and $T_{p'}$. This is a quotient of T. Let H be the kernel of the composite map $J_X(\mathbf{K}) \to T = J_X(\mathbf{K})/J_X(\mathbf{k}) \to G$. Let $\mathfrak{o}$ be the order of G. We note that

$$\#J_X(\mathbf{K})/J_X(\mathbf{k}) \geqslant (q-1)^{2g_X} / (\sqrt{q}+1)^{2g_X} = (\sqrt{q}-1)^{2g_X}$$

so

$$\mathfrak{o} \geqslant \sqrt{\sharp T} \geqslant (\sqrt{q}-1)^{g_X} \tag{23}$$

grows exponentially in g_X provided $q \geqslant 9$. Also G is a p-group or a p'-group. We find ourselves in the situation of Section 7.1. Let $Y_{\max}$ be the maximal unramified cover of X over $\mathbf{K}$ which is totally decomposed over $\mathbf{K}$ above P_1. Let Y be the quotient of $Y_{\max}$ by H. The fibers of

$$\tau : Y \to X$$

above the points $P_1, \ldots, P_n$ all split over $\mathbf{K}$. Let Q be the pullback of P by τ. This is a divisor on Y of degree

$$N = \mathfrak{o}.n.$$

We choose a real number ϱ such that

$$0 < \varrho < \frac{\sqrt{q}}{2} - 2. \tag{24}$$

Our goal is to correct up to $\varrho.\mathfrak{o}.g_X$ errors. Let D be a divisor on X that is disjoint from P and such that

$$\deg(D) = \lceil (\sqrt{q}-2-2\varrho)g_X \rfloor$$

the closest integer to $(\sqrt{q}-2-2\varrho)g_X$. Let E be the pullback of D by τ. We deduce from Equation (24) that condition (19) is met at least asymptotically. From X, Y, E, and Q the construction in Section 8.2 produces a code $C_{\mathcal{L}}$ over the field $\mathbf{K}$ with q^2 elements, having length

$$N = \mathfrak{o}.n \simeq (\sqrt{q}-1).\mathfrak{o}.g_X$$

and dimension

$$K = \mathfrak{o}.(\deg(D) - g_X + 1) \simeq (\sqrt{q}-3-2\varrho).\mathfrak{o}.g_X.$$

We set $k = K/\mathfrak{o}$ and deduce from Equation (23) that the lim inf of $(\log \mathfrak{o})/n$ and $(\log \mathfrak{o})/k$ are strictily positive. As explained in Remark 3, this implies that the code $C_{\mathcal{L}}$ can be encoded and parity-checked in quasi-linear deterministic time in its length N, and decoded with the same complexity when there are no errors. Using the basic decoding algorithm as in Section 8.3 one can decode in the presence of errors in quasi-quadratic probabilistic (Las Vegas) time up to the distance

$$d_{\text{basic}} = \frac{N - \deg(E) - 1 - g_Y}{2} \simeq \varrho.\mathfrak{o}.g_X$$

defined by Equation (22) as explained in Remark 4. We denote by δ_{basic} the relative distance d_{basic}/N. The existence of a divisor D_0 with all the properties required in Section 8.3 is granted because G is either a p-group or a p'-group. So we can apply Proposition 5 or Proposition 6 depending on the case. This finishes the proof of the theorem below.

THEOREM 3. *Let p be a prime integer and let q be a power of p. Assume that q is a square and*

$$q \geqslant 25. \tag{25}$$

Let ϱ be a real such that

$$0 < \varrho < \frac{\sqrt{q}}{2} - 2. \tag{26}$$

There exists a family of linear error correcting codes over the field with q^2 elements having length N tending to infinity and such that

(1) *the rate R satisfies*

$$\lim R = \frac{\sqrt{q} - 3 - 2\varrho}{\sqrt{q} - 1}$$

(2) *for each code there exists a straight-line program that encodes in quasi-linear time in the length N,*

(3) *for each code there exists a computation tree that decodes in quasi-quadratic probabilistic (Las Vegas) time in the length N up to the relative distance δ_{basic} and*

$$\lim \delta_{\text{basic}} = \frac{\varrho}{\sqrt{q} - 1}.$$

REMARK 5. The complexity statements in the theorem above are non-uniform in the sense that they bound the complexity of coding and decoding assuming that the code is given by its generating, parity-check and interpolation matrices having coefficients in the group algebra $\mathbf{K}[G]$. The theorem claims nothing about the complexity of finding these matrices. The example detailed in section 7.2 and remark 2 suggest that this complexity could be quasi-quadratic in the length N of the code. Proving such a complexity result would probably be quite heavy due to the relative sophistication of the methods used to find e.g. the interesting torsion points in the Picard group.

REMARK 6. A calculation similar to the one in [**30**, §7.3] shows that for any $q \geqslant 47^2$, some among the codes constructed above are excellent in the sense that the accumulation point $(2\delta_{\text{basic}}, R)$ stands above the Varshamov-Gilbert limit for codes over the field with q^2 elements. To our knowledge these are the first excellent codes that can be encoded in quasi-linear time and decoded in quasi-quadratic time. Recall that Reed–Solomon codes can be encoded and decoded in quasi-linear time but cannot be said to be asymptotically good because the length of the code is bounded by the size of the alphabet.

REMARK 7. We compare fast basic decoding of the codes in Theorem 3 as explained in Section 8.3 with the general purpose algorithm of Beelen, Rosenkilde, Solomatov [**5**]. Using the latter, one can decode up to half the Goppa designed minimum distance. Inequalities (25) and (26) are then replaced by

$$q \geqslant 16 \quad \text{and} \quad 0 < \varrho < \frac{\sqrt{q} - 3}{2},$$

and the limit of the rate becomes

$$\lim R = \frac{\sqrt{q} - 2 - 2\varrho}{\sqrt{q} - 1}.$$

However the complexity of decoding is then of order $\mu^{\omega-1}(N+g_Y)$ where N is the length of the code, μ is the gonality of Y, and ω is the exponent in the complexity of matrix multiplication. Curves with many points have large gonality. In particular $\mu \geqslant N/(q^2+1)$ in our situation, so that for fixed q, the complexity of this decoder is of order greater than N^ω. It is known [**1**] that $2 \leqslant \omega < 2.37286$ but it is not granted that $\omega = 2$.

References

[1] Josh Alman and Virginia Vassilevska Williams, *A refined laser method and faster matrix multiplication*, Proceedings of the 2021 ACM-SIAM Symposium on Discrete Algorithms (SODA), [Society for Industrial and Applied Mathematics (SIAM)], Philadelphia, PA, 2021, pp. 522–539, DOI 10.1137/1.9781611976465.32. MR4262465

[2] S. Ballet and D. Le Brigand, *On the existence of non-special divisors of degree g and $g-1$ in algebraic function fields over $\mathbb{F}_q$*, J. Number Theory **116** (2006), no. 2, 293–310, DOI 10.1016/j.jnt.2005.04.009. MR2195927

[3] S. Ballet and R. Rolland, *Multiplication algorithm in a finite field and tensor rank of the multiplication*, J. Algebra **272** (2004), no. 1, 173–185, DOI 10.1016/j.jalgebra.2003.09.031. MR2029030

[4] Stéphane Ballet, *Curves with many points and multiplication complexity in any extension of $\mathbf{F}_q$*, Finite Fields Appl. **5** (1999), no. 4, 364–377, DOI 10.1006/ffta.1999.0255. MR1711833

[5] Peter Beelen, Johan Rosenkilde, and Grigory Solomatov, *Fast decoding of AG codes*, IEEE Trans. Inform. Theory **68** (2022), no. 11, 7215–7232, DOI 10.1109/tit.2022.3188843. MR4524636

[6] Leo I. Bluestein, *A linear filtering approach to the computation of discrete Fourier transform*, IEEE Transactions on Audio and Electroacoustics **18** (1970), 451–455.

[7] Alin Bostan, Frédéric Chyzak, Marc Giusti, Romain Lebreton, Grégoire Lecerf, Bruno Salvy, and Éric Schost, *Algorithmes efficaces en calcul formel*, Frédéric Chyzak, August 2017, 686 pages, in French.

[8] N. Bourbaki, *Éléments de mathématique. Algèbre. Chapitre 8. Modules et anneaux semi-simples* (French), Springer, Berlin, 2012. Second revised edition of the 1958 edition [MR0098114], DOI 10.1007/978-3-540-35316-4. MR3027127

[9] Jean Chaumine, *Multiplication in small finite fields using elliptic curves*, Algebraic geometry and its applications, Ser. Number Theory Appl., vol. 5, World Sci. Publ., Hackensack, NJ, 2008, pp. 343–350, DOI 10.1142/9789812793430_0018. MR2484063

[10] D. V. Chudnovsky and G. V. Chudnovsky, *Algebraic complexities and algebraic curves over finite fields*, J. Complexity **4** (1988), no. 4, 285–316, DOI 10.1016/0885-064X(88)90012-X. MR974928

[11] Jean-Marc Couveignes and Tony Ezome, *The equivariant complexity of multiplication in finite field extensions*, J. Algebra **622** (2023), 694–720, DOI 10.1016/j.jalgebra.2023.01.022. MR4556885

[12] Jean-Marc Couveignes and Reynald Lercier, *Elliptic periods for finite fields*, Finite Fields Appl. **15** (2009), no. 1, 1–22, DOI 10.1016/j.ffa.2008.07.004. MR2468989

[13] Charles W. Curtis and Irving Reiner, *Representation theory of finite groups and associative algebras*, Pure and Applied Mathematics, Vol. XI, Interscience Publishers (a division of John Wiley & Sons, Inc.), New York-London, 1962. MR144979

[14] Bas Edixhoven and Jean-Marc Couveignes (eds.), *Computational aspects of modular forms and Galois representations*, Annals of Mathematics Studies, vol. 176, Princeton University Press, Princeton, NJ, 2011. MR2849700

[15] Arnaldo García and Henning Stichtenoth, *A tower of Artin-Schreier extensions of function fields attaining the Drinfel′d-Vlăduţ bound*, Invent. Math. **121** (1995), no. 1, 211–222, DOI 10.1007/BF01884295. MR1345289

[16] V. D. Goppa, *Codes on algebraic curves* (Russian), Dokl. Akad. Nauk SSSR **259** (1981), no. 6, 1289–1290. MR628795

[17] V. D. Goppa, *Algebraic-geometric codes* (Russian), Izv. Akad. Nauk SSSR Ser. Mat. **46** (1982), no. 4, 762–781, 896. MR670165

[18] V. D. Goppa, *Geometry and codes*, Mathematics and its Applications (Soviet Series), vol. 24, Kluwer Academic Publishers Group, Dordrecht, 1988. Translated from the Russian by N. G. Shartse, DOI 10.1007/978-94-015-6870-8. MR1029027

[19] Venkatesan Guruswami and Chaoping Xing, *Optimal rate list decoding over bounded alphabets using algebraic-geometric codes*, J. ACM **69** (2022), no. 2, Art. 10, 48, DOI 10.1145/3506668. MR4433318

[20] D. R. Heath-Brown, *Zero-free regions for Dirichlet L-functions, and the least prime in an arithmetic progression*, Proc. London Math. Soc. (3) **64** (1992), no. 2, 265–338, DOI 10.1112/plms/s3-64.2.265. MR1143227

[21] F. Hess, *Computing Riemann-Roch spaces in algebraic function fields and related topics*, J. Symbolic Comput. **33** (2002), no. 4, 425–445, DOI 10.1006/jsco.2001.0513. MR1890579

[22] Ming-Deh Huang and Doug Ierardi, *Efficient algorithms for the Riemann-Roch problem and for addition in the Jacobian of a curve*, J. Symbolic Comput. **18** (1994), no. 6, 519–539, DOI 10.1006/jsco.1994.1063. MR1334660

[23] Yasutaka Ihara, *Some remarks on the number of rational points of algebraic curves over finite fields*, J. Fac. Sci. Univ. Tokyo Sect. IA Math. **28** (1981), no. 3, 721–724 (1982). MR656048

[24] Jørn Justesen, Knud J. Larsen, H. Elbrønd Jensen, Allan Havemose, and Tom Høholdt, *Construction and decoding of a class of algebraic geometry codes*, IEEE Trans. Inform. Theory **35** (1989), no. 4, 811–821, DOI 10.1109/18.32157. MR1013689

[25] Erich Kaltofen and B. David Saunders, *On Wiedemann's method of solving sparse linear systems*, Applied algebra, algebraic algorithms and error-correcting codes (New Orleans, LA, 1991), Lecture Notes in Comput. Sci., vol. 539, Springer, Berlin, 1991, pp. 29–38, DOI 10.1007/3-540-54522-0_93. MR1229306

[26] Goro C. Kato and Saul Lubkin, *Zeta matrices of elliptic curves*, J. Number Theory **15** (1982), no. 3, 318–330, DOI 10.1016/0022-314X(82)90036-1. MR680536

[27] Kiran S. Kedlaya, *Counting points on hyperelliptic curves using Monsky-Washnitzer cohomology*, J. Ramanujan Math. Soc. **16** (2001), no. 4, 323–338. MR1877805

[28] Kamal Khuri-Makdisi, *Linear algebra algorithms for divisors on an algebraic curve*, Math. Comp. **73** (2004), no. 245, 333–357, DOI 10.1090/S0025-5718-03-01567-9. MR2034126

[29] Kamal Khuri-Makdisi, *Asymptotically fast group operations on Jacobians of general curves*, Math. Comp. **76** (2007), no. 260, 2213–2239, DOI 10.1090/S0025-5718-07-01989-8. MR2336292

[30] Gilles Lachaud, *Les codes géométriques de Goppa* (French), Astérisque **133-134** (1986), 189–207. Seminar Bourbaki, Vol. 1984/85. MR837220

[31] Alan G. B. Lauder and Daqing Wan, *Counting points on varieties over finite fields of small characteristic*, Algorithmic number theory: lattices, number fields, curves and cryptography, Math. Sci. Res. Inst. Publ., vol. 44, Cambridge Univ. Press, Cambridge, 2008, pp. 579–612. MR2467558

[32] Alfred J. Menezes, Tatsuaki Okamoto, and Scott A. Vanstone, *Reducing elliptic curve logarithms to logarithms in a finite field*, IEEE Trans. Inform. Theory **39** (1993), no. 5, 1639–1646, DOI 10.1109/18.259647. MR1281712

[33] J.-F. Mestre, *Lettre adressée à Gaudry et Harley*, `https://webusers.imj-prg.fr/~jean-francois.mestre/`, December 2010.

[34] Harald Niederreiter and Chaoping Xing, *A general method of constructing global function fields with many rational places*, Algorithmic number theory (Portland, OR, 1998), Lecture Notes in Comput. Sci., vol. 1423, Springer, Berlin, 1998, pp. 555–566, DOI 10.1007/BFb0054892. MR1726101

[35] Jonathan S. Pila, *Frobenius maps of Abelian varieties and finding roots of unity in finite fields*, ProQuest LLC, Ann Arbor, MI, 1988. Thesis (Ph.D.)–Stanford University. MR2637049

[36] Heinz-Georg Quebbemann, *Cyclotomic Goppa codes: Coding techniques and coding theory*, IEEE Trans. Inform. Theory **34** (1988), no. 5, 1317–1320, DOI 10.1109/18.21261. MR987676

[37] Lawrence R. Rabiner, Ronald W. Schafer, and Charles M. Rader, *The chirp z-transform algorithm and its application*, Bell System Tech. J. **48** (1969), 1249–1292, DOI 10.1002/j.1538-7305.1969.tb04268.x. MR243724

[38] Hugues Randriambololona, *Bilinear complexity of algebras and the Chudnovsky-Chudnovsky interpolation method*, J. Complexity **28** (2012), no. 4, 489–517, DOI 10.1016/j.jco.2012.02.005. MR2925903

[39] Michael Rosen, *The Hilbert class field in function fields*, Exposition. Math. **5** (1987), no. 4, 365–378. MR917350

[40] Takakazu Satoh, *The canonical lift of an ordinary elliptic curve over a finite field and its point counting*, J. Ramanujan Math. Soc. **15** (2000), no. 4, 247–270. MR1801221

[41] René Schoof, *Elliptic curves over finite fields and the computation of square roots mod p*, Math. Comp. **44** (1985), no. 170, 483–494, DOI 10.2307/2007968. MR777280

[42] Jean-Pierre Serre, *Sur la topologie des variétés algébriques en caractéristique p* (French), Symposium internacional de topología algebraica International symposium on algebraic topology, Universidad Nacional Autónoma de México and UNESCO, México, 1958, pp. 24–53. MR98097

[43] Jean-Pierre Serre, *Algebraic groups and class fields*, Graduate Texts in Mathematics, vol. 117, Springer-Verlag, New York, 1988. Translated from the French, DOI 10.1007/978-1-4612-1035-1. MR918564

[44] Jean-Pierre Serre, *Rational points on curves over finite fields*, Documents Mathématiques (Paris), vol. 18, Société Mathématique de France, Paris, 2020.

[45] Mohammad Amin Shokrollahi, *Optimal algorithms for multiplication in certain finite fields using elliptic curves*, SIAM J. Comput. **21** (1992), no. 6, 1193–1198, DOI 10.1137/0221071. MR1192302

[46] Igor E. Shparlinski, Michael A. Tsfasman, and Serge G. Vladut, *Curves with many points and multiplication in finite fields*, Coding theory and algebraic geometry (Luminy, 1991), Lecture Notes in Math., vol. 1518, Springer, Berlin, 1992, pp. 145–169, DOI 10.1007/BFb0087999. MR1186422

[47] Alexei N. Skorobogatov and Sergei G. Vlăduţ, *On the decoding of algebraic-geometric codes*, IEEE Trans. Inform. Theory **36** (1990), no. 5, 1051–1060, DOI 10.1109/18.57204. MR1066592

[48] The Sage Developers, *Sagemath, the Sage Mathematics Software System (Version 9.4)*, `https://www.sagemath.org`, 2021.

[49] M. A. Tsfasman and S. G. Vlăduţ, *Algebraic-geometric codes*, Mathematics and its Applications (Soviet Series), vol. 58, Kluwer Academic Publishers Group, Dordrecht, 1991. Translated from the Russian by the authors, DOI 10.1007/978-94-011-3810-9. MR1186841

[50] M. A. Tsfasman, S. G. Vlăduţ, and Th. Zink, *Modular curves, Shimura curves, and Goppa codes, better than Varshamov-Gilbert bound*, Math. Nachr. **109** (1982), 21–28, DOI 10.1002/mana.19821090103. MR705893

[51] Gerard van der Geer, *Hunting for curves with many points*, Coding and cryptology, Lecture Notes in Comput. Sci., vol. 5557, Springer, Berlin, 2009, pp. 82–96, DOI 10.1007/978-3-642-01877-0_9. MR2836234

[52] Emil J. Volcheck, *Computing in the Jacobian of a plane algebraic curve*, Algorithmic number theory (Ithaca, NY, 1994), Lecture Notes in Comput. Sci., vol. 877, Springer, Berlin, 1994, pp. 221–233, DOI 10.1007/3-540-58691-1_60. MR1322725

[53] Douglas H. Wiedemann, *Solving sparse linear equations over finite fields*, IEEE Trans. Inform. Theory **32** (1986), no. 1, 54–62, DOI 10.1109/TIT.1986.1057137. MR831560

[54] Triantafyllos Xylouris, *On the least prime in an arithmetic progression and estimates for the zeros of Dirichlet L-functions*, Acta Arith. **150** (2011), no. 1, 65–91, DOI 10.4064/aa150-1-4. MR2825574

Univ. Bordeaux, CNRS, INRIA, Bordeaux-INP, IMB, UMR 5251, F-33400 Talence, France

Email address: `jean-marc.couveignes@math.u-bordeaux.fr`

Univ. Bordeaux, CNRS, INRIA, Bordeaux-INP, IMB, UMR 5251, F-33400 Talence, France

Email address: `jean.gasnier@math.u-bordeaux.fr`

Contemporary Mathematics
Volume **832**, 2026
https://doi.org/10.1090/conm/832/16607

The relative class number one problem for function fields, II

Kiran S. Kedlaya

ABSTRACT. We establish that any finite extension of function fields of genus greater than 1 whose relative class group is trivial is Galois and cyclic. This depends on a result from a preceding paper which establishes a finite list of possible Weil polynomials for both fields. Given this list, we analyze most cases by computing options for the splittings of low-degree places in the extension, then consider the effect of these options on the Weil polynomials of certain isogeny factors of the Jacobian of the Galois closure. In one case, we use instead an analysis based on principal polarizations, modeled on an argument of Howe.

1. Introduction

This paper continues the work done in [**10**] on the *relative class number one problem* for function fields, building upon work of Leitzel–Madan [**13**] and Leitzel–Madan–Queen [**14**]. That is, we seek to identify finite extensions F'/F of function fields of curves over finite fields for which the two class numbers are equal. In this paper, we establish the following result; the argument relies heavily on results from [**10**] as described below. (Throughout, as in [**10**], we implicitly consider only separable extensions.)

THEOREM 1.1. *Let F'/F be a finite extension of function fields of genus greater than 1 with the same constant field and class number. Then F'/F is Galois and cyclic.*

Before continuing, we recall some notation from [**10**]. Given a finite extension F'/F of function fields, we write C, C' for the curves corresponding to F, F'; $q_F, q_{F'}$ for the orders of the base fields of C, C'; $g_F, g_{F'}$ for the genera of C, C'; and $h_F, h_{F'}$ for the class numbers of F, F'. In this paper we only consider the case where $q_F = q_{F'}$, in which case we say F'/F is a *purely geometric extension* (the other extreme is when $g_{F'} = g_F$, in which case we say F'/F is a *constant extension*). We recall here [**10**, Theorem 1.3(a)].

THEOREM 1.2. *Let F'/F be a finite purely geometric extension of function fields with $g_{F'} > g_F > 1$ and $h_{F'} = h_F$.*

2020 *Mathematics Subject Classification.* Primary 11G20.

Thanks to Everett Howe for helpful discussions, particularly about Lemma 10.1 and Lemma 10.2. Kedlaya was supported by NSF (grants DMS-1802161, DMS-2053473) and UC San Diego (Warschawski Professorship).

TABLE 1. Options for the triple $(d, g_F, g_{F'})$, as established in [**10**].

d	2										3				4			5	6	7
g_F	2			3		4		5	6	7	2		3	4	2		3	2	2	2
$g_{F'}$	3	4	5	5	6	7	8	9	11	13	4	6	7	10	5	6	9	6	7	8

(a) *If $q_F > 2$, then the tuple $(q_F, d, g_F, g_{F'}, F)$ is listed in* [**10**, Table 4]. *Moreover, F'/F is cyclic.*
(b) *If $q_F = 2$, then the triple $(d, g_F, g_{F'})$ is listed in Table* 1. *Moreover, the pair of Weil polynomials of C and C' belong to an explicit list of* 209 *triples (see below).*

For brevity we do not include the complete list of Weil polynomials in Theorem 1.2(b); instead, we catalog in §2 all of the specific features of this list that we need here (see especially Table 2). The full table can be found in an Excel spreadsheet in the repository [**12**].

To establish Theorem 1.1, it thus suffices to check the claim for F having one of the Weil polynomials allowed by Theorem 1.2(b). We organize this according to the value of d. (For a uniform description of the paradigm applicable to $d = 3, 4, 5, 6$, see Remark 4.6.)

- For $d = 2$ there is nothing to check, as F'/F is automatically Galois and cyclic.
- For $d = 3, 4$, we analyze the quadratic resolvent of F'/F by using the point counts of C and C' to constrain the splitting of places from C to C' (Lemma 6.1, Lemma 7.2). This can be seen as a natural continuation of the arguments of [**10**, §8].
- For $d = 5$, we make a similar argument, but now accounting more fully for the representation theory of A_5, its effect on the isogeny splitting of the Jacobian of the Galois closure (as in [**18**]), and the point counts on additional quotients of the Galois closure of degree 6 and 10 over C (Lemma 8.2).
- For $d = 6$, we make a similar argument, but using the quotient of the Galois closure of degree 6 over C arising from the outer automorphism of S_6, and also accounting for the possible Galois group $\mathrm{PGL}(2, 5)$ (Lemma 9.2).
- For $d = 7$, the similar argument becomes so complicated that we did not attempt to execute it. Fortunately, we only have to analyze one pair of Weil polynomials, which is known to occur for a cyclic cover. Even more fortunately, this example is amenable to an approach of Howe [**6**]: we analyze the possible principal polarizations of abelian varieties in the isogeny class of the Jacobian of C' to show that the cover is forced to be cyclic (Lemma 10.1).

In passing, we note some resemblance between these methods and the work of Rigato [**19**] classifying curves of low genus over $\mathbb{F}_2$ with the maximum number of $\mathbb{F}_2$-rational points. A pithy slogan for the overall strategy might be "the most radical [extreme] covers are radical [cyclic]": noncyclic covers have to meet the Weil polynomial constraint for multiple isogeny factors of the Galois closure, so their point counts have fewer degrees of freedom along which to vary to achieve extreme values.

As a side effect of the analysis for $d = 7$, we also establish another missing assertion from [**10**]: there is exactly one curve of genus 6 over $\mathbb{F}_2$ which occurs as a cover of a curve of genus 1 with relative class number 1 (Lemma 10.2). This confirms the corresponding entry in [**10**, Table 3].

In light of Theorem 1.1 and the results of [**10**], the relative class number one problem now reduces to the case of purely geometric quadratic extensions of function fields over $\mathbb{F}_2$. This requires identifying curves of genus 6 or 7 over $\mathbb{F}_2$ with some specified Weil polynomials; we leave this task to a separate paper [**11**].

Although it costs us some shortcuts to do so (Remark 3.4), it is possible to limit the use of the hypothesis that $h_{F'} = h_F$ to the list of pairs of Weil polynomials provided by Theorem 1.2. Consequently, our methods can in principle be adapted to the relative class number m problem for $m > 1$, though in that case one expects to find some cases where a noncyclic extension does exist, and therefore additional arguments are needed to find *all* extensions with the corresponding Weil polynomials. For $d = 3, 4, 5$ it may be feasible to use Bhargava's orbit parametrizations [**2**–**4**] for this purpose; for $d = 3$, an analogous computation for number fields has been implemented by Belabas [**1**].

As in [**10**], the arguments depend on a number of computations in SageMath and Magma; we have documented these in some Jupyter notebooks associated to this paper, which are available from a GitHub repository [**12**]. The computations can be reproduced in less than a minute on a single CPU (Intel i5-1135G7@2.40GHz). In a few places, we adopt the convention of [**10**] of referring to isogeny classes of abelian varieties over $\mathbb{F}_2$ via their labels in LMFDB [**15**], formatted as links to the corresponding webpages.

2. Numerical data

Definition 2.1. For the remainder of the paper, let F'/F be an extension of function fields of degree $d > 1$ with $q_F = q_{F'} = 2$, $g_{F'} > g_F > 1$, and $h_{F'/F} = 1$. We typically write g, g' instead of $g_F, g_{F'}$. Let $C' \to C$ be the corresponding cover of curves over $\mathbb{F}_2$. For i a positive integer, let $a_i(F)$ and $a_i(F')$ denote the number of degree-i places of F and F', respectively. Recall that by Riemann–Hurwitz, F'/F is everywhere unramified if and only if

$$g' = g + d(g-1). \tag{2.1.1}$$

The main purpose of this section is to record some partial information from [**10**, Theorem 1.3(b)] about the possible Weil polynomials of the curves C and C'. To begin with, we read off the following statement.

Lemma 2.2. *We have*

$$a_1(F) > \sum_{i=1}^{\lfloor (d-1)/2 \rfloor} a_i(F') + \begin{cases} \frac{1}{2} a_{d/2}(F') & \text{if (2.1.1) holds,} \\ a_{d/2}(F') & \text{otherwise,} \end{cases} \tag{2.2.1}$$

except in one case where

$$d = 5,\ g = 2,\ g' = 6,\ a_1(F) = 5,\ a_1(F') = 0,\ a_2(F') = 5,\ a_3(F') = 0. \tag{2.2.2}$$

While we can mostly ignore the case $d = 2$ because any quadratic extension is automatically cyclic, we will need a few features of that case in order to analyze larger d.

REMARK 2.3. In case $d = 2$, the following statements hold.

- For $(g, g') = (2, 3)$, we have
$$(\#C(\mathbb{F}_2), \#C(\mathbb{F}_4); \#C'(\mathbb{F}_2)) \in \{(2, 8; 0), (4, 8; 2)\};$$
moreover, by [**10**, Theorem 1.3(c)], both cases are possible.
- For $(g, g') = (3, 5)$, we have $\#C'(\mathbb{F}_2) \leq 2$. Moreover, if $\#C(\mathbb{F}_2) = 4$ then $\#C(\mathbb{F}_4) = 8$.
- For $(g, g') = (3, 6)$, we have $\#C(\mathbb{F}_2) \geq 4$.
- For $(g, g') = (4, 7)$, we have $\#C(\mathbb{F}_2) \geq 3$.
- For $(g, g') = (5, 9)$, we have $\#C(\mathbb{F}_2) \geq 2$. Moreover, if $\#C(\mathbb{F}_2) = 2$ then $\#C'(\mathbb{F}_4) = 4$.

REMARK 2.4. Remark 2.3 has the following specific consequence which will be of special interest. According to LMFDB (see 2.2.ab_c), there is exactly one curve C_1 of genus 2 over $\mathbb{F}_2$ with $(\#C_1(\mathbb{F}_{2^i}))_{i=1}^2 = (2, 8)$. If $C' \to C_1$ is a degree-2 étale covering, then either this covering or its relative quadratic twist has relative class number 1.

By the same token, there is exactly one curve C_2 of genus 2 over $\mathbb{F}_2$ with $(\#C_2(\mathbb{F}_{2^i}))_{i=1}^2 = (4, 8)$, namely the quadratic twist of C_1. If $C' \to C_2$ is a degree-2 étale covering, then again either this covering or its relative quadratic twist has relative class number 1.

Table 2 summarizes the possible point counts for C and C' when $d > 2$; we include only information needed in our proofs, leading to some gaps in the table.

3. Splitting types and splitting sequences

We now formalize the main strategy used in the arguments.

DEFINITION 3.1. For a place of F which does not ramify in F', the *splitting type* of this place is the partition of d corresponding to the Frobenius conjugacy class of the place in the symmetric group S_d; in other words, it records the relative degrees of the places of F' lying above the original place. When describing a splitting type, we write a^b to represent b copies of a in the partition.

We then define the *splitting sequence* of F'/F as the sequence $(s_1, s_2, \dots)$ in which s_i is the multiset of splitting types of degree-i places of F. When describing a splitting sequence, we write $(\times n)$ after a partition to indicate that it occurs with multiplicity $n > 1$.

As a first application of the strategy, we record the following consequence of Lemma 2.2.

LEMMA 3.2. *There exists at least one degree-*1 *place of* F *which lifts to a degree-*d *place of* F'.

PROOF. Suppose the contrary. Each degree-1 place of F which does (resp. does not) ramify in F' lifts either to at least one place of F' of degree strictly less than $d/2$ or to at least one place (resp. at least two places) of degree exactly $d/2$. However, this contradicts (2.2.1) save for the exceptional case of Lemma 2.2. In that case, (2.2.2) tells us that F'/F is everywhere unramified and F' has no places of degree 1 or 3; this leaves no possible splitting types for a degree-1 place other than a single degree-5 place. □

TABLE 2. Options for the point counts of C and C' in Theorem 1.2 when $d > 2$. We have omitted some data not used in the proof of Theorem 1.1.

	d	g	g'	$\#J(C)(\mathbb{F}_2)$	$\#J(C)(\mathbb{F}_4)$	$\#C(\mathbb{F}_{2^i})$	$\#C'(\mathbb{F}_{2^i})$
(1)	3	2	4	$3/9/7/13$			$0/1$
(2)	3	2	4			3	0
(3)	3	2	6			$3/4/5$	$0, 2$
(4)	3	3	7	$23/27$		4	
(5)	3	3	7			$2, 8, 14$	$0, 0$
(6)	3	3	7			$3, 7$	$0, 0$
(7)	3	3	7			$4, 6/8$	$0, 0$
(8)	3	3	7			$4, 12$	$0, 6$
(9)	3	3	7			5	$1, 1$
(10)	3	3	7			$5, 9$	$1, 3$
(11)	3	3	7			6	$2, 2, 14$
(12)	3	4	10			$6, 6$	$0, 0$
(13)	3	4	10			$7/8$	0
(14)	4	2	5				$1, *, 1$
(15)	4	2	5	4	40	$2, 8$	$0, 0$
(16)	4	2	5	8	16	$4, 4$	$0, 4$
(17)	4	2	5	10	40	$4, 8$	$0, 8$
(18)	4	2	6			5	$1, 1, *, 17$
(19)	4	3	9	36		$5, 9$	$0, 0, *, 28$
(20)	4	3	9	50	200	$6, 8$	$0, 0$
(21)	5	2	6			4	$0, 6, 0, 18, 0$
(22)	5	2	6	11		$4, 10, 7$	$0, 2, 15$
(23)	5	2	6	19		$6, 6$	$1, 3, 7$
(24)	5	2	6	15		$5, 9, 5$	$0, 6, 3$
(25)	5	2	6	5		$3, 5, 9, 33, 33$	$0, 0, 0, 20, 15$
(26)	5	2	6			$4, 8, 10, 24, 14$	$0, 0, 15, 20, 20$
(27)	5	2	6	15		$5, 9, 5, 17, 25$	$0, 10, 0, 10, 25$
(28)	5	2	6			$3, 7, 9, 31, 33, 43, 129$	$0, 0, 9, 8, 30, 33, 168$
(29)	6	2	7	13		$5, 5$	$0, 2$
(30)	6	2	7	15		5	$1, 1, 1$
(31)	6	2	7	19		$6, 6$	$0, 2/4$
(32)	6	2	7			$4, 8, 10, 24, 14, 56$	$0, 0, 6, 8, 30, 24$
(33)	6	2	7			$5, 7, 11, 15, 15$	$0, 2, 6, 10, 5$
(34)	6	2	7	13		$5, 5, 17, 9, 25, 65$	$0, 0, 12, 4, 15, 90$
(35)	7	2	8	14		$5, 7$	$0, 0, 0, 0, 0, 84, 133, 336$

REMARK 3.3. Given candidate tuples $(\#C(\mathbb{F}_{2^i}))_{i=1}^n, (\#C'(\mathbb{F}_{2^i}))_{i=1}^n$ for some n, identifying the splitting sequences compatible with these values is a combinatorial exercise akin to a pencil-and-paper logic puzzle. While this exercise is pleasant enough in each individual instance, given the number of instances involved it is more reliable to automate this process. We do this by a recursive procedure: given a candidate for the first $n-1$ terms of the splitting sequence, we iterate over possible n-th terms (i.e., $a_n(F)$-element multisets of partitions of d) to see which ones give the correct values of $\#C'(\mathbb{F}_{2^n})$. This completes all cases of the problem referenced throughout this paper in negligible time.

Since we are in the special situation where $h_{F'} = h_F$, we can make some additional arguments. These are not strictly necessary, and indeed are not be used in the uniform application of Remark 3.3 (Lemma 5.6); however, we will use them in some of the human-readable alternate calculations in order to shorten the arguments.

REMARK 3.4. Suppose that $\#J(C)(\mathbb{F}_2)$ is coprime to d. Then the composition

$$J(C)(\mathbb{F}_2) \to J(C')(\mathbb{F}_2) \to J(C)(\mathbb{F}_2),$$

in which the first map is pullback of divisors and the second map is pushforward of divisors, is multiplication by d and hence an isomorphism. Since we are requiring $h_{F'} = h_F$, the pullback map is an injection between two finite groups of the same order, and hence also an isomorphism. Consequently, any degree-0 divisor on C' which pushes forward to a principal divisor is itself principal.

This immediately implies that no degree-1 place of F can lift to more than one degree-1 place of F'. In a similar vein, if some degree-3 place of F lifts to 4 or more degree-3 places of F', then C' admits a g_3^r with $r \geq 2$, which implies $g' \leq 1$.

We can also use this logic in conjunction with the Castelnuovo–Severi inequality [**22**, Theorem 3.11.3]: if C' admits two maps to $\mathbf{P}^1$ which are incommensurable (in that they do not factor through a common map which is not an isomorphism) of degrees d_1 and d_2, then $g' \leq (d_1 - 1)(d_2 - 1)$. This implies that if there are two different degree-3 places of F, each of which lifts to 2 or more degree-3 places of F', then $g' \leq 4$.

4. Subfields of the Galois closure

Let F'' be the Galois closure of F'/F and let C'' be the associated curve. Let G be the Galois group of F''/F, viewed as a transitive subgroup of S_d. Building on [**10**, Lemma 8.2], it will be profitable to consider some other subfields of F'' and the covers of C corresponding to them. In doing so, we must keep in mind that these fields need not be purely geometric extensions of F; see for example Remark 4.4.

DEFINITION 4.1. Write q'' for $q_{F''}$. Let G_0 be the Galois group of F'' over $F \cdot \mathbb{F}_{q''}$. Then G_0 is a normal subgroup of G and $G/G_0 \cong \mathrm{Gal}(\mathbb{F}_{q''}/\mathbb{F}_q)$ is cyclic.

Let $\chi_0, \chi_1, \dots$ be the nontrivial irreducible characters of G over $\mathbb{Q}$, numbered so that χ_0 is the restriction of the standard representation of S_d. For each i, let d_i be the dimension of each irreducible *complex* representation of G_0 appearing in χ_i. Then writing Res for Weil restriction, we have an isogeny decomposition

$$\mathrm{Res}_{\mathbb{F}_{q''}/\mathbb{F}_q} J(C'') \sim J(C) \times \prod_{i \geq 0} B_i^{d_i} \tag{4.1.1}$$

of abelian varieties; when $C' \to C$ is étale, we have $\dim(B_i) = (\dim \chi_i)(g-1)$. In particular, B_0 is the (generalized) Prym variety A of the original cover $C' \to C$.

Following the convention of [**10**], we write $T_{*,q}$ for the q-Frobenius trace of $*$ (which could be either a curve or an abelian variety); for every positive integer n,

$$\#C'(\mathbb{F}_{q^n}) = \#C(\mathbb{F}_{q^n}) - T_{A,q^n}. \tag{4.1.2}$$

More generally, if H is a subgroup of G which surjects onto G/G_0, then

$$J(C''/H) \sim J(C) \times \prod_{i \geq 0} B_i^{c_i} \tag{4.1.3}$$

where c_i is the multiplicity of χ_i in the representation of G induced from the trivial representation of H. For every positive integer n,

$$\#(C''/H)(\mathbb{F}_{q^n}) = \#C(\mathbb{F}_{q^n}) - \sum_i c_i T_{B_i,q^n}. \tag{4.1.4}$$

DEFINITION 4.2. An important special case of Definition 4.1 occurs when $G_0 \not\subseteq A_d$ and $H = G_0 \cap A_d$. In this case, the function field of C''/H is the *quadratic resolvent* of F'/F. The decomposition (4.1.3) reduces to $J(C''/H) \sim J(C) \times B$ where B is the Prym variety of $C''/H \to C$; in the notation of Definition 4.1, B corresponds to the sign representation of G_0. An unramified splitting type for $C' \to C$ corresponds to the splitting type 1^2 or 2 for the quadratic resolvent according to whether it is the cycle type of an even or odd permutation (i.e., whether or not the number of parts in the partition has the same parity as d).

REMARK 4.3. If the quadratic extension is an *everywhere unramified* quadratic extension, then by class field theory,

$$\#J(C)(\mathbb{F}_2) \equiv 0 \pmod 2. \tag{4.3.1}$$

REMARK 4.4. We must account for the possibility that the quadratic resolvent is a constant extension rather than a purely geometric one. In this case, the Frobenius conjugacy class of a degree-i place of F must be odd if i is odd and even if i is even. That is, B is the quadratic twist of $J(C)$.

From Lemma 3.2 and Remark 4.4, we immediately deduce the following.

LEMMA 4.5. *The group G contains a d-cycle. Consequently:*

(a) *if d is even, then $G \not\subseteq A_d$;*
(b) *if d is odd, then either $G \subseteq A_d$ or the quadratic resolvent is purely geometric.*

REMARK 4.6. We can now describe the basic paradigm that we will use to address the cases $d = 3, 4, 5, 6$ of Theorem 1.1. We start with an option for the Weil polynomials of C and C'. We then use the process indicated in Remark 3.3 to compute possible values of the first few terms of the splitting sequence consistent with the point counts of C and C', any known limitations on G (which in some cases get stricter as we compute more terms), and the constraint that the traces of any abelian variety known to occur as isogeny factors of $J(C'')$ must come from a Weil polynomial of the appropriate degree (see below).

REMARK 4.7. The set of Weil polynomials corresponding to abelian varieties of dimension ≤ 6 over $\mathbb{F}_2$ can be recovered using SAGEMATH as in [**10**]. The results have also been tabulated in LMFDB; note that on the page for the isogeny class of a given abelian variety A, the "number of points on the curve" over $\mathbb{F}_q$ equals $q + 1 - T_{A,q}$ whether or not A is isogenous to a Jacobian.

For the benefit of the human reader, we spell out some of the most relevant constraints on the traces of an abelian variety A over $\mathbb{F}_2$.

- We have

$$|T_{A,2}| \leq 2\dim(A) \tag{4.7.1}$$

with equality iff A is isogenous to a power of an elliptic curve with trace ± 2 (e.g., by [**21**, Theorem 2.1.1]). When equality occurs, we have $T_{A,4} = 0$.

- For $\dim(A) = 2$,

$$T_{A,2} = -3 \Longrightarrow T_{A,4} \in \{-1, -3\}, \tag{4.7.2}$$

$$T_{A,2} = -2 \Longrightarrow T_{A,4} \in \{-6, -4, -2, 0\}. \tag{4.7.3}$$

5. Subfields by degree

We now make explicit some consequences of the discussion from §4 for $d = 3, 4, 5, 6$, then describe a unified calculation that addresses most cases.

REMARK 5.1. Suppose that $d = 3$ and $G_0 = G = S_3$. The equation (4.1.4) specializes to

$$\#C''(\mathbb{F}_q) = \#C(\mathbb{F}_q) - 2T_{A,q} - T_{B,q}. \tag{5.1.1}$$

In particular, if $\#C'(\mathbb{F}_q) = 0$, then $\#C''(\mathbb{F}_q) = 0$ and combining (4.1.2) with (5.1.1) yields

$$T_{A,q} = -T_{B,q} = \#C(\mathbb{F}_q). \tag{5.1.2}$$

REMARK 5.2. Suppose that $d = 4$ and $G = S_4$. The maximal constant subextension of F''/F is cyclic, so it must equal either F or the quadratic resolvent.

Suppose now that $C' \to C$ is étale. If the quadric resolvent is constant, then the cubic resolvent is a cyclic cubic étale cover of $C_{\mathbb{F}_4}$ which is not a base extension from C, forcing

$$\#J(C)(\mathbb{F}_4)/\#J(C)(\mathbb{F}_2) \equiv 0 \pmod 3. \tag{5.2.1}$$

Similarly, if the quadratic resolvent is purely geometric, then it admits a cyclic cubic étale cover which is not a base extension from C, forcing

$$\#B(\mathbb{F}_2) \equiv 0 \pmod 3. \tag{5.2.2}$$

REMARK 5.3. Suppose that $d = 5$, $G = A_5$, and $C' \to C$ is étale. Since A_5 is simple, this implies $G_0 = A_5$. We may number the characters in Definition 4.1 so that $\dim(B_1) = 5(g-1)$, $\dim(B_2) = 6(g-1)$. Then (4.1.4) specializes to

$$\#(C''/D_5)(\mathbb{F}_q) = \#C(\mathbb{F}_q) - T_{B_1,q} \tag{5.3.1}$$

$$\#(C''/A_3)(\mathbb{F}_q) = \#C(\mathbb{F}_q) - 2T_{A,q} - T_{B_1,q} - T_{B_2,q}. \tag{5.3.2}$$

The conversion of splitting types from C' to these quotients is as follows:

Type in C'	Type in C''/D_5	Type in C''/A_3
5	$5+1$	5^4
$3+1^2$	3^2	3^6+1^2
2^2+1	2^2+1^2	2^{10}
1^5	1^6	1^{20}

REMARK 5.4. Suppose that $d = 6$, $G = S_6$, and $C' \to C$ is étale. The maximal constant subextension of F''/F is cyclic, so it must equal either F or the quadratic resolvent. In either case, we may number the characters in Definition 4.1 so that χ_1 is the image of χ_0 under the action of an outer automorphism of S_6; then $\dim(B_1) = 5(g-1)$. The quotient by C'' by the image of S_5 under an outer automorphism is a cover of C whose Jacobian is isogenous to $J(C) \times B_1$; we call this the *sextic twin* of the original cover. Note that even if the quadratic resolvent

is constant, the outer automorphism of S_6 preserves A_6, so the sextic twin descends canonically to $\mathbb{F}_2$.

For reference, we list here the possible splitting types for unramified places and the effect of an outer automorphism on these types.

$$\begin{array}{llllll} \text{odd}: & 6 \leftrightarrow 3+2+1, & 4+1^2, & 2^3 \leftrightarrow 2+1^4, & & \\ \text{even}: & 5+1, & 4+2, & 3^2 \leftrightarrow 3+1^3, & 2^2+1^2, & 1^6. \end{array}$$

REMARK 5.5. Suppose that $d = 6$, $G = \mathrm{PGL}(2,5)$, and $C' \to C$ is étale. The splitting types $3+2+1$, $2+1^4$, $4+2$, and $3+1^3$ cannot occur, as these correspond via the outer automorphism to splitting types with no singletons.

Following the model of Remark 5.4, we may construct a sextic twin, but in this case it is reducible: it is the disjoint union of C with a degree-5 cover. The Jacobian of the sextic twin is isogenous to $J(C)^2 \times B_1'$ with $\dim(B_1') = 4(g-1)$.

Implementing Remark 4.6, we obtain the following via a unified calculation. We will also step through the cases individually in subsequent sections; this will help to illustrate why some of the conditions appear.

LEMMA 5.6. *Consider a candidate pair of point count sequences*

$$(\#C(\mathbb{F}_{2^i}))_{i=1}^{\infty}, \qquad (\#C'(\mathbb{F}_{2^i}))_{i=1}^{\infty}$$

from [**10**, Theorem 1.3(b)] *for which* $3 \leq d \leq 6$ *and* $g' = dg - d + 1$. *Choose* $G \subseteq S_d$ *such that*

$$(d, G) \in \{(3, S_3), (4, S_4), (5, A_5), (6, S_6), (6, \mathrm{PGL}(2,5))\}.$$

If d is odd, set $G_0 := G$; otherwise, choose $G_0 \in \{G, G \cap A_d\}$. Then there is no splitting sequence of length 7 compatible with the point counts and all of the following restrictions.

- *Restrictions on $J(C)$:*
  - *If $G_0 \not\subseteq A_d$, then the mod-2 congruence* (4.3.1) *holds.*
  - *If $d = 4$ and $G_0 = A_4$, then the mod-3 congruence* (5.2.1) *holds.*
- *Restrictions on splitting types:*
  - *If $G \neq G_0$, then Remark* 4.4 *applies.*
  - *If $(d, G) = (5, A_5)$, then the splitting types $4+1, 3+2, 2+1^3$ do not occur (Remark* 5.3*).*
  - *If $(d, G) = (6, \mathrm{PGL}(2,5))$, then the splitting types $4+2, 3+2+1, 3+1^3, 2+1^4$ do not occur (Remark* 5.5*).*
- *Compatibility with Weil polynomials:*
  - *If $G_0 \not\subseteq A_d$, then the traces of B are compatible with a Weil polynomial. Moreover, if $g = 2$ and $(\#C(\mathbb{F}_{2^i}))_{i-1}^2 \in \{(2,8),(4,8)\}$, then $T_{B,2} \in \{\pm 2\}$ (Remark* 2.4*).*
  - *If $d = 5$, then the traces of B_1 and B_2 are compatible with Weil polynomials (Remark* 5.3*).*
  - *If $d = 6$ and $G_0 = G$, then the traces of B_1 (if $G = S_6$) or B_1' (if $G = \mathrm{PGL}(2,5)$) are compatible with a Weil polynomial (Remark* 5.4*, Remark* 5.5*).*

In the remaining sections, we proceed through the values $d > 2$ allowed by Theorem 1.2. For ease of reference, when citing Table 2 we will specify rows individually or in short ranges using an abbreviated syntax: e.g., "Table 2(4–6)" refers to rows (4), (5), and (6) of the table.

6. Degree 3

For $d = 3$, we rule out $G = S_3$ using an analysis of quadratic extensions.

LEMMA 6.1. *If $d = 3$, then $C' \to C$ is Galois and cyclic.*

PROOF. Suppose to the contrary that $G = S_3$. By Theorem 1.2, we have $(g, g') \in \{(2,4),(2,6),(3,7),(4,10)\}$. By Lemma 4.5, the quadratic resolvent is purely geometric, so $G_0 = S_3$ and the mod-2 congruence (4.3.1) applies unless $(g, g') = (2, 6)$. Let $C''' := C''/A_3$ be the curve corresponding to the quadratic resolvent.

We first treat the case $(g, g') = (2, 6)$. Table 2(3) shows that

$$\#C(\mathbb{F}_2) \geq 3, \qquad \#C'(\mathbb{F}_2) = 0, \qquad \#C'(\mathbb{F}_4) = 2.$$

By Riemann–Hurwitz, $C' \to C$ is ramified at either one or two geometric points of C'. Since $\#C'(\mathbb{F}_2) = 0$, $C' \to C$ must ramify at the unique degree-2 place of F'; moreover, this must be a triple point and not a double point (as the latter would force $a_2(F') \geq 2$). It follows that $C''' \to C$ is étale, and so $\dim(B) = 1$ and $T_{B,2} = -\#C(\mathbb{F}_2) \leq -3$ by (5.1.2), violating (4.7.1)

In the remaining cases, $C' \to C$ is étale and we may appeal to the uniform calculation (Lemma 5.6). Alternatively, we may break the individual cases down as follows.

- For $(g, g') = (4, 10)$, we have $\dim(B) = 3$. From Table 2(12,13), we have
$$\#C(\mathbb{F}_2) \geq 7 \text{ or } (\#C(\mathbb{F}_{2^i}))_{i=1}^2 = (6,6), \qquad \#C'(\mathbb{F}_2) = 0.$$
By (5.1.2), $T_{B,2} \leq -7$ or $T_{B,2} = T_{B,4} = -6$, contradicting Remark 4.7.
- For $(g, g') = (3, 7)$, we have $\dim(B) = 2$.
  - Table 2(4) is ruled out by the mod-2 congruence (4.3.1).
  - If $\#C'(\mathbb{F}_2) > 1$, then from Table 2(11),
  $$\#C(\mathbb{F}_2) = 6, \qquad (\#C'(\mathbb{F}_{2^i}))_{i=1}^3 = (2, 2, 14).$$
  There are not enough places of F' of degree at most 3 to cover the degree-1 places of F.
  - If $\#C'(\mathbb{F}_2) = 1$, then some degree-1 place of F has splitting type $2 + 1$, so $\#C'(\mathbb{F}_4) \geq 3$. This rules out Table 2(9); from Table 2(10),
  $$(\#C(\mathbb{F}_{2^i}))_{i=1}^2 = (5, 9), \qquad (\#C'(\mathbb{F}_{2^i}))_{i=1}^2 = (1, 3).$$
  The splitting sequence begins $\{3(\times 4), 2+1\}, \{3(\times 2)\}$. This means that $(\#C'''(\mathbb{F}_{2^i}))_{i=1}^2 = (8, 18)$ and $(T_{B,2^i})_{i=1}^2 = (-3, -9)$, contradicting (4.7.2).
  - If $\#C'(\mathbb{F}_2) = 0$ and $\#C'(\mathbb{F}_4) > 0$, then from Table 2(8),
  $$(\#C(\mathbb{F}_{2^i}))_{i=1}^2 = (4, 12).$$
  By (5.1.2), $T_{B,2} = -4$. By Remark 4.7, B is isogenous to the square of an elliptic curve with trace -2, so the relative quadratic twist of C''' is a degree-2 étale cover of C with relative class number 1. However, $\#C(\mathbb{F}_4) = 12$ which violates Remark 2.3.
  - If $\#C'(\mathbb{F}_2) = \#C'(\mathbb{F}_4) = 0$, then from Table 2(5–7),
  $$(\#C(\mathbb{F}_{2^i}))_{i=1}^2 \in \{(2,8),(3,7),(4,2),(4,8)\}.$$
  By (5.1.2), $(T_{B,2^i})_{i=1}^2 = (-\#C(\mathbb{F}_{2^i}))_{i=1}^2$, contradicting Remark 4.7.

- For $(g, g') = (2, 4)$, we have $\dim(B) = 1$. The mod-2 congruence (4.3.1) rules out Table 2(1). From Table 2(2), we have

$$\#C(\mathbb{F}_2) = 3, \qquad \#C'(\mathbb{F}_2) = 0.$$

By (5.1.2), $T_{B,2} = -3$, contradicting Remark 4.3. □

7. Degree 4

For $d = 4$, we rule out $G = D_4$ using an analysis of quadratic extensions.

LEMMA 7.1. *If $d = 4$, then the Galois group of F'/F cannot equal D_4.*

PROOF. By Theorem 1.2, we have $(g_F, g_{F'}) \in \{(2,5), (2,6), (3,9)\}$. Suppose that F'/F contains an intermediate subfield E; then $g_E = 2g_F - 1$ and both E/F and F'/E are purely geometric quadratic extensions with relative class number 1. Applying Remark 2.3 to these extensions, we rule out the cases $(g_F, g_{F'}) = (2,6), (3,9)$. For $(g_F, g_{F'}) = (2,5)$, we see from Remark 2.3 that C has p-rank 1, as then does the intermediate curve C''' by the Deuring–Shafarevich formula [**10**, (7.2)]. In particular, $C_{\overline{\mathbb{F}}_2}$ and $C'''_{\overline{\mathbb{F}}_2}$ each admit only one étale double cover, which forces F'/F to be cyclic. □

LEMMA 7.2. *If $d = 4$, then $C' \to C$ is Galois and cyclic.*

PROOF. Suppose to the contrary that $G \neq C_4$. By Theorem 1.2, we have $(g, g') \in \{(2,5), (2,6), (3,9)\}$. We have $G \neq D_4$ by Lemma 7.1 and $G \not\subseteq A_4$ by Lemma 4.5, so $G = S_4$. The maximal constant subextension of F''/F is cyclic, and so must equal either F or the quadratic resolvent. In the former case, let $C''' := C''/A_4$ be the curve corresponding to the quadratic resolvent and let F''' be its function field.

We first treat the case $(g, g') = (2, 6)$. The map $C' \to C$ ramifies at one geometric point, which must be $\mathbb{F}_2$-rational. By Table 2(18), we have $a_2(F') = 0$, so the ramified place of F' must be alone in its fiber, violating Riemann–Hurwitz.

In the remaining cases, $C' \to C$ is étale and we may appeal to the uniform calculation (Lemma 5.6). Alternatively, we may break the individual cases down as follows. If $(g, g') = (2, 5)$ and $\#C'(\mathbb{F}_2) = 1$, then by Table 2(14), $\#C'(\mathbb{F}_8) = 1$ and there is no way to accommodate the degree-1 place of F' in a fiber. To handle the remaining cases, suppose first that the quadratic resolvent is constant.

- For $(g, g') = (3, 9)$, the mod-3 congruence (5.2.1) rules out Table 2(20). From Table 2(19), we have

$$(\#C(\mathbb{F}_{2^i}))_{i=1}^2 = (5, 9), \qquad (\#C'(\mathbb{F}_{2^i}))_{i=1}^4 = (0, 0, 0, 28).$$

By Remark 4.4, the splitting sequence begins $\{4(\times 5)\}, \{2^2(\times 2)\}$, but this creates too many degree-4 places of F'.
- For $(g, g') = (2, 5)$, the mod-3 congruence (5.2.1) rules out Table 2(15–17).

Suppose next that the quadratic resolvent is purely geometric.

- For $(g, g') = (3, 9)$, we have $\dim(B) = 2$. From Table 2(19,20) we have

$$\#C(\mathbb{F}_2) \in \{5, 6\}, \qquad \#C'(\mathbb{F}_2) = \#C'(\mathbb{F}_4) = 0.$$

Consequently, each degree-1 place of F lifts to a degree-4 place of F', and hence to a degree-2 place of F'''. Since $0 = \#C'''(\mathbb{F}_2) = \#C(\mathbb{F}_2) - T_{B,2}$ by (4.1.4), we have $T_{B,2} = \#C(\mathbb{F}_2) \geq 5$, contradicting (4.7.1).

- For $(g, g') = (2, 5)$, we have $\dim(B) = 1$. From Table 2(15–17) we have

$$(\#C(\mathbb{F}_{2^i}))_{i=1}^2 \in \{(2,8), (4,4), (4,8)\} \tag{7.2.1}$$

and $\#J(C)(\mathbb{F}_2) \not\equiv 0 \pmod 3$. By the mod-3 congruence (5.2.2), $\#B(\mathbb{F}_2) \equiv 0 \pmod 3$; by (4.7.1) this forces $T_{B,2} = 0$. By this plus Remark 2.4 and (7.2.1), we must have $(\#C(\mathbb{F}_{2^i}))_{i=1}^2 = (4,4)$; from Table 2(16), we have $(\#C'(\mathbb{F}_{2^i}))_{i=1}^2 = (0,4)$. Now $\#C'''(\mathbb{F}_2) = \#C(\mathbb{F}_2)$ because $T_{B,2} = 0$, so two of the four degree-1 places of F have even splitting types in F'. Since $\#C'(\mathbb{F}_2) = 0$, the splitting type 2^2 must occur twice, but this forces the contradiction $\#C'(\mathbb{F}_4) \geq 8$. □

8. Degree 5

For $d = 5$, we rule out $G \in \{D_5, S_5\}$ using an analysis of quadratic extensions.

LEMMA 8.1. *If $d = 5$, then the Galois group of F'/F cannot equal D_5 or S_5.*

PROOF. By Theorem 1.2, $g = 2$ and $C' \to C$ is étale. By Lemma 4.5, G contains a 5-cycle, so if $G = S_5$ then the quadratic resolvent $C''' := C''/A_5$ is purely geometric, while if $G = D_5$ then $C''' := C''/C_5$ is a purely geometric degree-2 étale cover of C. In both cases, the mod-2 congruence (4.3.1) holds, which rules out Table 2(22–25,27).

At this point, by Table 2(21,26,28), $\#C(\mathbb{F}_2) \geq 3$ and every degree-1 point of C lifts to a degree-5 point of C'. Let B' be the Prym of $C''' \to C'$, so that $\dim(B') = 1$. If $G = S_5$, then $\#C'''(\mathbb{F}_2) = 0$ and so by (4.1.4), $T_{B',2} = \#C(\mathbb{F}_2) - \#C'''(\mathbb{F}_2) \geq 3$; if $G = D_5$, then $\#C'''(\mathbb{F}_2) = 2\#C(\mathbb{F}_2)$ and so by (4.1.4), $T_{B',2} = \#C(\mathbb{F}_2) - \#C'''(\mathbb{F}_2) \leq -3$. In both cases, we violate (4.7.1). □

LEMMA 8.2. *If $d = 5$, then $C' \to C$ is Galois and cyclic.*

PROOF. Suppose by way of contradiction that $G \neq C_5$. By Theorem 1.2, $g = 2$ and $C' \to C$ is étale; by Lemma 8.1, $G = A_5$. We may thus appeal to the uniform calculation (Lemma 5.6); alternatively, we may break down Table 2(21–28) as follows. We start with some rows for which computing the splitting sequence already yields a contradiction (even if the cover is cyclic).

- From Table 2(21), we have

$$\#C(\mathbb{F}_2) = 4, \qquad (\#C'(\mathbb{F}_{2^i}))_{i=1}^5 = (0,6,0,18,0).$$

There are not enough places of F' of degree at most 5 to cover the degree-1 places of F.

- From Table 2(22), we have

$$(\#C(\mathbb{F}_{2^i}))_{i=1}^3 = (4,10,7), \qquad (\#C'(\mathbb{F}_{2^i}))_{i=1}^3 = (0,2,15).$$

The splitting sequence begins $\{5(\times 3)\}, \{5(\times 2), 2^2+1\}, \{1^5\}$. Since $\#J(C)(\mathbb{F}_2) = 11$ is coprime to 5, the degree-3 places create a contradiction as per Remark 3.4.

- From Table 2(23), we have

$$(\#C(\mathbb{F}_{2^i}))_{i=1}^2 = (6,6), \qquad (\#C'(\mathbb{F}_{2^i}))_{i=1}^2 = (1,3).$$

There is no way to include the degree-2 place of F' in a fiber.

- From Table 2(24), we have
$$(\#C(\mathbb{F}_{2^i}))_{i=1}^3 = (5,9,5), \qquad (\#C'(\mathbb{F}_{2^i}))_{i=1}^3 = (0,6,3).$$
There is no way to include the degree-3 place of F' in a fiber.

For the remaining rows of Table 2, let B_1, B_2 be the abelian varieties described in Remark 5.3, so that $\dim(B_1) = 5$, $\dim(B_2) = 6$. We combine the analysis of splitting sequences with the point counts of B_1 and/or B_2 from (4.1.2), (5.3.1), and (5.3.2), then compare to Weil polynomial data (see Remark 8.3).

- From Table 2(25), we have
$$(\#C(\mathbb{F}_{2^i}))_{i=1}^5 = (3,5,9,33,33), \qquad (\#C'(\mathbb{F}_{2^i}))_{i=1}^5 = (0,0,0,20,15).$$
The splitting sequence begins $\{5(\times 3)\}, \{5\}, \{5(\times 2)\}, ?, \{5(\times 6)\}$ and
$$(T_{B_2,2^i})_{i=1}^5 = (-3,-5,-9,?,-48),$$
but the latter is inconsistent with Remark 8.3.
- From Table 2(26), we have
$$(\#C(\mathbb{F}_{2^i}))_{i=1}^3 = (4,8,10), \qquad (\#C'(\mathbb{F}_{2^i}))_{i=1}^3 = (0,0,15).$$
The splitting sequence begins $\{5(\times 4)\}, \{5(\times 2)\}, \{5, 1^5\}$ and
$$(T_{B_2,2^i})_{i=1}^3 = (-4,-8,-25),$$
but the latter is inconsistent with Remark 8.3.
- From Table 2(27), we have
$$(\#C(\mathbb{F}_{2^i}))_{i=1}^3 = (5,9,5), \qquad (\#C'(\mathbb{F}_{2^i}))_{i=1}^3 = (0,10,0).$$
The splitting sequence begins $\{5(\times 5)\}, \{5, 1^5\}, \emptyset$ and
$$(T_{B_2,2^i})_{i=1}^3 = (-5,-19,-5),$$
but the latter is inconsistent with Remark 8.3.
- From Table 2(28), we have
$$(\#C(\mathbb{F}_{2^i}))_{i=1}^7 = (3,7,9,31,33,43,129), \; (\#C'(\mathbb{F}_{2^i}))_{i=1}^7 = (0,0,9,8,30,33,168).$$
The splitting sequence begins
$$\{5(\times 3)\}, \qquad \{5(\times 2)\}, \qquad \{3+1^2, 2^2+1\},$$
$$\{5(\times 5), 3+1^2\} \text{ or } \{5(\times 4), 2^2+1(\times 2)\},$$
$$\{5(\times 4), 3+1^2, 2^2+1\} \text{ or } \{5(\times 3), 2^2+1(\times 3)\},$$
$$\{5(\times 4), 3+1^2\} \text{ or } \{5(\times 3), 2^2+1(\times 2)\}$$
and $(T_{B_2,2^i})_{i=1}^6 = (-3,-7,3,-27/-7,-28/-3,-49/-19)$. By comparing with Remark 8.3, we deduce that
$$(T_{B_2,2^i})_{i=1}^7 = (-3,-7,3,-7,-3,-19,25);$$
the splitting sequence continues
$$\{5(\times 4), 3+1^2(\times 2), 2^2+1(\times 10), 1^5(\times 2)\}$$
$$\text{or } \{5(\times 4), 3+1^2(\times 6), 2^2+1(\times 7), 1^5\} \text{ or } \{5(\times 4), 3+1^2(\times 10), 2^2+1(\times 4)\}$$
and we obtain $(T_{B_1,2^i})_{i=1}^7 = (0,0,0,-8,-30,-24,-126/-42/42)$. However, the Weil polynomial of B_1 is uniquely determined by $(T_{B_1,2^i})_{i=1}^5$ and forces $T_{B_1,2^7} = 0$. □

REMARK 8.3. We summarize here the Weil polynomial data used in the proof of Lemma 8.2. For $\dim(A) = 6$,

$$(T_{A,2^i})_{i=1}^3 = (-3,-5,-9) \Longrightarrow T_{A,2^5} \geq -13$$
$$(T_{A,2^i})_{i=1}^2 = (-4,-8) \Longrightarrow T_{A,2^3} \geq -16$$
$$(T_{A,2^i})_{i=1}^2 = (-5,-19) \Longrightarrow T_{A,2^3} = 25$$
$$(T_{A,2^i})_{i=1}^4 = (-3,-7,3,-27) \Longrightarrow T_{A,2^5} \geq 17$$
$$(T_{A,2^i})_{i=1}^5 = (-3,-7,3,-7,-28) \Longrightarrow T_{A,2^6} = 41$$
$$(T_{A,2^i})_{i=1}^5 = (-3,-7,3,-7,-3) \Longrightarrow T_{A,2^6} \geq -19.$$

9. Degree 6

For $d = 6$, we have a number of Galois groups to worry about. We deal with most of these by analyzing intermediate subfields.

LEMMA 9.1. *If $d = 6$, then F'/F does not contain an intermediate subfield.*

PROOF. Suppose to the contrary that E is an intermediate subfield; then both E/F and F'/E are extensions with relative class number 1. By Theorem 1.2, we have $(g_F, g_{F'}) = (2,7)$.

In case $[E : F] = 3$, from Table 2(1,2), the intermediate curve has at most one $\mathbb{F}_2$-rational point. Remark 2.3 then shows that no extension F'/E can exist.

In case $[E : F] = 2$, from Table 2(29–34) we have $\#C(\mathbb{F}_2) \geq 4$. Comparing with Remark 2.3, we deduce that the intermediate curve has 2 $\mathbb{F}_2$-rational points. From Table 2(4–11), we see that the cover F'/E corresponds to Table 2(5), so the Jacobian of the intermediate curve has isogeny class 3.2.ab_c_a which does not appear in [**10**, Table 5]. Hence by [**10**, Theorem 1.3], F'/E cannot be cyclic, contradicting Lemma 6.1. □

LEMMA 9.2. *If $d = 6$, then $C' \to C$ is Galois and cyclic.*

PROOF. Suppose to the contrary that $G \neq C_6$. By Theorem 1.2, $g = 2$ and $C' \to C$ is étale. By Lemma 4.5, G contains a 6-cycle, so $G \not\subseteq A_6$. By Lemma 9.1 and Lemma 6.1, F'/F has no intermediate subfields; consequently, G must be either $\mathrm{PGL}(2,5) \cong S_5$ or S_6. We may thus appeal to the uniform calculation (Lemma 5.6); alternatively, we may break the individual cases down as follows.

Suppose first that the quadratic resolvent is constant. Among Table 2(29–34), for some rows we obtain a contradiction directly from the splitting sequence by keeping in mind Remark 4.4.

- From Table 2(29), we have
$$(\#C(\mathbb{F}_{2^i}))_{i=1}^2 = (5,5), \qquad (\#C'(\mathbb{F}_{2^i}))_{i=1}^2 = (0,2).$$
The degree-2 place of F' maps to a degree-1 place of F, forcing the splitting type $4+2$ in degree 1.
- From Table 2(30), we have
$$\#C(\mathbb{F}_2) = 5, \qquad (\#C'(\mathbb{F}_{2^i}))_{i=1}^3 = (1,1,1).$$
The degree-1 place of F' forces the splitting type $5+1$ in degree 1.

- From Table 2(31), we have
$$(\#C(\mathbb{F}_{2^i}))_{i=1}^2 = (6,6), \qquad (\#C'(\mathbb{F}_{2^i}))_{i=1}^2 = (0,2/4).$$
The degree-2 places of F' must map to *distinct* degree-1 places of F, forcing the splitting type $4+2$ in degree 1.
- From Table 2(32), we have
$$(\#C(\mathbb{F}_{2^i}))_{i=1}^6 = (4,8,10,24,14,56), \qquad (\#C'(\mathbb{F}_{2^i}))_{i=1}^6 = (0,0,6,8,30,24).$$
There are not enough degree-6 places of F' to cover the degree-1 places of F, forcing the splitting type $3+3$ in degree 1.

For the remaining rows (still assuming constant quadratic resolvent), the splitting sequence begins $\{6(\times 5)\}$ and the splitting type $3+2+1$ is forced to occur in degree 5 for parity reasons, yielding $G = S_6$. We now argue in terms of the sextic twin (Remark 5.4), keeping in mind that $\dim(B_1) = 5$.

- From Table 2(33), we have
$$(\#C(\mathbb{F}_{2^i}))_{i=1}^5 = (5,7,11,15,15), \qquad (\#C'(\mathbb{F}_{2^i}))_{i=1}^5 = (0,2,6,10,5).$$
The splitting sequence continues
$$\{5+1\}, \{6, 4+1^2\} \text{ or } \{4+1^2, 2^3\} \text{ or } \{3+2+1(\times 2)\},$$
$$\{5+1(\times 2)\} \text{ or } \{4+2, 2^2+1^2\} \text{ or } \{3^2, 2^2+1^2\};$$
this yields $(T_{B_1,2^i})_{i=1}^4 = (0,-10,-27/-18/-9,-22/-10)$, contradicting Remark 9.3.
- From Table 2(34), we have
$$(\#C(\mathbb{F}_{2^i}))_{i=1}^6 = (5,5,17,9,25,65), \qquad (\#C'(\mathbb{F}_{2^i}))_{i=1}^6 = (0,0,12,4,15,90).$$
Since $\#J(C)(\mathbb{F}_2) = 13$ is coprime to 6, Remark 3.4 implies that in degree 3 we cannot use the splitting type $2+1^4$ at all, or the splitting type $4+1^2$ more than once. The splitting sequence thus continues
$$\emptyset, \{6, 4+1^2, 3+2+1(\times 2)\} \text{ or}$$
$$\{4+1^2, 3+2+1(\times 2), 2^3\} \text{ or } \{3+2+1(\times 4)\}, \{5+1\}$$
and
$$(T_{B_1,2^i})_{i=1}^4 = (0,-10,-3/-12/-21,-10).$$
Combining this with Remark 9.3 yields
$$(T_{B_1,2^i})_{i=1}^6 = (0,-10,-3,-10,10/15,23).$$
The splitting sequence now begins
$$\{6(\times 5)\}, \emptyset, \{3+2+1(\times 4)\}, \{5+1\}, \{6, 3+2+1(\times 3)\},$$
but there is no possible extension matching $T_{B_1,2^6}$.

Suppose next that the quadratic resolvent is purely geometric. The mod-2 congruence (4.3.1) rules out Table 2(29–31,34). For Table 2(32,33) we must allow both $G = \mathrm{PGL}(2,5)$ and $G = S_6$; we treat these cases in parallel, listing the numerical values that arise in Table 3.

- We first use the facts that $\dim(B) = 1$ and $|T_{B,2}| \leq 2$ to determine the first term of the splitting sequence.

TABLE 3. Numerics from the proof of Lemma 9.2.

$(\#C(\mathbb{F}_{2^i}))_{i=1}^5$	(4,8,10,24,14)	(5,7,11,15,15)
$(\#C'(\mathbb{F}_{2^i}))_{i=1}^5$	(0,0,6,8,30)	(0,2,6,10,5)
Degree 1 splittings	$\{6(\times 3), 3^2\}$	$\{6(\times 3), 4+2, 3^2\}$
$(T_{B,2^i})_{i=1}^5$	$(2, 0, -4, -8, -8)$	$(1, -3, -5, 1, 11)$
Degree 2 splittings	$\{(6/2^3)(\times 2)\}$	$\{6/2^3\}$
Degree 3 splittings	$\{(4+2/3^2)(\times 2)\}$	$\{(4+2/3^2)(\times 2)\}$
$(T_{B_1,2^i})_{i=1}^3$ (splittings)	$(-2, -20/-14/-8, -26/-17/-8)$	$(-1, -15/-9, -25/-16/-7)$
$(T_{B_1,2^i})_{i=1}^4$ (Weil polys)	$(-2, -8, -8, -4/\cdots/16)$	$(-1, -9, -7, -13/\cdots/19)$
Degree 4 splittings	$\{4+2(\times 2), 3+2+1(\times 2)\}$	$\{6, 3+2+1\}$
$T_{B_1,2^4}$ from splittings	0	-13
$T_{B_1,2^5}$ from Weil polys	33	49
$(T_{B_1',2^i})_{i=1}^3$ (splittings)	$(2, -12/-6/0, -16)$	$(4, -8/-2, -14)$
$(T_{B_1',2^i})_{i=1}^4$ (Weil polys)	$(2, 0, -16, -8/-4)$	none
Degree 4 splittings	$\{6, 5+1(\times 2), 2^3\}/\{6(\times 2), 3^2, 2^2+1^2\}$	none
$T_{B_1',2^4}$ from splittings	-4	none
$T_{B_1',2^5}$ from Weil polys	2	none

- Let $C''' := C''/(G \cap A_6)$ be the curve corresponding to the quadratic resolvent. We then compute $T_{B,4}, T_{B,8}$ and then $\#C'''(\mathbb{F}_4), \#C'''(\mathbb{F}_8)$. These values imply that the splitting types must all be odd in degree 2 and even in degree 3.
- We then compute candidates for the second and third terms of the splitting sequence compatible with B, keeping in mind the splitting types that cannot be used if $G = \mathrm{PGL}(2,5)$; derive from these the possible values of $(T_{B_1,2^i})_{i=1}^3$ or $(T_{B_1',2^i})_{i=1}^3$; and identify Weil polynomials for B_1 or B_1' matching these values. In each case, we either reach a contradiction or find a unique candidate for the first three terms of the splitting sequence (although not yet a unique Weil polynomial).
- We then repeat the previous step for the fourth term of the splitting sequence. In each remaining case, we determine at most one possible value for $T_{B_1,2^4}$ or $T_{B_1',2^4}$.
- We now find that there are no options for the fifth term of the splitting sequence consistent with both the point counts and the Weil polynomial constraints. □

REMARK 9.3. We summarize here the Weil polynomial data used in the proof of Lemma 9.2. For $\dim(A) = 4$,

$$(T_{A,2^i})_{i=1}^2 = (2,-12) \Longrightarrow T_{A,2^3} \geq -10$$
$$(T_{A,2^i})_{i=1}^2 = (2,-6) \Longrightarrow T_{A,2^3} \geq -10$$
$$(T_{A,2^i})_{i=1}^3 = (2,0,-16) \Longrightarrow T_{A,2^4} \in \{-8,-4\}$$
$$(T_{A,2^i})_{i=1}^2 = (4,-8) \Longrightarrow T_{A,2^3} \geq -11$$
$$(T_{A,2^i})_{i=1}^2 = (4,-2) \Longrightarrow T_{A,2^3} \geq -8.$$

For $\dim(A) = 5$,

$$(T_{A,2^i})_{i=1}^{3} = (0, -10) \Longrightarrow T_{A,2^3} \geq -18$$
$$(T_{A,2^i})_{i=1}^{3} = (0, -10, -18) \Longrightarrow T_{A,2^4} = 18$$
$$(T_{A,2^i})_{i=1}^{3} = (0, -10, -9) \Longrightarrow T_{A,2^4} \geq -6$$
$$(T_{A,2^i})_{i=1}^{3} = (0, -10, -12) \Longrightarrow T_{A,2^4} \geq 2$$
$$(T_{A,2^i})_{i=1}^{4} = (0, -10, -3, -10) \Longrightarrow T_{A,2^5} \in \{10, 15\}, T_{A,2^6} = 23$$
$$T_{A,2} = -2 \Longrightarrow T_{A,2^2} \geq -18$$
$$(T_{A,2^i})_{i=1}^{2} = (-2, -14) \Longrightarrow T_{A,2^3} \geq -2$$
$$(T_{A,2^i})_{i=1}^{2} = (-2, -8) \Longrightarrow T_{A,2^3} \geq -11$$
$$(T_{A,2^i})_{i=1}^{3} = (-2, -8, -8) \Longrightarrow -4 \leq T_{A,2^4} \leq 16$$
$$(T_{A,2^i})_{i=1}^{2} = (-1, -15) \Longrightarrow T_{A,2^3} \geq -1$$
$$(T_{A,2^i})_{i=1}^{2} = (-1, -9) \Longrightarrow T_{A,2^3} \geq -13$$
$$(T_{A,2^i})_{i=1}^{3} = (-1, -9, -7) \Longrightarrow -13 \leq T_{A,2^4} \leq 19.$$

REMARK 9.4. The astute reader will have noticed that Lemma 9.1 also rules out $G = C_6$, so Lemma 9.2 can be upgraded to say that the case $d = 6$ cannot occur at all! This is consistent with [**10**, Table 5]. By contrast, the proof of Lemma 7.1 must work around the fact that $d = 4$ does occur in [**10**, Table 5].

10. Degree 7

For $d = 7$, the complexity of the representation theory of S_7 and the presence of additional transitive subgroups (notably PSL(2, 7)) together make it quite difficult to follow the paradigm of Remark 4.6. Instead, we make a detailed analysis of polarizations in the style of [**6**].

LEMMA 10.1. *If $d = 7$, then $C' \to C$ is Galois and cyclic.*

PROOF. By Theorem 1.2, we have $g = 2$. From Table 2, we may read off the Weil polynomials of C and A: they are P_1 and P_1P_2 for

$$P_1(T) = T^2 + 2T - 1, \qquad P_2(T) = T^6 - 5T^5 - 3T^4 + 43T^3 - 33T^2 - 59T + 43.$$

We analyze this case following [**6**, Theorem 4.5]. The polynomials P_1, P_2 are both irreducible; let $K_1 = \mathbb{Q}[\pi_1]/(P_1(\pi_1)), K_2 = \mathbb{Q}[\pi_2]/(P_2(\pi_2))$ be the number fields defined by P_1, P_2. Using MAGMA, we compute that $\mathbb{Z}[\pi_1, \overline{\pi}_1], \mathbb{Z}[\pi_2, \overline{\pi}_2]$ are the maximal orders of K_1, K_2, and both orders have class number 1. It follows that up to isomorphism there are unique abelian varieties A_1, A_2 over $\mathbb{F}_2$ with respective Weil polynomials P_1, P_2; in particular, $A_1 \cong J(C)$.

By [**10**, Lemma 9.1], there is an exact sequence

$$0 \to \Delta \to J(C) \times_{\mathbb{F}_2} A_2 \to J(C') \to 0$$

in which Δ is a nontrivial finite flat group scheme killed by 7. In K_2, the rational prime 7 decomposes as $\mathfrak{p}^6$ where $\mathfrak{p} = (\zeta_7 - 1)$ for some nontrivial seventh root of unity $\zeta_7 \in K_2$. Consequently, Δ must be isomorphic to the kernel of $\zeta_7 - 1$ acting on A_2.

We now compare with the proof of [**6**, Proposition 4.2]. The principal polarization on $J(C')$ pulls back to a polarization (λ_1, λ_2) on $J(C) \times_{\mathbb{F}_2} A_2$ of degree

$(\#\Delta)^2$. The automorphism $1 \times \zeta_7$ acts on $J(C) \times_{\mathbb{F}_2} A_2$ with its polarization and acts trivially on the image of Δ, so $J(C')$ admits an automorphism of order 7 that is compatible with its polarization and equivariant with respect to $J(C) \to J(C')$. By Torelli [**17**, Theorem 12.1], C' admits an automorphism of order 7 over C, proving the claim. □

As promised, we adapt the argument to cover a related statement that was asserted without proof in [**10**, Remark 6.2].

LEMMA 10.2. *Let C_0 be the curve $y^2 + y = x^5$ over $\mathbb{F}_2$. Let C be a curve of genus 6 over $\mathbb{F}_2$ such that $(\#C(\mathbb{F}_{2^i}))_{i=1}^6 = (0, 0, 0, 20, 15, 90)$. Then there exists a cyclic étale morphism $C \to C_0$ of degree 5. In particular, C is unique up to isomorphism.*

PROOF. The real Weil polynomial of C factors as

$$(T - 2)(T + 2)(T^4 - 3T^3 - 6T^2 + 18T + 1).$$

Let E_1, E_2 be elliptic curves over $\mathbb{F}_2$ with $\#E_1(\mathbb{F}_2) = 1, \#E_2(\mathbb{F}_2) = 5$. By [**6**, Theorem 4.2], there exist morphisms $f_1 \colon C \to E_1$, $f_2 \colon C \to E_2$ of degrees dividing $2^2 \times 5$, $2^2 \times 19$ respectively. By [**6**, Theorem 4.2] again, there exist morphisms $g_1 \colon C_0 \to E_1$, $g_2 \colon C_0 \to E_2$ of degrees dividing 4. The composition $f_{i*} \circ f_i^*$ equals the isogeny $[\deg(f_i)]$ on $J(E_i)$; similarly, the composition $g_{i*} \circ g_i^*$ equals the isogeny $[\deg(g_i)]$ on $J(E_i)$.

The composition of $g_1^* \times g_2^* : E_1 \times E_2 \to J(C_0)$ followed by $g_{1*} \times g_{2*} : J(C_0) \to E_1 \times E_2$ is multiplication by a power of 2 on $E_1 \times E_2$, which in particular is an isogeny. Since $E_1 \times E_2$ and $J(C_0)$ are of the same dimension, it follows that both maps in the composition are also isogenies, and their composition in the opposite order is also multiplication by a power of 2 (the same one in fact).

Let $f^* \colon J(C_0) \to J(C)$ and $f_* \colon J(C) \to J(C_0)$ be the respective compositions

$$J(C_0) \xrightarrow{g_{1*} \times g_{2*}} E_1 \times E_2 \xrightarrow{f_1^* \times f_2^*} J(C), \qquad J(C) \xrightarrow{f_{1*} \times f_{2*}} E_1 \times E_2 \xrightarrow{g_1^* \times g_2^*} J(C).$$

The composition $f_* \circ f^* \colon J(C_0) \to J(C_0)$ is multiplication by an integer of the form $2^n 5^i 19^j$ with $n \geq 0$ and $i, j \in \{0, 1\}$. Let A be the reduced closed subscheme of the identity component of $\ker(f_*)$. Then as in [**10**, Lemma 9.3], we have an exact sequence

$$0 \to \Delta \to J(C_0) \times_{\mathbb{F}_2} A \to J(C) \to 0$$

in which the map $J(C_0) \to J(C)$ is f^* and Δ is a finite flat group scheme over $\mathbb{F}_2$ killed by $2^n \times 5 \times 19$ for some $n \geq 0$. In fact we must have $n = 0$ because $J(C_0)$ is supersingular (so any 2-power-torsion group subscheme of it is biconnected) whereas A_2 is ordinary (so any 2-power-torsion group subscheme of it has trivial biconnected constituent).

Let P be the Weil polynomial of A and put $K_2 = \mathbb{Q}[\pi]/(P(\pi))$. Using MAGMA, we compute that $\mathbb{Z}[\pi, \overline{\pi}]$ is the maximal order of K and its class number is 1. Consequently, A is the unique abelian variety with its Weil polynomial. Since $\zeta_5 \in K$, we deduce that A admits an automorphism α of order 5.

Since $19 \not\equiv 1 \pmod 5$, $\alpha - 1$ kills the 19-part of Δ. In the field K, the rational prime 5 decomposes as $\mathfrak{p}_1^4 \mathfrak{p}_2^4$ and $\zeta_5 - 1$ has positive valuation with respect to both $\mathfrak{p}_1$ and $\mathfrak{p}_2$; hence $\alpha - 1$ also kills the 5-part of Δ. We conclude that $\alpha - 1$ kills all of Δ; as in the proof of [**6**, Proposition 4.2], we conclude that $J(C)$ admits an automorphism of order 5 preserving its principal polarization and fixing $J(C_0)$. By

Torelli, C admits an automorphism of order 5, the quotient by which is a curve whose Jacobian is isogenous to $J(C_0)$. From LMFDB we see that $J(C_0)$ is the unique Jacobian in its isogeny class, so C is in fact a cyclic degree-5 cover of C_0, which by Riemann–Hurwitz must be étale. (Note that *a posteriori* the 19-part of Δ is trivial.) The map $C \to C_0$ defines an extension of function fields of relative class number 1; by the uniqueness aspect of [**10**, Theorem 1.3(c)], this leaves only one possible isomorphism class for C. □

REMARK 10.3. Note that the proof of Lemma 10.1 does not rule out the existence of a curve with the point counts ascribed to C', but which does not cover C. By contrast, the proof of Lemma 10.2 does rule out the existence of a curve with the point counts ascribed to C, but which does not cover C_0; this situation is more typical of applications of this strategy as described in [**6**].

REMARK 10.4. It is now possible to give an alternate brute-force proof of the final assertion of Lemma 10.1. To wit, the isogeny class of $J(C)$ is 6.2.ad_c_a_f_am_q, whose LMFDB entry includes data from a census of curves of genus 6 over $\mathbb{F}_2$ [**7**] which was unavailable at the time of writing of [**10**]; this data shows that C is unique.

Appendix A. Errata for [10]

There are some numbering discrepancies between the arXiv and published versions of [**10**]. This does not affect any of the citations in these errata.

As noted in the introduction, there is a missing running assumption in [**10**] (and [**11**]) that we only consider finite *separable* extensions of function fields, as otherwise a purely inseparable extension always has relative class number 1. Thanks to Jean-Pierre Serre for bring this to our attention.

In [**10**, Theorem 1.3(b)], the number of pairs should be 209 rather than 208. For the extra pair, $(q_F, g_F, g_{F'}) = (2, 4, 7)$ and there are no corresponding extensions, so no other results in [**10**] or this article or [**11**] are affected.

The proof of [**10**, Lemma 3.2] is incomplete in two aspects:

- The reduction to the case of d prime does not suffice to exclude the case $(q, d, g) = (2, 4, 1)$. One must also observe that in the case $(q, d, g) = (4, 2, 1)$, the isogeny class for $J(C)$ must be 1.4.ae and this is not a base extension from $\mathbb{F}_2$. Alternatively, one may rule this case out using [**8**, Corollary 3.4] which does not require d to be prime.
- The application of [**10**, Lemma 3.1] implies that if $d > 2$ is prime, then $q = 2, d = 3$; but it does not by itself imply that $g = 1$. To correct this, note that [**10**, Lemma 3.1] does imply that $J(C)$ has Weil polynomial $(T^2 + T + 2)^i (T^2 + 2T + 2)^{g-i}$, so

$$T_{C,2} = -i - 2(g - i), \qquad T_{C,8} = 5i + 4(g - i)$$

and

$$0 \leq \#C(\mathbb{F}_8) - \#C(\mathbb{F}_2) = (8 + 1 - T_{C,8}) - (2 + 1 - T_{C,2}) = 6 - 6g.$$

In [**10**, Table 1], the row label $q = 2^3$ should read $q = 2^4$.

In (8.1), the definition of c was omitted: it equals 1 if q is even and 2 if q is odd.

The proof of [**10**, Lemma 10.1] is incomplete: the proof of the inequality

$$\begin{aligned} T_{A,q} &+ 0.3366(T_{A,q} + T_{A,q^2} - t) + 0.1137(T_{A,q^3} - T_{A,q}) + 0.0537(T_{A,q^4} - T_{A,q^2}) \\ &\leq a_1 + 0.3366(2a_2) + 0.1137(3a_3) + 0.0537(4a_4) \end{aligned}$$

includes some faulty algebra. To correct it, note that (5.2) implies a stronger version of (5.3): for all i and j, $\#C'(\mathbb{F}_{q^{ij}}) - \#C'(\mathbb{F}_{q^i}) \geq 0$ and so

$$\#C(\mathbb{F}_{q^{ij}}) - \#C(\mathbb{F}_{q^i}) \geq T_{A,q^{ij}} - T_{A,q^i}.$$

Applying (5.3), (7.7), and this new inequality yields

$$\begin{aligned} T_{A,q} &+ 0.3366(T_{A,q} + T_{A,q^2} - t) + 0.1137(T_{A,q^3} - T_{A,q}) + 0.0537(T_{A,q^4} - T_{A,q^2}) \\ &\leq \#C(\mathbb{F}_q) + 0.3366(\#C(\mathbb{F}_{q^2}) - \#C(\mathbb{F}_q)) \\ &\quad + 0.1137(\#C(\mathbb{F}_{q^3}) - \#C(\mathbb{F}_q)) + 0.0537(\#C(\mathbb{F}_{q^4}) - \#C(\mathbb{F}_{q^2})) \\ &= a_1 + 0.3366(2a_2) + 0.1137(3a_3) + 0.0537(4a_4). \end{aligned}$$

References

[1] K. Belabas, CUBIC version 1.3, `https://www.math.u-bordeaux.fr/~kbelabas/research/cubic.html`.

[2] Manjul Bhargava, *Higher composition laws. II. On cubic analogues of Gauss composition*, Ann. of Math. (2) **159** (2004), no. 2, 865–886, DOI 10.4007/annals.2004.159.865. MR2081442

[3] Manjul Bhargava, *Higher composition laws. III. The parametrization of quartic rings*, Ann. of Math. (2) **159** (2004), no. 3, 1329–1360, DOI 10.4007/annals.2004.159.1329. MR2113024

[4] Manjul Bhargava, *Higher composition laws. IV. The parametrization of quintic rings*, Ann. of Math. (2) **167** (2008), no. 1, 53–94, DOI 10.4007/annals.2008.167.53. MR2373152

[5] Everett W. Howe and Kristin E. Lauter, *New methods for bounding the number of points on curves over finite fields*, Geometry and arithmetic, EMS Ser. Congr. Rep., Eur. Math. Soc., Zürich, 2012, pp. 173–212, DOI 10.4171/119-1/12. MR2987661

[6] Everett W. Howe, *Deducing information about curves over finite fields from their Weil polynomials* (English, with English and French summaries), Curves over finite fields—past, present and future, Panor. Synthèses, vol. 60, Soc. Math. France, Paris, 2023, pp. 1–36. MR4727503

[7] Y. Huang, K. S. Kedlaya, and J. B. Lau, *A census of genus 6 curves over* $\mathbb{F}_2$, Preprint, `arXiv:2402.00716v1` (2024).

[8] Borys Kadets, *Estimates for the number of rational points on simple abelian varieties over finite fields*, Math. Z. **297** (2021), no. 1-2, 465–473, DOI 10.1007/s00209-020-02520-w. MR4204701

[9] Kiran S. Kedlaya, *Search techniques for root-unitary polynomials*, Computational arithmetic geometry, Contemp. Math., vol. 463, Amer. Math. Soc., Providence, RI, 2008, pp. 71–81, DOI 10.1090/conm/463/09047. MR2459990

[10] Kiran S. Kedlaya, *The relative class number one problem for function fields, I*, Res. Number Theory **8** (2022), no. 4, Paper No. 79, 21, DOI 10.1007/s40993-022-00369-y. MR4493405

[11] Kiran S. Kedlaya, *The relative class number one problem for function fields, III*, LuCaNT: LMFDB, computation, and number theory, Contemp. Math., vol. 796, Amer. Math. Soc., [Providence], RI, [2024] ©2024, pp. 55–73, DOI 10.1090/conm/796/15997. MR4732683

[12] K. S. Kedlaya, GitHub repository, `https://github.com/kedlaya/same-class-number`.

[13] James R. C. Leitzel and Manohar L. Madan, *Algebraic function fields with equal class number*, Acta Arith. **30** (1976), no. 2, 169–177, DOI 10.4064/aa-30-2-169-177. MR414523

[14] James R. C. Leitzel, Manohar L. Madan, and Clifford S. Queen, *Algebraic function fields with small class number*, J. Number Theory **7** (1975), 11–27, DOI 10.1016/0022-314X(75)90004-9. MR369326

[15] The LMFDB Collaboration, *L-functions and modular forms database*, `https://lmfdb.org`.

[16] The Magma Group (Univ. of Sydney), MAGMA version 2.27-1, 2022, `http://magma.maths.usyd.edu.au`. Accessed November 2025.

[17] J. S. Milne, *Jacobian varieties*, Arithmetic geometry (Storrs, Conn., 1984), Springer, New York, 1986, pp. 167–212. MR861976

[18] Jennifer Paulhus, *Decomposing Jacobians of curves with extra automorphisms*, Acta Arith. **132** (2008), no. 3, 231–244, DOI 10.4064/aa132-3-3. MR2403651

[19] Alessandra Rigato, *Uniqueness of low genus optimal curves over* $\mathbb{F}_2$, Arithmetic, geometry, cryptography and coding theory 2009, Contemp. Math., vol. 521, Amer. Math. Soc., Providence, RI, 2010, pp. 87–105, DOI 10.1090/conm/521/10276. MR2744036

[20] The Sage Developers, SAGEMATH version 9.7, 2022, `https://www.sagemath.org`.

[21] Jean-Pierre Serre, *Rational points on curves over finite fields*, Documents Mathématiques (Paris) [Mathematical Documents (Paris)], vol. 18, Société Mathématique de France, Paris, [2020] ©2020. With contributions by Everett Howe, Joseph Oesterlé and Christophe Ritzenthaler; Edited by Alp Bassa, Elisa Lorenzo García, Christophe Ritzenthaler and René Schoof. MR4242817

[22] Henning Stichtenoth, *Algebraic function fields and codes*, 2nd ed., Graduate Texts in Mathematics, vol. 254, Springer-Verlag, Berlin, 2009. MR2464941

DEPARTMENT OF MATHEMATICS, UNIVERSITY OF CALIFORNIA, SAN DIEGO, CALIFORNIA 92093

Contemporary Mathematics
Volume **832**, 2026
https://doi.org/10.1090/conm/832/16608

The refined Humbert invariant for an automorphism group of a genus 2 curve

Harun Kır

ABSTRACT. The purpose of this paper is to list the refined Humbert invariants for a given automorphism group of a smooth curve C/K of genus 2 over an algebraically closed field K with $\mathrm{char}(K) = 0$. This invariant is an algebraic generalization of the (usual) *Humbert invariant*. It is a positive definite quadratic form associated to the curve C, and it encodes many geometric properties of the curve. The paper has a special interest in the cases where $\mathrm{Aut}(C) \simeq D_4$ or D_6. In these cases, several applications of the main results are discussed including the curves with elliptic subcovers of a given degree.

1. Introduction

The smooth curves of genus 2 with a group of automorphisms have been considered by many authors, e.g. see the references in §1.5 of [**1**], [**14**], and [**6**]. The classification of these curves for a given automorphism group is achieved in different ways; see [**1**], [**6**], [**34**]. In the present article, we also describe these curves for a given automorphism group by examining their associated quadratic forms. In doing so, we obtain interesting applications to the curves of genus 2; see §5 and 6 below.

Let C/K be a (smooth) curve of genus 2 over an algebraically closed field K of characteristic 0. Kani [**15**] introduced a positive definite quadratic form q_C attached to the curve C; see [**18**] or §2 below. This form is called the *refined Humbert invariant of C*. This invariant is a useful ingredient since it translates geometric problems into arithmetic problems. An elegant illustration of making use of the refined Humbert invariant can be found in Kani [**18**], [**19**], [**24**], and in the author's Ph.D. thesis.

Kani showed that the automorphism group $\mathrm{Aut}(C)$ of a genus 2 curve C can be determined from its associated refined Humbert invariant q_C; see Theorem 29 of [**24**] (or Table 1 below). Moreover, in certain cases, he gave a list of the refined Humbert invariants q_C encoding the information $\mathrm{Aut}(C)$; see Proposition 33 of [**24**]. By extending the idea of the argument in the proof of his result, we give a complete list of the possible ternary quadratic forms.

2020 *Mathematics Subject Classification*. Primary 14H37.

Key words and phrases. Curves of genus 2, automorphism groups, ternary quadratic forms, elliptic curves with complex multiplication, product surfaces, Jacobians, Humbert invariant, Humbert surfaces, elliptic subcovers.

THEOREM 1.1. *Let q_C be a ternary quadratic form. Then the Jacobian J_C is isogenous to a product $E \times E$ of an elliptic curve E/K with complex multiplication. Then* $|\operatorname{Aut}(C)| > 2$ *if and only if the refined Humbert invariant q_C is equivalent to one of the ternary quadratic forms listed in* Propositions 4.4, 4.7, 4.8 and 4.9 *below.*

In addition, if q is one of the forms listed in Propositions 4.4, 4.7, 4.8 and 4.9(i) *below, then there is a curve C/K of genus 2 such that q_C is* $(\mathrm{GL}_3(\mathbb{Z}))$*-equivalent to q.*

We have actually finer results than what was stated in Theorem 1.1. More precisely, Propositions 4.4, 4.7, 4.8 and 4.9 below explicitly show which ternary (quadratic) forms belong to a specific automorphism group; see Table 5 below.

One can apply the list provided by Theorem 1.1 to the interesting topics discussed in Cardona and Quer [**6**] and Kani [**20**], [**23**]. More precisely, they characterized in different ways the genus 2 curves whose automorphism group contains (up to isomorphism) the dihedral group D_n of order $2n$, for $n = 6$ and 4. To illustrate, Kani proved in Theorem 4(a) of [**20**] that:

(1.1)
$$q_C \text{ primitively represents the binary form } 4x^2 + 4xy + 4y^2 \;\Leftrightarrow\; D_6 \leqslant \operatorname{Aut}(C),$$

(1.2) $$q_C \text{ primitively represents the binary form } 4x^2 + 4y^2 \;\Leftrightarrow\; D_4 \leqslant \operatorname{Aut}(C).$$

In the special case of Theorem 1.1 that $|\operatorname{Aut}(C)| \geq 12$, we have a simple explicit list for the characterization of (1.1):

THEOREM 1.2. *Let C/K be a curve of genus 2. Then* $\operatorname{Aut}(C)$ *contains a subgroup isomorphic to the dihedral group D_6 if and only if the refined Humbert invariant q_C is equivalent to $4x^2 + 4xy + 4y^2$ or $4x^2 + 4y^2 + cz^2 + 4yz + 4xz + 4xy$ or $4x^2 + 4y^2 + cz^2 - 4xy$, where $c \equiv 0, 1 \pmod 4$, and $c > 1$.*

In addition, if q is one of the forms listed above, then there is a curve C/K of genus 2 such that q_C is equivalent to q.

The first assertion of this result has already been proved for odd numbers c; see Proposition 33 of [**24**] (and (1.1)). We also obtain a similar explicit list (but more complicated) in the case that $\operatorname{Aut}(C)$ contains a subgroup D_4 up to isomorphism; see Theorem 5.6 below.

We will see in Sections 5 and 6 that Theorem 1.2 (and Theorem 5.6 below) lead to some interesting results regarding the curves of genus 2 having *elliptic subcovers* of a given degree. To state them more precisely, we recall that Kani in [**19**, p. 25] introduced the sets $H(q)$ to determine how the curves of genus 2 which lie on some product abelian surface are distributed in the moduli space $\mathcal{M}_2$ of genus 2 curves. This set $H(q)$ is called a *generalized Humbert set* which is associated with a given quadratic form q. This set is defined by using the refined Humbert invariant, and it is a subset of the moduli space $\mathcal{A}_2$ (which classifies isomorphism classes $\langle A, \theta \rangle$ of principally polarized abelian surfaces). This set is a generalization of a *Humbert surface H_n of invariant n*; see §3 of [**19**] or §5 below (also, see van der Geer [**39**], Ch. IX for a contemporary treatment of Humbert surfaces). Moreover, we have from Theorem 7.1 of [**23**] that if q (primitively) represents n^2, for some n, then $H(q)$ is a closed subset of the moduli space $\mathcal{A}_2$, in which case it is called a *generalized Humbert scheme*.

It is interesting to understand the intersections of the generalized Humbert schemes and/or Humbert surfaces; see [**23**]. For example, McMullen [**28**] posed a question regarding the description of the intersection of Humbert surfaces $H_N \cap H_M$. He solved the problem for the intersection $H_N \cap H_1$. In particular, Kani [**23**], Proposition 5.2(d), gave a formula/algorithm for the intersection $H_n \cap H_m$ of two Humbert surfaces. This formula involves the $H(q)$'s for certain binary forms q's; see Example 5.8 below. The results above can be used to give a formula for the intersection $H(q) \cap H_m$ for a binary form q and integer m in certain cases; see Corollaries 5.5 and 5.7 below. This formula involves the $H(q)$'s for certain ternary forms q's.

The list provided in Theorem 1.1 has another application concerning the quantity

$$k(q_C) := \frac{|\operatorname{Aut}(q_C)|}{|\operatorname{Aut}(C)|},$$

which is an integer by Theorem 4 of [**24**] (see also (1.5) and (2.3) below). If $|\operatorname{Aut}(C)| > 2$, where q_C is a ternary form, then

$$k(q_C) \leq 2, \text{ with the exception of one class of forms for which } k(q_C) = 4,$$

see Proposition 4.10 below. Rather than an upper bound, we have actually explicit information for the formula $k(q_C)$ in virtually all cases; see Tables 5 and 6 below. The significance of this quantity will be discussed later.

To obtain the results for the formula $k(q_C)$, we list the ternary quadratic forms according to their automorphism groups. Section 3 is mostly dedicated to doing that because it is also of independent interest. In a special case of Theorem 3.8 below, we have:

THEOREM 1.3. *Let $q(x, y, z) = ax^2+by^2+cz^2+2ryz+2sxz+2txy$ be an Eisenstein reduced positive definite ternary quadratic form. Let us put $u = |\operatorname{Aut}^+(q)|$. If $3 \mid u$ and $u > 6$, then $u = 12$ or 24. Moreover, we have*

(i) $u = 12 \Leftrightarrow a = b = -2t,\ r = s = 0$ or $b = c = -2r,\ s = t = 0$.

(ii) $u = 24 \Leftrightarrow a = b = c,\ r = s = t = \kappa$, *where* $\kappa = 0,\ a/2$ *and* $-a/3$.

We present similar results for other cases; see Theorem 3.8, Proposition 3.14 and Theorem 3.16 below. While we prove these results, the tables of Theorem 105 of Dickson [**9**] will be a cornerstone, and so these tables are explained with a suitable modification; see Tables 2 and 3 below.

Let

$$(1.3) \qquad R_n(q) := \{(x_1, \ldots, x_r) \in \mathbb{Z}^r : q(x_1, \ldots, x_r) = n\}$$

denote the set of the r-tuples of the integers of representations of an integer n by a quadratic form q. Let $r_n(q) = |R_n(q)|$ denote the cardinality of the set $R_n(q)$. For a positive definite ternary form q, let

$$(1.4) \qquad a(q) = \max(1, r_4(q), 3r_4(q) - 12).$$

This quantity is the other cornerstone in this article because it has a very nice connection with $\operatorname{Aut}(C)$ by Theorem 4 of [**24**] (see Remark 2.1 below) as follows.

$$(1.5) \qquad 2a(q_C) = |\operatorname{Aut}(C)|.$$

By equation (1.5), we have that $1/k(q_C) = a(q_C)/|\operatorname{Aut}^+(q_C)|$, which plays an important role in [**24**] because it is a part of the formula of the number of isomorphism classes of curves C of genus 2 lying on a given abelian surface A.

Kani computed $a(q_C)$ when q_C is a primitive ternary form (see Proposition 3.9 below). He mentioned in Remark 36 of [**24**] that the computation of $a(q_C)$ is more complicated when q_C is not primitive. Our list makes explicit the formula $a(q_C)$ for any imprimitive ternary form q_C; see Table 5.

By making use of the properties of the refined Humbert invariant q_C, one can study the elliptic subcovers of degree n of a genus 2 curve C; see Theorem 4.5 of [**15**] (or see (6.1) below). In Section 6, we apply our results to this research area, and describe the curves C of having an elliptic subcover of degree n for a given $\mathrm{Aut}(C)$ in some cases.

2. The refined Humbert invariant

In this section, we recall the concept of the refined Humbert invariant which was introduced by Kani [**15**], [**19**]. It is the main ingredient of the present article. We also discuss many useful properties of the refined Humbert invariant as derived in [**19**] and [**18**]. This invariant is a quadratic form that is intrinsically attached to a principally polarized abelian surface.

Let A/K be an abelian surface over an algebraically closed field K of char $K = 0$. If $\lambda : A \xrightarrow{\sim} \hat{A}$ is a principal polarization of A, then $\lambda = \phi_\theta$ for a (unique) $\theta \in \mathrm{NS}(A) = \mathrm{Div}(A)/\equiv$, where $\equiv$ denotes numerical equivalence; (see [**31**, p. 60] for the discussion of ϕ).

Let $\mathcal{P}(A) \subset \mathrm{NS}(A)$ denote the set of principal polarizations on A, and let $\theta \in \mathcal{P}(A)$. The quadratic form $\tilde{q}_{(A,\theta)}$ of a principally polarized abelian surface (A, θ) on $\mathrm{NS}(A)$ is defined by the formula

$$\tilde{q}_{(A,\theta)}(D) \; = \; (D.\theta)^2 - 2(D.D), \quad \text{for } D \in \mathrm{NS}(A), \tag{2.1}$$

where (.) denotes the intersection number of divisors. By [**15**, p. 200], the form $\tilde{q}_{(A,\theta)}$ induces a positive definite quadratic form $q_{(A,\theta)}$ on the quotient module

$$\mathrm{NS}(A, \theta) \; = \; \mathrm{NS}(A)/\mathbb{Z}\theta.$$

The quadratic form $q_{(A,\theta)}$ or, more correctly, the quadratic module

$$(\mathrm{NS}(A, \theta), q_{(A,\theta)})$$

is called the *refined Humbert invariant* of the principally polarized abelian surface (A, θ); as defined in [**19**], [**18**]. Thus $q_{(A,\theta)}$ gives rise to an equivalence class of integral quadratic forms in $\rho - 1$ variables since $\mathrm{NS}(A, \theta) \simeq \mathbb{Z}^{\rho-1}$, where ρ is the Picard number of A. Hence, when $\rho = 4$, the (corresponding) refined Humbert invariant is actually a ternary quadratic form.

Recall from [**30**] that if C is a curve of genus 2, then its Jacobian J_C is an abelian surface and there is a divisor θ_C on J_C, called the *theta-divisor* such that θ_C is a principal polarization in $\mathrm{NS}(J_C)$. We then write

$$q_C \; := \; q_{(J_C,\theta_C)}$$

for its associated refined Humbert invariant.

One key property of the refined Humbert invariant $q_{(A,\theta)}$ is that it can be used to determine whether or not (A, θ) is a Jacobian. By Proposition 6 of [**19**], we have that

$$(A, \theta) \simeq (J_C, \theta_C), \text{ for some curve } C \;\Leftrightarrow\; q_{(A,\theta)}(D) \neq 1, \; \forall D \in \mathrm{NS}(A, \theta). \tag{2.2}$$

When A/K is a *CM (complex multiplication) abelian product surface*, i.e., $A \simeq E_1 \times E_2$ for some CM elliptic curves $E_1/K, E_2/K$ with $E_1 \sim E_2$, we have that $q := q_{(A,\theta)}$ is a ternary quadratic form because the Picard number $\rho(A) = 4$ in this case. If $(A, \theta) \simeq (J_C, \theta_C)$, for some curve C, then q does not represent 1 by equation (2.2), and hence, we can apply Theorem 3 and Proposition 17 of [**24**] to the form q. Then we obtain from these results that

$$a(q) \mid |\operatorname{Aut}^+(q)|, \tag{2.3}$$

where $\operatorname{Aut}^+(q) = \{\alpha \in \operatorname{Aut}(q) : \det(\alpha) = 1\}$ (and $a(q)$ is as in the introduction).

Another key property of the refined Humbert invariant q_C is that it possesses information about $\operatorname{Aut}(C)$. This property is the key ingredient in this article, and so we now give it as a table explicitly from Theorem 29 of [**24**]:

Let C be a curve of genus 2 such that q_C is a binary or a ternary form. Let C_n denote the cyclic group of order n. Then we have Table 1.

TABLE 1. Automorphism groups of a curve C of genus 2 with $r_4(q_C)$

$\operatorname{Aut}(C)$	$a(q_C)$	$r_4(q_C)$
C_2	1	0
C_{10}	1	0
$C_2 \times C_2$	2	2
D_4	4	4
D_6	6	6
$C_3 \rtimes D_4$	12	8
$\mathrm{GL}_2(3)$	24	12

Since this table plays a key role in the present article, we outline the main ideas of its proof. The first column of this table gives the complete list of possibilities of $\operatorname{Aut}(C)$; see Theorem 2 of [**34**]. To verify the third column, if $i(\operatorname{Aut}(C))$ denotes the number of involutions of the group $\operatorname{Aut}(C)$, then we have from Theorem 20 of [**21**] and Proposition 35 of [**20**] that

$$i(\operatorname{Aut}(C)) - 1 = r_4(q_C)$$

(see Proposition 26 of [**24**]). Hence, one can easily verify the third column by the group theory, and the second column follows from the third column by the definition (see (1.4)).

REMARK 2.1. If a refined Humbert invariant $q_{(A,\theta)}$ is a ternary form, then $\rho(A) = 4$. By the structure theorems for $\operatorname{End}(A)$ (see Proposition IX.1.2 of [**39**]), it follows that $A \sim E \times E$, for some CM elliptic curve E. By Shioda and Mitani's Theorem [**35**] (or by Theorem 2 of [**17**]), it follows that A is a CM abelian product surface.

Therefore, if q_C is a ternary form for some curve C of genus 2, then $\operatorname{Aut}(C)$ cannot be isomorphic to C_{10} by the fact that J_C is simple when $\operatorname{Aut}(C) \simeq C_{10}$; see [**14**, p. 648]. Thus, we can eliminate the second line of Table 1 while we list the forms q_C, for a given automorphism group $\operatorname{Aut}(C)$. Hence, we are just left with different integers $a(q_C)$ (and $r_4(q_C)$) for each case in the table, and so it follows that $2a(q_C) = |\operatorname{Aut}(C)|$, in this case as stated in (1.5).

3. Ternary quadratic forms

Since the form q_C encodes the information of $\mathrm{Aut}(C)$ (by Table 1), we aim to list the ternary quadratic forms q_C with given $\mathrm{Aut}(C)$. For this, since $a(q_C) \mid |\mathrm{Aut}^+(q_C)|$ by equation (2.3), we first determine the ternary quadratic forms f according to the group $\mathrm{Aut}^+(f)$ in this section.

For simplicity, we use the abbreviation

$$[a, b, c, r, s, t] := ax^2 + by^2 + cz^2 + ryz + sxz + txy = f(x, y, z)$$

to denote an *integral ternary quadratic form* $f(x, y, z)$ with $a, b, c, r, s, t \in \mathbb{Z}$. Note that if all non-diagonal coefficients, namely r, s, t, are even, then f is also an *integral ternary quadratic form in Dickson's sense* [**9**].

The properties of an *Eisenstein reduced* form are often used below, so let us recall it here. Firstly, by equivalent ternary forms q_1 and q_2, we mean $\mathrm{GL}_3(\mathbb{Z})$-equivalence, and we denote it by $q_1 \sim q_2$ as in [**40**, p. 5]. We say that $f = [a, b, c, 2r, 2s, 2t]$ is an *Eisenstein reduced form* if the following are true, where $\epsilon = a + b + 2r + 2s + 2t$:

(3.1) r, s, t are all positive or all are non-negative.

(3.2) $a \le b \le c$, $\epsilon \ge 0$.

(3.3) $a \ge |2s|$, $a \ge |2t|$, $b \ge |2r|$.

(3.4) If $a = b$, $|r| \le |s|$; if $b = c$, $|s| \le |t|$; If $\epsilon = 0$, $a + 2s + t \le 0$.

(3.5) If $a = -2t$, $s = 0$; if $a = -2s$, $t = 0$; if $b = -2r$, $t = 0$.

(3.6) If $a = 2t$, $s \le 2r$; if $a = 2s$, $t \le 2r$; if $b = 2r$, $t \le 2s$.

The condition (3.5) is valid for r, s, $t \le 0$, and the condition (3.6) is valid for $r, s, t > 0$. The main point of introducing the definition of an Eisenstein reduced form is as follows:

THEOREM 3.1 (Theorems 103 and 104 of [**9**]). *Every positive definite integral ternary quadratic form in Dickson's sense is equivalent to a unique Eisenstein reduced form.*

There are many useful results on the reduced forms. In particular, it is easy to see the minima of the reduced forms by the following result.

THEOREM 3.2 (Theorem 101 of [**9**]). *Let $f = [a, b, c, 2r, 2s, 2t]$ be a positive definite Eisenstein reduced ternary quadratic form. Then a is the minimum value of f and b is the second minimum value of f.*

We defined an Eisenstein reduced form for positive definite ternary quadratic forms. The following proposition suggests that we can relax this hypothesis considerably. Note that by $\mathrm{disc}(f)$, we mean the *discriminant* of a quadratic form f as defined in [**40**, p. 2] (or [**3**, p. 335]).

PROPOSITION 3.3. *Let $f = [a, b, c, 2r, 2s, 2t]$ be an integral ternary quadratic form. If $a > 0$, and if f satisfies the conditions* (3.1)–(3.3), *then f is positive definite.*

PROOF. We use Theorem 8 of [**9**] to prove this statement. We have by (3.2) that $ab - t^2 \geq a^2 - t^2$. We also have from (3.3) that $a^2 - t^2 > a^2 - 4t \geq 0$, if $t \neq 0$, and otherwise $a^2 - t^2 = a^2 > 0$. Thus, it follows that $ab - t^2 > 0$, and hence, it suffices to show that $\operatorname{disc}(f) < 0$.

For this, we first see that $ar^2 \leq abc$, $bs^2 \leq abc$ and $ct^2 \leq abc$ by (3.2) and (3.3) because $ar^2 \leq 4ar^2 \leq ab^2 \leq abc$, $bs^2 \leq 4bs^2 \leq ba^2 \leq abc$, and $ct^2 \leq 4ct^2 \leq ca^2 \leq abc$. Since $0 < a \leq b \leq c$ by (3.2) (and by hypothesis), we have that

$$\begin{aligned} \operatorname{disc}(f) &= -4abc - rst + ar^2 + bs^2 + ct^2 \\ &\leq -4abc - rst + abc + abc + abc = -abc - rst. \end{aligned}$$

By (3.1), we know that $r, s, t > 0$ or $r, s, t \leq 0$. In the first situation, it is clear that $\operatorname{disc}(f) \leq -abc - rst < 0$. In the second case, it follows from (3.3) that $abc \geq |2s|\,|2t|\,|2r| = -8rst$, which implies that $abc + 8rst \geq 0$. If $rst = 0$, then it is clear that $\operatorname{disc}(f) \leq -abc - rst = -abc < 0$. If $rst \neq 0$, then $abc + rst > abc + 8rst \geq 0$, which implies that $\operatorname{disc}(f) \leq -(abc + rst) < 0$. Thus the assertion follows from Theorem 8 of [**9**]. □

To list the ternary forms q according to $\operatorname{Aut}^+(q)$, we employ the tables of Theorem 105 of Dickson [**9**]. There, he listed the existence of an automorph[1] of a positive definite (Eisenstein) reduced ternary quadratic form according to certain relations of the coefficients. Calculating the orders of the elements in those tables, we can obtain two useful tables (see Tables 2 and 3 below). These tables make our arguments below more readable and simplify the proofs considerably.

REMARK 3.4. If $q = [a, b, c, 2r, 2s, 2t]$ is a reduced ternary form such that some of the nondiagonal coefficients $2r, 2s, 2t$ are zero, then there might be automorphs of $\operatorname{Aut}^+(q)$ not listed in the table of [**9**, p. 180] (or in Table 3 below). More precisely, we have by the footnote of [**9**, p. 180] that if $r = s = 0$, then new automorphs (not listed there) are derived from each one listed in the table by changing the signs of x and y. Similarly, we can derive new automorphs from each one listed in the table when $r = t = 0$ or $s = t = 0$. Also, it follows from the same footnote that if $r = s = t = 0$, then we can derive three new automorphs from each one listed in the table.

In the second columns of the following two tables, we listed all conditions on the coefficients of a reduced ternary form $q = [a, b, c, 2r, 2s, 2t]$ as done in the tables of Theorem 105 of [**9**]. For any case in each line, Dickson listed automorphs as an element of $\mathrm{SL}_3(\mathbb{Z})$, and so it is easy to find their orders. Regarding these listed automorphs, let us introduce the following notation:

For a given reduced ternary form $q = [a, b, c, 2r, 2s, 2t]$ and an integer $n > 0$, let

$$\operatorname{Aut}^+(q)_n = \{\alpha \in \operatorname{Aut}^+(q) : |\alpha| = n, \text{ where } \alpha \text{ is listed in Theorem 105 of [9]}\}$$

denote the set of automorphs in $\operatorname{Aut}^+(q)$ listed in Theorem 105 of [**9**] whose order is n. Let $u_n = u(q)_n := |\operatorname{Aut}^+(q)_n|$ be the cardinality of the set $\operatorname{Aut}^+(q)_n$. Hence, we find the number u_n of automorphs of order n in each line of the tables of Theorem 105 of [**9**], and we fill out the third, fourth, fifth, and sixth columns in the following tables accordingly.

[1] Please note that an automorph is an automorphism. We are using this (older) terminology, which was employed in Dickson's book [**9**], our primary reference for this discussion.

In some situations, if one case of a line holds in a table, then it might be true that some other cases of lines in the same table hold. In the seventh column of the tables, we listed all such cases. By doing so, we obtain a lower bound for $|\operatorname{Aut}^+(q)| = u$, which is written in the last column of the tables. Note that if at least two of r, s, t are zero, then we apply Remark 3.4 to obtain this lower bound. Let u_L denote this lower bound, i.e., u_L be the number of automorphs listed in 3-7 columns plus the identity as well as derived ones by Remark 3.4 (when this remark is applicable).

TABLE 2. Number of automorphs when $r, s, t > 0$

	Conditions on form	u_2	u_3	u_4	u_6	Extra Automorphs	u_L
1	$a = 2s = 2t$	1					2
2	$a = 2s, t = 2r$	1					2
3	$a = 2t, s = 2r$	1					2
4	$b = 2r, t = 2s$	1					2
5	$a = b, r = s$	1					2
6	$b = c, s = t$	1					2
7	$b = c, a = 2s = 2t$	1				1, 6	4
8	$b = c, a = 2s = 2t = 4r$			2		1-3, 6-7	8
9	$a = b = 2r = 2s = 2t$	1	2			1, 5	6
10	$a = b = c, r = s = t$	1	2			5, 6	6
11	$a = b = c = 2r = 2s = 2t$	3	4	6		1, 5-7, 9-10	24

PROOF OF TABLE 2. As was mentioned above, we listed in the second column of Table 2 the conditions on the coefficients of a reduced ternary form $q = [a, b, c, 2r, 2s, 2t]$ as in Dickson's Table of [**9**, p. 179]. By determining the orders of the automorphs listed in Dickson's table, we derive the values u_n, which correspond to columns 3-6 of Table 2. For example, in the case of line 1 of Table 2, we see that there is only one automorph in Dickson's table, namely $(x + y + z, -y, -z)$ in the notation defined there. This notation corresponds to the matrix $\left(\begin{smallmatrix} 1 & 1 & 1 \\ 0 & -1 & 0 \\ 0 & 0 & -1 \end{smallmatrix}\right)$, which has order 2, and thus $u_2 = 1$.

As was mentioned above, if one case of a line holds in Table 2, then some other cases of lines in the same table might hold as well. For example, if the case of line 7 holds, i.e., $b = c$ and $a = 2s = 2t$, then this implies that the cases of lines 1 and 6 also hold as stated in the seventh column. Thus, this means that we have extra automorphs not counted in the columns 3-6 of line 7. In the case that there are extra automorphs, we wrote all line numbers in the seventh column.

Finally, we can find a lower bound for $u = |\operatorname{Aut}^+(q)|$ by counting the number of the automorphs in the columns 3-7. For example, in the case of line 10 of this table, we have that $u_2 + u_3 + u_4 + u_6 = 1 + 2 + 0 + 0 = 3$. In the seventh column, we see that there are extra automorphs in this case, which are listed in lines 5 and 6. In both lines, we see that $u_2 = 1$ and $u_3 = u_4 = u_6 = 0$. Therefore, it follows that there are at least $3 + 1 + 1 = 5$ automorphs except the identity in line 10, and so $u_L = 6$, as stated in the last column. Thus all columns can be verified in this way. □

TABLE 3. Number of automorphs when $r, s, t \leq 0$

	Conditions on form	u_2	u_3	u_4	u_6	Extra automorphs	u_L
1	$a = -2s, t = 0$	1					2
2	$a = -2t, s = 0$	1					2
3	$b = -2r, t = 0$	1					2
4	$a + 2s + t = b + 2r + t = 0$	1					2
5	$a = b, r = s$	1					2
6	$b = c, s = t$	1					2
7	$a = b, \epsilon = 0$	1					2
8	$a = b = -2r = -2s, t = 0$			2		1, 3-5, 7	8
9	$a = b = -2t, r = s = 0$	1	1		1	2, 5	12
10	$b = c = -2r, s = t = 0$	1	1		1	3, 6	12
11	$b = c, s = t, b + 2r = -s, a = -3s$	1	2			4, 6	6
12	$a = b = c, r = s = t$	1	2			5, 6	6
13	$a = b = c, \epsilon = 0$	2				7	4
14	$a = b = c, \epsilon = 0, r = s$			2		4, 5, 7, 13	8
15	$a = b = c, \epsilon = 0, s = t$	1		2		6, 7, 13	8
16	$a = b = c = -3s, r = s = t$		4	2		4-7, 11-15	24

Notation: Here, $\epsilon = a + b + 2r + 2s + 2t$.

PROOF OF TABLE 3. As in Table 2, we listed in the second column the conditions on the coefficients of a reduced ternary form $q = [a, b, c, 2r, 2s, 2t]$ as in Table of [**9**, p. 180]. The verification of columns 3-7 follows the same approach as in the proof of Table 2.

To verify the last column, note first that if at least two of r, s, t are not zero, then the last column follows as in the proof of Table 2. Otherwise, we need to use Remark 3.4. For example, if the case of line 9 of Table 3 holds, then there are a total of $u_2 + u_3 + u_4 + u_6 = 1 + 1 + 0 + 1 = 3$ automorphs coming from 3-6 columns. Also, the seventh column tells us that the cases of lines 2 and 5 hold, and so we have 2 more automorphs in these lines. Since the identity is always an automorph, we have at least $3 + 2 + 1 = 6$ automorphs listed in the table. By Remark 3.4, we derive new automorphs for each one since $r = s = 0$ (and $t \neq 0$), and hence there are at least $6 \times 2 = 12$ automorphs in this situation as indicated in the last column. Therefore, the table can be verified in this way. □

REMARK 3.5. There are only two lines 9 and 10 of Table 2 and two lines 11 and 12 of Table 3 having automorphs of order 3 and no automorphs of orders 4 or 6.

LEMMA 3.6. *Let $q = [a, b, c, 2r, 2s, 2t]$ be a positive definite ternary quadratic form. Let $G = \mathrm{Aut}^+(q)$, and let $u = |G|$. Then $u \leq 24$, and also, if $3 \mid u$, then $u = 24, 12$ or 6. Moreover, if $u = 24, 12$ or $u = 6$, then G is isomorphic to S_4, D_6 or S_3, respectively.*

PROOF. Recall that G is a finite subgroup of $\mathrm{SL}_3(\mathbb{Z})$ since q is positive definite. Hence, we can use Tahara's results in [**38**]. Then it directly follows from the table of [**38**, p. 170] that $u \leq 24$, and also that if $3 \mid u$, then $u = 3, 6, 12, 24$.

By Theorem 3.1, we may suppose that q is reduced. We then see from Tables 2 and 3 that if there is an automorph of order 3 in G, then there is always an automorph of order 2 or 4 in G simultaneously, and so $u \neq 3$. This proves the first assertion.

If $u = 24$, then the assertion directly follows from the Corollary of Proposition 9 of [**38**] because this corollary tells us that there is only one subgroup of order 24 in $\mathrm{SL}_3(\mathbb{Z})$, up to isomorphism, which is the symmetric group S_4.

If $u = 12$, then it follows from the Corollary of Proposition 7 of [**38**] that G is isomorphic to D_6 or alternating group A_4. We show that the latter is not possible. Let us assume that $G \simeq A_4$, and thus there is no $g \in G$ such that its order $|g| = 4$ or 6, and also there are 8 elements of G whose order is 3 by group theory. (This follows from the fact that G has 4 Sylow-3 subgroups).

Now, if $r, s, t > 0$, then by Remark 3.5, there are two cases of lines 9 and 10 of Table 2. Moreover, if both cases hold, then the last line of this table also holds, which immediately implies that $u \geq 24$ by its last column. Thus, both cases cannot be true simultaneously. However, if only one of the cases is true, then we could have 4 automorphs of order 3, which is a contradiction.

If $r, s, t \leq 0$, then we have similarly two cases of lines 11 and 12 of Table 3 by Remark 3.5. Also, both cases cannot hold simultaneously since then the last line of this table would hold, which means that $u \geq 24$ by its last column. Note that if line 11 holds, then $rst \neq 0$, and we cannot derive new automorphs from the ones listed in this table by changing the signs of x, y, z (see Remark 3.4). Moreover, if the line 12 holds, then $rst \neq 0$; otherwise, $u \geq 6 \times 4 = 24$ by Remark 3.4 and by the last column of the table. Therefore, we conclude that if one of the lines 11 and 12 holds, then there are 4 elements of order 3, which is again not possible. Hence, $G \simeq D_6$, as stated.

Suppose now that $u = 6$. We know by group theory that there exist exactly 2 groups of order 6, up to isomorphism, namely C_6 and S_3. If $G \simeq C_6$, then there is an element $g \in G$ such that its order $|g| = 6$, and thus, the case of lines 9 or 10 of Table 3 holds. However, in both cases, we know from the last column of the table that $u \geq 12$, which is impossible, and thus $G \simeq S_3$. □

For later use, we now prove the possible values of the cardinality $|\operatorname{Aut}^+(q)|$, where q is a positive definite ternary form.

PROPOSITION 3.7. *If q is as in Lemma 3.6, then we have that* $|\operatorname{Aut}^+(q)| \in \{1, 2, 4, 6, 8, 12, 24\}$.

PROOF. Since $\operatorname{Aut}^+(q) \leqslant \mathrm{SL}_3(\mathbb{Z})$, we can use Tahara's results in [**38**] as in the proof of Lemma 3.6. Then we have by Table of [**38**, p. 170] that $|\operatorname{Aut}^+(q)| \in \{1, 2, 3, 4, 6, 8, 12, 24\}$. Since we see by Lemma 3.6 that $|\operatorname{Aut}^+(q)| \neq 3$, the assertion follows. □

Recall that Smith [**37**, p. 278] and Dickson [**9**, p. 179] stated (without proving) Proposition 3.7.

The following theorem lists the positive definite ternary quadratic forms q when $3 \mid |\operatorname{Aut}^+(q)|$ which is a generalization of Theorem 1.3.

THEOREM 3.8. *Let* $q = [a, b, c, 2r, 2s, 2t]$ *be an (Eisenstein) reduced positive definite ternary quadratic form. Let us put* $u = |\operatorname{Aut}^+(q)|$. *If* $3 \mid u$, *then* $u = 6, 12, 24$. *Moreover, we have*

(i) $u = 24 \Leftrightarrow q = [-3r, -3r, -3r, 2r, 2r, 2r]$ *or* $q = [a, a, a, \kappa a, \kappa a, \kappa a]$, *where* $\kappa = 0, 1$.

(ii) $u = 12 \Leftrightarrow q = [a, a, c, 0, 0, -a]$ *or* $q = [a, b, b, -b, 0, 0]$.

(iii) $u = 6 \Leftrightarrow q = [a, a, c, a, a, a]$, *with* $a \neq c$ *or* $q = [-3s, b, b, -b - s, 2s, 2s]$, *with* $b \neq -3s$ *or* $q = [a, a, a, 2r, 2r, 2r]$, *with* $r \neq 0$, $a \neq \epsilon r$, *where* $\epsilon = 2, -3$.

PROOF. Recall first from Lemma 3.6 that $u \leq 24$, and also that if $3 \mid u$, then $u = 6, 12, 24$.

(i) Suppose first that $u = 24$, and so we know from Lemma 3.6 that $\operatorname{Aut}^+(q) \simeq S_4$. By group theory, there exist elements of order 3 and 4 in this group. Thus, if $(r, s, t) \neq (0, 0, 0)$, then the last line of Table 2 or 3 holds, i.e., $a = b = c = 2r = 2s = 2t$ or $a = b = c = -3s$, $r = s = t$ respectively. This means that $q = [a, a, a, a, a, a]$ and $q = [-3r, -3r, -3r, 2r, 2r, 2r]$. In addition, if $r = s = t = 0$, then it follows by the footnote of [**9**, p. 180] (or by Remark 3.4 and Table 3) that $u = 24 \Leftrightarrow a = b = c$. Hence, $q = [a, a, a, 0, 0, 0]$ in this case. This verifies the implication ($\Rightarrow$) of the assertion (i). The other implication easily follows from the last columns of Tables 2 and 3 (and from the fact that $u \leq 24$).

(ii) Suppose next that $u = 12$, and so $\operatorname{Aut}^+(q) \simeq D_6$ by Lemma 3.6. Since this group has an element of order 6, we have the two cases given in lines 9 and 10 of Table 3. Then this implies that $a = b = -2t$, $r = s = 0$ or $b = c = -2r$, $s = t = 0$, respectively, i.e., $q = [a, a, c, 0, 0, -a]$ or $q = [a, b, b, -b, 0, 0]$, which proves the implication ($\Rightarrow$). The other implication ($\Leftarrow$) follows from the last column of the table. Indeed, if $q = [a, a, c, 0, 0, -a]$, then we are just left with the cases of lines 2, 5, and 9 as indicated in the table, and this means that $u = 12$ by the last column of the table since no other cases occur. Similarly, if $q = [a, b, b, -b, 0, 0]$, then we are left with the cases of lines 3, 6, and 10, and we obtain that $u = 12$, which proves (ii).

(iii) Suppose now that $u = 6$. By Lemma 3.6, $\operatorname{Aut}^+(q) \simeq S_3$, and so there is an automorph of order 3 but not of order 4 and 6. If $r, s, t > 0$, then it follows by Remark 3.5 that only the cases of lines 9 or 10 of Table 2 are true. Thus, we have that $a = b = 2r = 2s = 2t$ or that $a = b = c$, $r = s = t$. Note that if both hold simultaneously, then $u = 24$ by (i), which is impossible. So we respectively obtain that $q = [a, a, c, a, a, a]$, with $a \neq c$ or that $q = [a, a, a, 2r, 2r, 2r]$, with $r > 0$ and $a \neq 2r$.

If $r, s, t \leq 0$, then we are left with the cases of lines 11 and 12 of Table 3 by Remark 3.5. Thus, it follows that $b = c$, $s = t$, $b + 2r = -s$, $a = -3s$ or that $a = b = c$, $r = s = t$, respectively. If both hold, then $u = 24$ by (i), and so both cannot hold simultaneously. Now, if the first case holds, then $b \neq -3s$ and $q = [-3s, b, b, -b - s, 2s, 2s]$ as stated. If the second case holds, then $a \neq -3s$ and $q = [a, a, a, 2r, 2r, 2r]$. Note that $r \neq 0$; otherwise, $u = 24$ by (i). Therefore, the part ($\Rightarrow$) of this assertion follows. The other part can be easily verified by Tables 2 and 3 as above, and hence all assertions have been verified. □

One can easily deduce Theorem 1.3 from Theorem 3.8.

Let $\mathrm{cont}(q)$ denote the greatest common divisor (gcd) of the coefficients of an integral quadratic form q. When $q_{(A,\theta)}$ is a ternary form, we have that

$$\text{either } \mathrm{cont}(q_{(A,\theta)}) = 1 \text{ or } \mathrm{cont}(q_{(A,\theta)}) = 4, \tag{3.7}$$

by Theorem 20 of [**18**] and Proposition 19 of [**25**] (cf. Proposition 15(ii) of [**25**]). Moreover, if $q \sim q_C$, for some curve C of genus 2, we have from [**19**, p. 24] that

$$q(x,y,z) \equiv 0,1 \pmod 4, \text{ for all } x,y,z \in \mathbb{Z}. \tag{3.8}$$

Recall from the introduction that $r_4(q) = |R_4(q)|$, where $R_4(q)$ is as in (1.3). Therefore, it suffices to determine $r_4(q)$, for Eisenstein reduced ternary forms q in the case that q satisfies the conditions in equations (3.7) and (3.8). It is useful to know that this is done in the primitive case, i.e., when $\mathrm{cont}(q) = 1$ by Kani [**24**]. To state this, recall from (1.4) the definition of $a(q)$.

PROPOSITION 3.9 (Corollary 35 of [**24**]). *Let q_C be a primitive ternary form. Then $q_C \sim q$, where q is Eisenstein reduced. Moreover, if we write*

$$q = [a, b, c, 2r, 2s, 2t],$$

then we have that

$$(a(q_C), r_4(q_C)) = \begin{cases} (1,0) & \text{if } a \neq 4 \\ (2,2) & \text{if } a = 4 \text{ and } b \neq a \\ (6,6) & \text{if } a = b = 4 \text{ and } (r,s,t) = (2,2,2) \text{ or } (0,0,-2) \\ (4,4) & \text{otherwise.} \end{cases}$$

Thus, we want to determine $r_4(q)$, for imprimitive ternary forms satisfying the condition in (3.7). However, the situation is more complicated in this case, and so it needs considerable work.

NOTATION. Let $q_{a,b,c}$ denote the diagonal ternary quadratic form $ax^2 + by^2 + cz^2$. Let

$$q_{1,c} = [4,4,c,4,4,4], \quad q_{2,c} = [4,4,c,0,0-4], \quad q_{3,c} = [4,4,c,-4,-4,0], \tag{3.9}$$

$$q_{4,c} = [4,c,c,-4,0,0], \quad q_{5,c} = [4,c,c,-c,0,0], \quad q_{6,c} = [4,4,c,0,-4,0]. \tag{3.10}$$

REMARK 3.10. It is clear that $q_{a,b,c}$ satisfy the conditions (3.1)–(3.6) if $0 < a \leq b \leq c$. In addition, it is straightforward to check that all the ternary forms $q_{i,c}$ satisfy the conditions (3.1)–(3.6), for $i = 1,2$ provided $c \geq 4$. Similarly, it is a simple calculation to see that all the ternary forms $q_{i,c}$ satisfy the conditions (3.1)–(3.6), for $3 \leq i \leq 6$, provided $c > 4$. It follows from Proposition 3.3 that all these forms are positive definite, and since all satisfy the conditions (3.1)–(3.6), all are reduced.

If we apply Theorem 3.8 to the ternary quadratic forms having stated there, then we obtain the following:

COROLLARY 3.11. *Let q be a positive definite ternary quadratic form such that* $\mathrm{cont}(q) = 4$ *and $r_4(q) > 0$. Let $u = |\operatorname{Aut}^+(q)|$.*

(i) *If $u = 24$, then $q \sim q_{1,4}$ or $q \sim q_{4,4,4}$.*

(ii) *If $u = 12$, then $q \sim q_{2,c}$ with $4 \mid c$, or $q \sim q_{5,c}$, with $4 \mid c$ and $c > 4$.*

(iii) *If $u = 6$, then $q \sim q_{1,c}$, with $4 \mid c > 4$.*

PROOF. Since $\operatorname{cont}(q) = 4$, it follows that q is an integral quadratic form in the sense of Dickson, and by replacing q by an equivalent form, we may assume that q is an Eisenstein reduced form by Theorem 3.1. Let $q = [a, b, c, 2r, 2s, 2t]$. Since $\min(q) = a$ by Theorem 3.2, and since $\operatorname{cont}(q) = 4$ and $r_4(q) > 0$, it follows that $a = 4$.

(i) If $u = 24$, then it follows from Theorem 3.8(i) that $q = [4, 4, 4, 4, 4, 4]$ or $q = [4, 4, 4, 0, 0, 0]$, which means that $q = q_{1,4}$ or $q = q_{4,4,4}$ respectively, and so (i) follows.

(ii) If $u = 12$, then it follows by Theorem 3.8(ii) that $q = [4, 4, c, 0, 0, -4]$ or $q = [4, c, c, -c, 0, 0]$, with $4 \mid c$ (since $\operatorname{cont}(q) = 4$). This means that $q = q_{2,c}$ or $q = q_{5,c}$, with $4 \mid c$. Note that $c > 4$ (by (3.4)) since q is reduced, and thus, (ii) follows.

(iii) If $u = 6$, then since $4 \neq -3s$, we have two possible cases by Theorem 3.8(iii): $q = [4, 4, c, 4, 4, 4]$, $c > 4$ and $q = [4, 4, 4, 2r, 2r, 2r]$, $r \neq 0$, $r \neq 2$. If the latter case holds, then since $\operatorname{cont}(q) = 4$ and since $4 \geq |2r|$ by (3.3), we see that $r = -2$, but this contradicts (3.5). Thus, only the former case is left, which means that $q \sim q_{1,c}$, for $c > 4$. Thus, the last assertion (iii) follows. □

We now find the number $r_4(q)$ for some ternary forms q to use Table 1.

LEMMA 3.12. *We have the following values of* $r_4(q)$ *for special ternary forms* q.

(i) $r_4(q_{4,4,4}) = 6$, $r_4(q_{1,4}) = 12$, *and* $r_4(q_{1,c}) = 6$ *for* $c > 4$ *and* $c \equiv 0, 1 \pmod 4$.

(ii) $r_4(q_{2,4}) = 8$ and $r_4(q_{2,c}) = 6$, *where* $c > 4$.

(iii) $r_4(q_{3,c}) = 4$, *for* $c > 4$ *and* $4 \mid c$.

(iv) $r_4(q_{6,c}) = 4$, *for* $c > 4$.

PROOF.

(i) We first see that $R_4(q_{4,4,4}) = \{(\pm1, 0, 0), (0, \pm1, 0), (0, 0, \pm1)\}$, and so $r_4(q_{4,4,4}) = 6$. We next note that $q_{1,4} = 2[(x + y)^2 + (y + z)^2 + (x + z)^2]$, so it is easy to see that

$$R_4(q_{1,4}) = \{(\pm1, 0, 0), (0, \pm1, 0), (0, 0, \pm1), \pm(-1, 1, 0), \pm(-1, 0, 1), \pm(0, -1, 1)\},$$

which means that $r_4(q_{1,4}) = 12$. So, the first two equations in the assertion (i) follow.

To prove the last equation in this assertion, we note that when $c \equiv 1 \pmod 4$, this was already proven in the proof of Proposition 33 of [**24**]. With the same argument, we prove the equation when $4 \mid c$. Indeed, we observe that

$$q_{1,c} = 2[(x + y)^2 + (y + z)^2 + (x + z)^2] + (c - 4)z^2. \tag{3.11}$$

Also, if $(x, y, z) \in R_4(q_{1,c})$ and $z \neq 0$, then $z = \pm1$ and $x + y = y + z = x + z = 0$ since $c \geq 8$. But, this would imply that $x = -y$ and $x = y$, and so $x = y = 0$, which is impossible since $y + z = 0$ and $z \neq 0$. Hence, if $(x, y, z) \in R_4(q_{1,c})$, then $z = 0$, and thus $r_4(q_{1,c}) = r_1(x^2 + xy + y^2)$. Also, it is known that $r_1(x^2 + xy + y^2) = 6$, namely $\{\pm(0, 1), \pm(1, 0), \pm(1, -1)\}$, and thus $r_4(q_{1,c}) = 6$, so (i) follows.

(ii) First consider $q_{2,4} = 4x^2 + 4y^2 + 4z^2 - 4xy = 2(x - y)^2 + 2x^2 + 2y^2 + 4z^2$. If $(x, y, z) \in R_4(q_{2,4})$ and $z \neq 0$, then it follows that $(x, y, z) = \pm(0, 0, 1)$. Suppose now that $z = 0$, and so we should find the number $r_1(x^2 - xy + y^2)$, but it is known that $x^2 - xy + y^2 = 1$ has 6 solutions (as in the first part). Therefore, we have totally 8 solutions, and so $r_4(q_{2,4}) = 8$.

Next, suppose that $c \geq 5$. Observe that

$$q_{2,c} = 4x^2 + 4y^2 + cz^2 - 4xy = 3x^2 + cz^2 + (x - 2y)^2. \tag{3.12}$$

Also, if $(x, y, z) \in R_4(q_{2,c})$, then $z = 0$ because $c \geq 5$. Then, we can see that $r_4(q_{2,c}) = r_1(x^2 - xy + y^2) = 6$ as was mentioned above. More precisely, we have $R_4(q_{2,c}) = \{\pm(0, 1, 0), \pm(1, 0, 0), \pm(1, 1, 0)\}$, and so $r_4(q_{2,c}) = 6$, which proves this assertion.

(iii) Observe that

$$q_{3,c} = 4x^2 + 4y^2 + cz^2 - 4yz - 4xz = 2[(x - z)^2 + (y - z)^2 + x^2 + y^2] + (c - 4)z^2.$$

Also, if $(x, y, z) \in R_4(q_{3,c})$, then we see that $z = 0$ as in part (i), because $c \geq 8$ by hypothesis. Then, we see that $R_4(q_{3,c}) = \{\pm(0, 1, 0), \pm(1, 0, 0)\}$, and this proves that $r_4(q_{3,c}) = 4$.

(iv) Observe that

$$q_{6,c}(x, y, z) = 4x^2 + 4y^2 + cz^2 - 4xz = 2(x - z)^2 + 2x^2 + 4y^2 + (c - 2)z^2.$$

If $(x, y, z) \in R_4(q_{6,c})$, then $z = 0$, because $c - 2 > 4$. Hence $R_4(q_{6,c}) = \{(x, y, 0) : (x, y) \in R_1(x^2 + y^2)\}$. Thus, (iv) follows easily since $R_1(x^2 + y^2) = \{(\pm 1, 0), (0, \pm 1)\}$. □

Before we list the ternary forms q when $|\operatorname{Aut}^+(q)|$ is 2, 4, or 8, we introduce more notations which will be often used below.

Notation. Assume that $4 \leq b \leq c$ and $b \geq |2r|$ below.

$$\begin{aligned} q_1 =& [4, b, c, 0, 0, -4] \text{ with } b > 4, \quad q_2 = [4, b, b, 2r, 4, 4] \text{ with } b > 4,\ r > 1, \\ q_3 =& [4, b, c, 0, -4, 0] \text{ with } b \neq c, \quad q_4 = [4, b, c, -b, -4, 0] \text{ with } b > 4,\ b \neq c, \\ q_5 =& [4, b, c, 2r, 4, 4] \text{ with } 4 < b \neq c,\ r > 1,\ q_6 = [4, b, c, 2r, 0, -4] \text{ with} \\ & 4 < b \neq -2r,\ r < 0,\ q_7 = [4, b, c, 2r, -4, 0] \text{ with } b \neq c,\ b \neq -2r,\ r < 0. \end{aligned}$$

The following lemma is a straightforward verification.

Lemma 3.13. *All forms q_i are Eisenstein reduced forms, for $1 \leq i \leq 7$.*

Proof. We first prove that all forms are positive definite by using Proposition 3.3. For this, note first that the first coefficients of the q_i are positive.

To show that these forms are positive definite (and reduced), we use the notations in the definition of the Eisenstein reduced form; see (3.1)–(3.6). So, by the nondiagonal coefficients, we mean r, s, t. It is clear that (3.1), (3.3) and the first inequality of (3.2) hold for all forms. It is also clear that $\epsilon > 0$, for the forms q_i, for $1 \leq i \leq 5$. Since $b \geq |2r|$, this also holds for the forms q_i, for $i = 6, 7$. Hence, the second inequality of (3.2) holds for all forms. Thus, it follows from Proposition 3.3 that all forms q_i are positive definite.

To show that these forms are reduced, we also check the conditions (3.4)–(3.6). Since $\epsilon > 0$, the last inequality of (3.4) does not apply in our case. To check the other inequalities of (3.4), we first suppose that $b = 4$. Then we are left the q_i, for $i = 3$, 5 or 7 to check. It is clear that the other inequalities of (3.4) hold for q_3. For the forms q_5 and q_7, those hold since $4 = b \geq |2r|$. We next suppose that $b \neq 4$. Then, the rest cases are q_i, for $i = 1$, 2 or 6. But, we clearly have that the second inequality of (3.4) holds.

We lastly check the forms q_i, for $i = 1$, 3, 4, 6 or 7 to show that (3.5) holds and the forms q_i, for $i = 2$ or 5 to show that (3.6) holds. For the first one, we easily see that if $t = -2$, then $s = 0$ in all cases which were mentioned above. Also, we see that if $s = -2$, then $t = 0$ in those cases. We now show that if $b = -2r$, then $t = 0$. Note first that $t = 0$ in the case for the q_i, for $i = 3, 4$. So, the forms q_3 and q_4 trivially satisfy this condition. Since the forms q_6 and q_7 do not satisfy that $b = -2r$, these forms vacuously satisfy that condition. Also, q_1 satisfies this condition trivially since $r = 0$. Therefore, (3.5) holds for the forms q_i, for $i = 1$, 3, 4, 6 or 7.

The property (3.6) holds for the forms q_2 and q_5 since $2r \geq 4$. Hence, we proved that all forms satisfy all properties of the reduced forms, and so the assertion follows. □

We now give the classification of the ternary quadratic forms q with $|\operatorname{Aut}^+(q)| = 8$ as in Theorem 3.8.

PROPOSITION 3.14. *Let $q = [a, b, c, 2r, 2s, 2t]$ be an (Eisenstein) reduced positive definite ternary quadratic form. Then $|\operatorname{Aut}^+(q)| = 8$ if and only if*

$$q = q_{a,a,c} \text{ with } a \neq c \text{ or } q = q_{a,b,b} \text{ with } a \neq b \text{ or } q = [a, b, b, a/2, a, a] \text{ or}$$
$$q = [a, a, c, -a, -a, 0] \text{ or } q = [a, a, a, 2r, 2r, -2(a + 2r)] \text{ with } a \neq -3r \text{ or}$$
$$q = [a, a, a, -2(a + 2s), 2s, 2s] \text{ with } a \neq -3s.$$

PROOF. Assume first that $r = s = t = 0$. Let $u := |\operatorname{Aut}^+(q)|$. Since $r = s = t = 0$, there are left lines 5, 6, and 12 of Table 3, i.e., $a = b$ or $b = c$ or $a = b = c$ respectively. Then, in the last situation, we know that $u = 24$ by Theorem 3.8(i). Moreover, if $q = q_{a,a,c}$, with $a \neq c$ or $q = q_{a,b,b}$, with $a \neq b$, then we obtain that $u = 8$ by Table 3 and Remark 3.4 (or directly by the footnote of [**9**, p. 180]). Therefore, it follows that $u = 8 \Leftrightarrow q = q_{a,a,c}$, with $a \neq c$ or $q = q_{a,b,b}$, with $a \neq b$ in this case, as stated.

Suppose now that $(r, s, t) \neq (0, 0, 0)$. If $u = 8$, then we observe that $\operatorname{Aut}^+(q) \simeq D_4$ because $\operatorname{Aut}^+(q) \leqslant \operatorname{SL}_3(\mathbb{Z})$. Indeed, there is only one subgroup of order 8 in $\operatorname{SL}_3(\mathbb{Z})$, up to isomorphism, namely D_4, by the Corollary of Proposition 6 of [**38**]. This means that $\operatorname{Aut}^+(q)$ has an automorph of order 4 but not one of order 3. Therefore, there is one case given in line 8 of Table 2 (when $r, s, t > 0$) and there are three cases given in lines 8, 14, and 15 of Table 3 (when $r, s, t \leq 0$). In these four cases, we see that $q = [a, b, b, a/2, a, a]$, $q = [a, a, c, -a, -a, 0]$, $q = [a, a, a, 2r, 2r, -2(a + 2r)]$ and $q = [a, a, a, -2(a + 2s), s, s]$, respectively. In the last two situations, we have that $a \neq -3r$ and $a \neq -3s$; otherwise, $u = 24$ by Theorem 3.8(i). This proves the implication ($\Rightarrow$) for the remaining cases.

For the other implication, observe that if the case of line 8 of Table 2 or 3 holds, then all possible automorphs are already listed in the seventh column of these tables, and so no other ones cannot occur. This means that $u = 8$ in this situation as was indicated in the last column of these tables. If the case of line 14 of Table 3 holds and $a \neq -3r$, then we similarly see that all possible automorphs are the ones listed in the seventh column of this table, and so $u = 8$ as above. Similarly, if the case of line 15 of this table holds and $a \neq -3s$, then we obtain that $u = 8$, which completes the proof in all cases. □

We give the list of the ternary forms q having some special properties in the case that $|\operatorname{Aut}^+(q)| = 2$, 4 or 8 in the following two results, namely Corollary 3.15 and Theorem 3.16 below.

COROLLARY 3.15. *Let* $q = [4, b, c, 2r, 2s, 2t]$ *be an Eisenstein reduced positive definite ternary quadratic form with* $q \equiv 0, 1 \pmod 4$. *Then* $|\operatorname{Aut}^+(q)| = 8$ *if and only if* $q = q_{4,4,c}$, *with* $c \neq 4$ *or* $q = q_{4,b,b}$, *with* $b \neq 4$ *or* $q = [4, b, b, 2, 4, 4]$ *or* $q = q_{3,c}$.

PROOF. This assertion follows from Proposition 3.14 by showing that the last two ternary forms listed in this proposition cannot occur here. Indeed, suppose that the first one of these forms occurs, i.e., that $q = [4, 4, 4, 2r, 2r, -2(4 + 2r)]$. Then all nondiagonal coefficients are non-positive by (3.1) since $4 + 2r \geq 0$ by (3.3). Thus, since $|2r| \leq |-8 - 4r| \leq 4$ by (3.3), it follows that $-2r \leq 8 + 4r \leq 4$, and so $r = -1$. But this implies that $4 = 2(4 + 2r)$, and so $r = 0$ by (3.5), which is impossible. Similarly, if $q = [4, 4, 4, -2(4 + 2s), 2s, 2s]$, then we see that $s \leq 0$ by (3.1) and (3.3), and then it follows from (3.3) and (3.4) that $|2s| \leq 4$ and $|-(4 + 2s)| \leq |s|$, respectively. Since $s \leq 0$, this means that $s = -2$ (because $q \equiv 0, 1 \pmod 4$). Then since $4 = -2s$, it would give that $s = 0$ by (3.5) again, and this is impossible. Therefore, the assertion follows from Proposition 3.14. □

THEOREM 3.16. *Let* $q = [4, b, c, 2r, 2s, 2t]$ *be an Eisenstein reduced positive definite ternary quadratic form with* $q \equiv 0, 1 \pmod 4$. *Let* $u = |\operatorname{Aut}^+(q)|$. *Then we have that*

(i) $u = 2 \Leftrightarrow q = [4, b, c, 2r, 0, -4]$ *with* $r \neq 0$ *or* $q = [4, b, c, 2r, -4, 0]$ *with* $b \neq -2r$ *or* $q = [4, b, c, 2r, 4, 4]$, *with* $b \neq 4$, $b \neq c$, $r \neq 1$.

(ii) $u = 4 \Leftrightarrow q = q_{4,b,c}$, *with* $b \neq 4$, $b \neq c$ *or* $q = [4, b, c, -b, 0, 0]$, *with* $b \neq c$, *or* $q = [4, b, b, 2r, 0, 0]$, *with* $b \neq -2r$ *or* $q = [4, b, c, 0, 0, -4]$ *with* $b \neq 4$ *or* $q = [4, b, b, 2r, 4, 4]$ *with* $b \neq 4$, $r \neq 1$ *or* $q = [4, b, c, 0, -4, 0]$ *or* $q = [4, b, c, -b, -4, 0]$ *with* $b \neq 4$ *or* $q = [4, b, c, 2, 4, 4]$, *with* $b \neq c$.

PROOF. We see from (3.3) that $4 \geq |2s|, |2t|$ since q is reduced. Furthermore, s and t are even integers since $q \equiv 0, 1 \pmod 4$, and thus $s, t \in \{0, \pm 2\}$. Moreover, we have by (3.5) that $s = -2 \Rightarrow t = 0$ and $t = -2 \Rightarrow s = 0$. Hence, there are four possible cases for the pairs s, t because $(s, t) \neq (2, 0)$ or $(0, 2)$ by (3.1). To verify the assertions, we use Tables 2 and 3 as was done above. We first show that many cases of those tables can be discarded.

Suppose now that $2 \leq u \leq 4$, and so $u = 2$ or 4 by Lemma 3.6. Since $u \geq 2$, there is at least one nonidentity automorph of q, and so at least one case of Tables 2 and 3 holds. We analyze below all possible situations of the coefficients of the form q such that $u \leq 4$, and we exactly determine u in each case, and this will verify all assertions.

The fact that $u \leq 4$ immediately leaves the cases of lines 1–7 of Table 2 and those of lines 1–7 and 13 of Table 3 by the last columns of these tables. We also observe that line 13 of this table cannot hold. Indeed, if $a = b = c$, then since q is reduced, it follows that $|r| \leq |s| \leq |t|$ (by (3.4)). This implies that $r = 0$ because we know that at least one of the s and t is zero by what was mentioned above about the possible pairs (s, t). Hence, $8 + 2s + 2t \geq 4$, and so the condition of line 13 does not hold. Therefore, we are thus left with the cases of lines 1–7 of both tables to analyze below.

CASE 1. Assume that $t = -2$ and $s = 0$.

If the condition of line 7 of Table 3 were to hold, then it would follow that $b + 2r = 0$, which implies that $t = 0$ by (3.5), and thus, this case does not hold. Moreover, if the case of line 5 were to hold, then line 9 would hold, which is not possible. Since lines 1, 3, 4, and 6 clearly do not hold, we are only left with line 2. Therefore, if $r \neq 0$, then $u = 2$ (by the last column of the table), and $q = [4, b, c, 2r, 0, -4]$. Also if $r = 0$, then it follows by Remark 3.4 that $u = 4$ and $q = [4, b, c, 0, 0, -4]$ with $b \neq 4$. (If $b = 4$, then line 5 holds, which is a contradiction).

CASE 2. Assume that $s = -2$ and $t = 0$.

The cases of lines 2 and 6 of Table 3 are not possible here. If line 5 or 7 were to hold, then also line 8 would hold, which is not possible, and hence we are thus left with the cases of lines 1, 3, and 4. Clearly, line 1 always holds. Moreover, we have from the table that lines 3 and 4 hold $\Leftrightarrow b = -2r \Leftrightarrow u = 4$. Thus, if $u = 2$, then $q = [4, b, c, 2r, -4, 0]$, with $b \neq -2r$ and if $u = 4$, then $q = [4, b, c, -b, -4, 0]$, with $b \neq 4$. (If $b = 4$, then line 5 holds, which is a contradiction). If $r = 0$, then we are only left with the case of line 1, and thus $u = 4$ by Remark 3.4 and by the last column of the table. We find that $q = [4, b, c, 0, -4, 0]$ in this case, as stated.

CASE 3. Suppose that $s = t = 2$.

We discard the case of line 5 of Table 2 since it would imply the case of line 9, which is impossible. Also, if $b = c$ and $r = 1$, then line 8 holds, which is a contradiction.

If $b \neq c$ and $r \neq 1$, then only line 1 applies, and so $u = 2$, i.e., $q = [4, b, c, 2r, 4, 4]$, with $b \neq 4$, $b \neq c$ and $r \neq 1$. (If $b = 4$, then $r = 2$ by (3.4), and so line 5 holds, which is a contradiction).

If $b = c$ and $r \neq 1$, then we are left with the cases of lines 1, 6, and 7, which implies that $u = 4$ by the table, and that $q = [4, b, b, 2r, 4, 4]$, with $r \neq 1, b \neq 4$.

If $b \neq c$ and $r = 1$, then we are thus left with the cases of lines 1, 2, 3, and so $u = 4$, i.e., $q = [4, b, c, 2, 4, 4]$, with $b \neq c$.

CASE 4. Suppose that $s = t = 0$.

If $r = 0$, then we obtain that $b \neq 4$ and $b \neq c$; otherwise, we would have $u = 8$ by Proposition 3.14. Then, we are thus left with the case of line 1 of Table 3, and so $u = 4$ by Remark 3.4 and by the table. We see that $q = q_{4,b,c}$, with $b \neq 4$ and $b \neq c$ in this case.

Suppose now that $r \neq 0$, and thus the cases of lines 1–2 and 4–5 of Table 3 are impossible. We eliminate the case of line 7 because $\epsilon = 4 + b + 2r \neq 0$ by (3.4). We are thus left with the possible cases of lines 3 and 6. Note that if both lines hold at the same time, then the case of line 10 also holds, which is not possible. Thus, we conclude that if only the case of line 3 holds, i.e., $b = -2r$ and $b \neq c$, then $q = [4, b, c, -b, 0, 0]$, with $b \neq c$, and thus $u = 4$ by Remark 3.4 and by the table. We also conclude that if only the case of 6 holds, i.e., $b = c$ and $b \neq -2r$, then $q = [4, b, b, 2r, 0, 0]$, with $b \neq -2r$, and thus $u = 4$ again for the same reason.

Since we have analyzed all possible cases of the forms with $u \leq 4$, the assertions follow. □

REMARK 3.17. We restate Theorem 3.16 by using the notations q_i. More precisely, we can restate Theorem 3.16(i) as follows.

$$u = 2 \Leftrightarrow q \in \{q_5, q_6, q_7\},$$

and similarly, we can restate Theorem 3.16(ii) as follows.

$$\begin{aligned} u = 4 \Leftrightarrow q &\in \{q_1, q_2, q_3, q_4\} \text{ or } q = q_{4,b,c} \text{ with } b \neq 4,\ b \neq c \text{ or} \\ q &= [4, b, c, -b, 0, 0] \text{ with } b \neq c, \text{ or } q = [4, b, b, 2r, 0, 0] \text{ with } b \neq -2r \text{ or} \\ q &= [4, b, c, 2, 4, 4] \text{ with } b \neq c. \end{aligned}$$

4. Lists of the ternary quadratic forms

The problem of determining which integral quadratic forms q admit a curve C/K of genus 2 such that $q \sim q_C$ has been extensively studied by Kani and the present author. In particular, this question was completely resolved in [**22**] and [**25**] when q is ternary. The main results of these papers which give a solution to this problem will be used frequently below.

In the previous section, we have determined the ternary quadratic forms based on the order of their automorphism groups. By using the main results in [**25**] and [**22**], we now determine which of those forms is equivalent to a refined Humbert invariant. While this task is quite easy in the imprimitive case, it is highly complicated in the primitive case. In the latter situation, our method benefits from the theory of *genus equivalence* for ternary quadratic forms. Then, we present a proof of Theorem 1.1 by employing the results in Section 3.

PROPOSITION 4.1. *Suppose that* $4 \mid c$ *and* $c > 0$. *Then*

(i) $q_{5,c} \sim q_C$, *for some curve* C *of genus* 2 *if and only if* $c = 4$,

(ii) $q_{i,c} \sim q_C$, *for some curve* C *of genus* 2, *for* $1 \leq i \leq 4$ *and* $i = 6$,

(iii) *if* $\mathrm{cont}(q_j) = 4$, *for* $1 \leq j \leq 7$, *then* $q_j \sim q_C$, *for some curve* C *of genus* 2.

PROOF. To prove (i), we first observe that $q_{5,c}$ never represents 1 since its content is 4. Also, we have that $\frac{1}{2}q_{5,c}$ is improperly primitive if and only if $c = 4$. Therefore, since $q_{5,c}$ is a positive definite form representing 4, this assertion follows from Theorem 2 of [**25**] and (2.2).

To prove assertions (ii) and (iii), we first note that all ternary quadratic forms in both statements are positive definite by Remark 3.10 and Lemma 3.13. Second, all those forms have a nondiagonal coefficient 4 or -4. Since the content of those forms is 4, we thus obtain that the forms $\frac{1}{2}q_{i,c}$ and the forms $\frac{1}{2}q_j$ are improperly primitive, and that all forms never represent 1. Then, since the first coefficient of all forms is 4, both assertions follow from Theorem 2 of [**25**] and (2.2) again. □

We continue to determine which of the forms discussed in the previous sections are equivalent to a refined Humbert invariant when they are primitive. In this case, its proof is more complicated, and it needs arguments related to the theory of ternary forms. To this end, we recall some basic concepts here. Let f be a positive definite primitive ternary quadratic form. Recall that f comes equipped with two *basic genus invariants* $I_1(f)$ and $I_2(f)$ introduced by Brandt [**2**, p. 316], [**3**, p. 336]. Moreover, recall that f has *reciprocal* as defined by Brandt [**3**, p. 336]; this is denoted by F_f. Depending on the basic genus invariants, we determine the set of the *assigned characters* $X(f)$ and $X(F_f)$ of f and F_f, respectively; see [**3**]. Let us put

$$R(f) = \{f(x) > 0 : x \in \mathbb{Z}^3\}.$$

Let us also put

$$\chi_\ell(x) = \left(\frac{x}{\ell}\right),$$

for x prime to ℓ, where ℓ is an odd prime, and let us define

$$\chi_{-4}(x) = \left(\frac{-4}{x}\right) = (-1)^{(x-1)/2} \quad \text{and} \quad \chi_8(x) = \left(\frac{8}{x}\right) = (-1)^{(x^2-1)/8},$$

for odd x, as in [**3**]. Here, the symbol $\left(\frac{\cdot}{\cdot}\right)$ is the the Kronecker-Jacobi symbol.

PROPOSITION 4.2. *We have that* $f_1 := q_{1,c}$, $f_2 := q_{2,c}$, $f_3 := q_{3,c'}$, $f_6 := q_{6,c}$, *and* $f_4 := q_{4,4,c}$ *are equivalent to a form* q_C, *for some genus* 2 *curve* C, *for all* $c, c' > 1$, $c \equiv 1 \pmod 4$ *and* $c' \equiv 1 \pmod 8$.

PROOF. To prove this assertion we show that the f_i satisfy the conditions in Theorem 20(iii) of [**22**]. Then, the assertion follows from this result and Theorem 1 of [**22**] (together with (2.2)) because those results give the classification of primitive ternary refined Humbert invariants. The f_i are primitive positive definite by Remark 3.10, and also they clearly satisfy the condition (3.8), for any c, c'. We claim that the following conditions (which are the ones in Theorem 20(iii) of [**22**]) hold for the f_i.

$$\chi(f_i) = 1, \text{ for every assigned character } \chi \in X(f_i), \quad \text{and} \tag{4.1}$$

$$\left(\frac{I_2(f_i)}{m}\right) = 1, \text{ for some } m \in R(F_{f_i}) \text{ with } \gcd(m, 2I_2(f_i)) = 1. \tag{4.2}$$

In order to prove these equations for f_i's, we find their (basic) genus invariants and one suitable element in $R(F_{f_i})$. In this regard, the table below gives all the necessary information on the forms.

Regarding the following table, when $3 \mid c$, we denote f_2 by $f_{2''}$, otherwise, we denote it by $f_{2'}$. Also, the second column of the following table shows the *adjoint forms* $|I_1(f_i)|F_{f_i}$, and the third and the fifth columns are the absolute value of the genus invariant I_1 and the genus invariant I_2 respectively. The fourth column shows the discriminants and the last column is an integer represented by the reciprocal forms.

TABLE 4. Invariants of some ternary quadratic forms

i	$\lvert I_1(f_i)\rvert F_{f_i}$	$\lvert I_1(f_i)\rvert$	$\mathrm{disc}(f_i)$	$I_2(f_i)$	$n \in R(F_{f_i})$
1	$16[c-1, c-1, 3, 1, 1, c-2]$	16	$16(4-3c)$	$4-3c$	3
$2'$	$16[c, c, 3, 0, 0, c]$	16	$-48c$	$-3c$	$4c+3$
$2''$	$48[c/3, c/3, 1, 0, 0, c/3]$	48	$-48c$	$-c/3$	1
3	$32[\frac{c'-1}{2}, \frac{c'-1}{2}, 2, 2, 2, 1]$	32	$64(2-c')$	$2-c'$	c'
4	$16[c, c, 4, 0, 0, 0]$	16	$-64c$	$-4c$	$c+4$
6	$16[c, c-1, 4, 0, 4, 0]$	16	$64(1-c)$	$4(1-c)$	c

To verify the table, recall first that adjoint $\mathrm{adj}(f)$ of f is defined by the formula:

$$M(\mathrm{adj}(f)) = -2\,\mathrm{adj}(M(f)),$$

where $M(f)$ is the *coefficient matrix* of f as defined in [**40**, p. 2] ($\mathrm{adj}(M)$ is the usual adjoint of a matrix M). Thus, the second column is just an easy matrix calculation. We then easily see from the second column that $|I_1(f_i)| = 16$, for $i = 1, 2', 4, 6$ and $|I_1(f_3)| = 32$ since $|I_1(f)|$ is equal to the content of the adjoint form of f (and since $c \equiv 1 \pmod 4$). Also, we see that $|I_1(f_{2''})| = 48$, and so the third column is verified.

Since we have a relation

$$I_1^2(f)I_2(f) = 16\,\mathrm{disc}(f) \tag{4.3}$$

by [**3**, p. 336], it suffices to find the discriminants of the forms f_i to determine the second genus invariant I_2. By the straightforward matrix calculation, we obtain the fourth column. Thus, the fifth column follows from equation (4.3). For the last column, we observe that $F_{f_1}(0,0,1) = 3$, $F_{f_{2'}}(2,0,1) = 4c+3$, $F_{f_{2''}}(0,0,1) = 1$, $F_{f_3}(1,1,0) = c'$, $F_{f_4}(1,0,1) = c+4$ and $F_{f_6}(1,0,0) = c$.

By Brandt [**3**], §19, the set of the assigned characters $X(f_i)$ of the f_i's are as follows: $X(f_i) = \{\chi_{-4}\}$, for $i = 1$, 4 and 6, $X(f_3) = \{\chi_{-4}, \chi_8\}$, and $X(f_{2''}) = \{\chi_{-4}, \chi_3\}$ and $X(f_{2'}) = \{\chi_{-4}\}$. Since the f_i clearly represent an integer $n \equiv 1 \pmod 4$, we have that $\chi_{-4}(f_i) = 1$. Moreover, since $f_3(0,0,1) = c' \equiv 1 \pmod 8$, it follows that $\chi_8(f_3) = 1$. We see that $\chi_3(f_{2''}) = \chi_3(4) = 1$. Thus, (4.1) holds for all forms f_i.

Observe that $\gcd(3, 2I_2(f_1)) = \gcd(3, 8-6c) = 1$. Since $4-3c \equiv 1 \pmod 3$, we have that $\left(\frac{4-3c}{3}\right) = 1$. Observe also that $\gcd(c, 8-8c) = 1$. Moreover, we see that $4-4c \equiv 2^2 \pmod c$, and thus $\left(\frac{4-4c}{c}\right) = 1$. Next, note that $\gcd(c', 4-2c') = 1$. We know that $\left(\frac{2}{c'}\right) = 1$, for all $c' \equiv 1 \pmod 8$, and thus $\left(\frac{2-c'}{c'}\right) = 1$. We lastly check that $\gcd(c+4, -8c) = 1$. Then we see that $-4c \equiv -4c + c(c+4) \equiv c^2 \pmod{c+4}$, which proves that $\left(\frac{-4c}{c+4}\right) = 1$.

We are thus left with the case f_2 to verify (4.2). For this, first suppose that $3 \nmid c$, and so $I_2(f_{2'}) = -3c$. We see that $4c+3$ is relatively prime to $2I_2(f_2) = -6c$ since $3 \nmid c$. Then we see that $-3c \equiv -3c + c(4c+3) \equiv (2c)^2 \pmod{4c+3}$, which proves that $\left(\frac{-3c}{4c+3}\right) = 1$. We next suppose that $3 \mid c$, and now we have that $I_2(f_{2''}) = -c/3$. Since $F_{f_{2''}}$ represents 1 in this case, (4.2) holds trivially. Therefore, (4.2) holds for all the forms f_i's.

We thus conclude that the f_i are equivalent to a refined Humbert invariant by Theorems 1 and 20 of [**22**]. Moreover, they do not represent 1 since they are reduced and their minimum value is their first coefficient (by Theorem 3.2). Thus, we obtain from equation (2.2) that $f_i \sim q_C$, for some genus 2 curve C, for the f_i's, so the assertion follows. □

We derive the following corollary from the above results, which will be used below.

COROLLARY 4.3. *Suppose that q_C is a ternary form with $r_4(q_C) > 0$. If $|\mathrm{Aut}^+(q_C)| \geq 12$, then $q_C \sim q_{1,4}$ or $q_{2,c}$, with $c \equiv 0, 1 \pmod 4$, $c \geq 4$. Moreover, if this is the case, then $r_4(q_C) \geq 6$.*

PROOF. Let us put $u := |\mathrm{Aut}^+(q_C)|$. By Proposition 3.7, $u = 12$ or 24. If q_C is imprimitive, then $\mathrm{cont}(q_C) = 4$ by (3.7), and so it follows from Corollary 3.11(i)–(ii) that $q_C \sim q_{1,4}, q_{4,4,4}, q_{2,c}$, with $4 \mid c$ or $q_C \sim q_{5,c}$, with $4 \mid c > 4$. Then, since $\frac{1}{2}q_{4,4,4}$ is not improperly imprimitive, q_C is not equivalent to $q_{4,4,4}$ by Theorem 2 of [**25**]. Also, q_C is not equivalent to $q_{5,c}$, with $c > 4$ by Proposition 4.1(i), and so the first assertion follows in this case.

Suppose now that q_C is primitive. By Proposition 3.9, $q_C \sim q = [4, b, c, 2r, 2s, 2t]$, where q is a reduced form. Since q is primitive, we see from Theorem 3.8 that $u \neq 24$; otherwise, we would have from this theorem that $b = c = 4$, which contradicts $\operatorname{cont}(q) = 1$. If $u = 12$, then $q = [4, 4, c, 0, 0, -4] = q_{2,c}$ or $q = [4, c, c, -c, 0, 0] = q_{5,c}$ by the same theorem. Note that $c \equiv 1 \pmod 4$ by (3.8) in both cases since q is primitive. But then $q_{5,c}(1, -1, 1) = 4 + 3c \equiv 3 \pmod 4$, which is not possible by (3.8). Hence the first assertion follows. Then the second assertion directly follows from Lemma 3.12(i)–(ii). □

We are now ready to start to list the ternary forms q_C based on the information $\operatorname{Aut}(C)$, i.e., to prove the main results stated in the introduction.

We have the following two results from [**34**], pages 711 and 712, and Theorem 2 (also from Igusa [**14**], pages 644 and 645). The first one is that there is only one genus 2 curve C up to isomorphism such that $\operatorname{Aut}(C) \simeq C_3 \rtimes D_4$, and it has a model $y^2 = x^6 - 1$. The second one is that there is a unique genus 2 curve C' up to isomorphism such that $\operatorname{Aut}(C') \simeq \mathrm{GL}_2(3)$, and it has a model $y^2 = x(x^4 - 1)$ (known as *Burnside's curve*; see [**1**, p. 141]). We first find the ternary forms q_C and $q_{C'}$ such that $\operatorname{Aut}(C) \simeq \mathrm{GL}_2(3)$ and $\operatorname{Aut}(C') \simeq C_3 \rtimes D_4$.

PROPOSITION 4.4. *Let q be a ternary form. Then there exists a genus 2 curve C with $\operatorname{Aut}(C) \simeq \mathrm{GL}_2(3)$ such that $q \sim q_C$ if and only if $q_C \sim q_{1,4}$. Also, there exists a genus 2 curve C' with $\operatorname{Aut}(C') \simeq C_3 \rtimes D_4$ if and only if $q_{C'} \sim q_{2,4}$.*

PROOF. We prove both assertions together since a similar argument works for both.

($\Leftarrow$) Note first from Proposition 4.1(iii) that $q_{1,4} \sim q_C$ and $q_{2,4} \sim q_{C'}$ for some genus 2 curves C, C'. In addition, we know that $r_4(q_C) = r_4(q_{1,4}) = 12$ and $r_4(q_{C'}) = r_4(q_{2,4}) = 8$ by Lemma 3.12(i)-(ii). Thus, this implication of the assertions follows by Table 1.

($\Rightarrow$) When $\operatorname{Aut}(C) \simeq \mathrm{GL}_2(3)$, there is a unique genus 2 curve C up to isomorphism as was stated above. Suppose that q_C is not equivalent to $q_{1,4}$. We know by Proposition 4.1(iii) that $q_{1,4} \sim q_{C_1}$, for some curve C_1 of genus 2. It follows from the previous implication that $\operatorname{Aut}(C_1) \simeq \mathrm{GL}_2(3)$, which contradicts the uniqueness of C, because $C_1 \not\simeq C$. Thus it follows that $q_{1,4} \sim q_C$.

Similarly, when $\operatorname{Aut}(C') \simeq C_3 \rtimes D_4$, there is a unique genus 2 curve C' up to isomorphism. Therefore, this implication follows from the previous implication in a similar way. □

The following lemma, which is easy to verify will be used below.

LEMMA 4.5. *Let q_C be a ternary form, and let q be its reduced form. Then:*

(i) *If $a(q) = 1$ or $r_4(q) = 0$, then $q(1, 0, 0) \geq 5$. If, in addition, q is imprimitive, then $4 \mid q(1, 0, 0) \geq 8$.*

(ii) *If $a(q) \neq 1$ or $r_4(q) > 0$, then $q(1, 0, 0) = 4$.*

PROOF. Recall first from (3.8) that $q \equiv 0, 1 \pmod 4$, and also from (2.2) that q does not represent 1.

It is easy to observe that $r_4(q) = 0$ if and only if $a(q) = 1$. Theorem 3.2 shows that $\min(q) = q(1, 0, 0)$. Since q does not represent 1 and $q \equiv 0, 1 \pmod 4$, it follows that $\min(q) = q(1, 0, 0) \geq 4$. Therefore, if $r_4(q) = 0$, then $q(1, 0, 0) \geq 5$. In addition, if q is imprimitive, then we know by (3.7) that $\operatorname{cont}(q) = 4$, and so $4 \mid q(1, 0, 0) \geq 8$, which proves the assertion (i). Also, if $r_4(q) > 0$, then $\min(q) = q(1,0,0) = 4$ since $\min(q) \geq 4$, and so (ii) follows. □

The following simple fact is often used below.

LEMMA 4.6. *Let q_C be a ternary form with $r_4(q_C) > 0$. Then there exists an Eisenstein reduced form of the form $q = [4, b, c, 2r, 2s, 2t]$ such that $q \sim q_C$. Moreover, we have that*

$$(4.4) \qquad r_4(q) = 2 \quad \Leftrightarrow \quad b \neq 4.$$

PROOF. If q_C is primitive, then both assertions are direct consequences of Proposition 3.9. If q_C is imprimitive, i.e., $\mathrm{cont}(q) = 4$ (by (3.7)), then we have from Theorem 3.1 that $q_C \sim q$, for a reduced form $q = [a, b, c, 2r, 2s, 2t]$, and hence, $a = 4$ by Lemma 4.5. Thus if q_C is imprimitive, the first assertion also follows.

To show that (4.4) holds, we first see that if $r_4(q) = 2$, then $b \neq 4$; otherwise, we would have $q(\pm 1, 0, 0) = 4$ and $q(0, \pm 1, 0) = 4$, which would imply that $r_4(q) > 2$. For the converse of (4.4), if $b \neq 4$, then $b > 4$ by (3.2) because q is reduced. Also, recall from Theorem 3.2 that b is the second minimum. Since $r_1(q) = 0$, we have that any solution for which $q = 4$ has to be primitive. Since the second minimum b is strictly larger than 4, all primitive solutions of $q = 4$ are $(\pm 1, 0, 0)$. Otherwise, any extra primitive solution of $q = 4$ would imply that $b = 4$, which is not possible. So the assertion follows. □

We are ready to prove more interesting parts of the assertions in Theorem 1.1.

PROPOSITION 4.7. *Let q be a ternary form. Then there exists a genus 2 curve C with $\mathrm{Aut}(C) \simeq D_6$ such that $q \sim q_C$ if and only if there exists $c > 4$ with $c \equiv 0, 1 \pmod 4$ such that $q \sim q_{1,c}$ or $q \sim q_{2,c}$.*

PROOF. ($\Rightarrow$) Since $\mathrm{Aut}(C) \simeq D_6$, we have that $a(q) = r_4(q) = 6$ by Table 1. If q is imprimitive, then $\mathrm{cont}(q) = 4$ by (3.7), and so we may assume Theorem 3.1 that q is an Eisenstein reduced form. If q is primitive, we may assume by Proposition 3.9 that q is an Eisenstein reduced form. So let us put $q = [a, b, c, 2r, 2s, 2t]$, which is a reduced form.

If q is a primitive form, it follows from Proposition 3.9 that $a = b = 4$ and $(r, s, t) = (2, 2, 2)$ or $(0, 0, -2)$, which means that $q = q_{1,c}$ or $q = q_{2,c}$. Since q is primitive and since $q \equiv 0, 1 \pmod 4$ by (3.8), we see that $c \equiv 0, 1 \pmod 4$, and so $c > 4$ by (3.2) because q is reduced. Note that this case was proven by Kani [**24**], Proposition 33.

We henceforth suppose that q is imprimitive, and so $\mathrm{cont}(q) = 4$ by (3.7). Thus it follows that $a = b = 4$ by Lemma 4.6. Moreover, since $a(q) = 6$ and since $a(q) \mid |\mathrm{Aut}^+(q)|$ by (2.3), we have from Lemma 3.6 that $u := |\mathrm{Aut}^+(q)| \in \{6, 12, 24\}$.

If $u \geq 12$, then we have by Corollary 4.3 that $q = q_{1,4}$ or $q = q_{2,c}$, with $4 \mid c$. But, we know from Lemma 3.12 that $r_4(q_{1,4}) = 12$ and $r_4(q_{2,4}) = 8$, which is a contradiction, and so $q = q_{2,c}$, with $c > 4$, as stated.

If $u = 6$, then by Corollary 3.11, it follows that there exists $c > 4$, $4 \mid c$, such that $q = q_{1,c}$, so this part follows.

($\Leftarrow$) Let $c > 4$. If c is 0 or 1 in modulo 4, then we see from Propositions 4.1(ii) and 4.2 respectively that there exists a genus 2 curve C such that $q \sim q_C$. By Lemma 3.12(i)-(ii), we have that $r_4(q_{1,c}) = r_4(q_{2,c}) = 6$, and thus it follows by Table 1 that $\mathrm{Aut}(q_C) \simeq D_6$, which proves the assertion. □

NOTATION. Let $Q_c = \{q_{4,4,c}, q_{6,c}, q_{3,c}\}$, if $c \equiv 1 \pmod 8$, and let $Q_c = \{q_{4,4,c}, q_{6,c}\}$, if $c \equiv 5 \pmod 8$, and let $Q_c = \{q_{3,c}, q_{6,c}\}$, if $c \equiv 0 \pmod 4$, with $c > 4$, and let $Q_4 = \{q_{1,4}, q_{2,4}\}$.

PROPOSITION 4.8. *Let q be a ternary form. Then there exists a genus 2 curve C with $\mathrm{Aut}(C) \simeq D_4$ such that $q \sim q_C$ if and only if there exists $c > 4$ with $c \equiv 0, 1 \pmod 4$ such that q is equivalent to a form in the set Q_c.*

PROOF. We first claim that

$$q \sim q_C, \text{ with } \mathrm{Aut}(C) \simeq D_4 \Rightarrow u := |\,\mathrm{Aut}^+(q)| = 4 \quad \text{or} \quad 8. \tag{4.5}$$

To prove the claim, assume that $q \sim q_C$, with $\mathrm{Aut}(C) \simeq D_4$. In this case, we know from Table 1 that $a(q) = r_4(q) = 4$, which means that $u := |\,\mathrm{Aut}^+(q)| \in \{4, 8, 12, 24\}$ by (2.3) and Proposition 3.7. By Corollary 4.3, if $u \geq 12$, then $r_4(q) \geq 6$, which is a contradiction, and so $u < 12$, which implies (4.5).

($\Rightarrow$) By Lemma 4.6, we see that $q \sim [4, 4, c, 2r, 2s, 2t] =: q'$, where q' is a reduced form. Also, we have from (4.5) that $u = 8$ or 4.

We first suppose that q' is imprimitive. If $u = 8$, then since q' is a positive definite reduced form and $\mathrm{cont}(q') = 4$, we can apply Corollary 3.15 to obtain that $q' = q_{3,c}$ or that $q' = q_{4,4,c}$, where $4 \mid c > 4$. But, since $\frac{1}{2}q_{4,4,c}$ is not improperly primitive, $q_{4,4,c}$ cannot be equivalent to a refined Humbert invariant by Theorem 2 of [**25**], and so this implication follows in this case.

If $u = 4$, then we can apply Theorem 3.16(ii). Since $q'(0, 1, 0) = 4$, we can immediately eliminate the forms having the condition that $b > 4$ in this result, and we obtain precisely one form (whose content is 4), namely $q_{6,c}$, with $4 \mid c > 4$. Thus this implication follows in this case.

We next assume that q is primitive, and so, $q \sim q' = [4, 4, c, 2r, 2s, 2t]$, where c is odd. Since $q' \sim q_C$, we have that $q' \equiv 0, 1 \pmod 4$ by (3.8), and thus Corollary 3.15 and Theorem 3.16 can be applied.

If $u = 8$, then by Corollary 3.15, we have that $q' = q_{3,c}$, or $q' = q_{4,4,c}$, where $c \equiv 1 \pmod 4$, $c > 4$, since the other forms in this result cannot be primitive in our case. If $u = 4$, then by Theorem 3.16(ii), we see (by a similar reason which was discussed in the previous part) that $q' = q_{6,c}$, where $c \equiv 1 \pmod 4$, $c > 4$. Hence, this implication follows.

($\Leftarrow$) Let q be equivalent to a from in the set Q_c, for $c > 4$ and $c \equiv 0, 1 \pmod 4$. If $4 \mid c$, then we see from Proposition 4.1(ii) that there exists a genus 2 curve C such that $q \sim q_C$. For c, $c' > 1$, with $c \equiv 1 \pmod 4$ and $c' \equiv 1 \pmod 8$, it follows from Proposition 4.2 that there exists genus 2 curve C such that $q \sim q_C$. We know from Lemma 3.12 that $r_4(q_{3,c}) = r_4(q_{6,c}) = 4$, for $c > 4$. Also, it is clear that $r_4(q_{4,4,c}) = 4$, for $c > 4$. Hence, we obtain that $\mathrm{Aut}(q_C) \simeq D_4$ by Table 1, and the assertions follow. □

PROPOSITION 4.9. *Let q be a positive definite ternary form.*

(i) *If q is imprimitive, then there exists a genus 2 curve C with $\mathrm{Aut}(C) \simeq C_2 \times C_2$ such that $q \sim q_C$ if and only if $q \sim q_{4,c}$, with $4 \mid c > 4$ or $q \sim q_i$, for $1 \leq i \leq 7$ provided that $q_i(0, 1, 0) \neq 4$, for all i.*

(ii) *If q is primitive, and if $q \sim q_C$, for some curve C of genus 2 with $\mathrm{Aut}(C) \simeq C_2 \times C_2$, then q is equivalent to $[4, b, b, 2, 4, 4]$, with $b > 1$, $b \equiv 1 \pmod 4$ or q is equivalent to one of the forms listed in* Theorem 3.16(ii) and (iii) *with $b \neq 4$.*

PROOF. As in the proof of Proposition 4.8, we first claim that

$$q \sim q_C, \text{ with } \mathrm{Aut}(C) \simeq C_2 \times C_2 \;\Rightarrow\; u := |\,\mathrm{Aut}^+(q)| \;=\; 2, 4 \text{ or } 8. \tag{4.6}$$

To prove the claim, assume that $q \sim q_C$, with $\mathrm{Aut}(C) \simeq C_2 \times C_2$. By Table 1, we have that $a(q) = r_4(q) = 2$, and then we see that $u := |\,\mathrm{Aut}^+(q)| \in \{2, 4, 6, 8, 12, 24\}$ by Proposition 3.7 and (2.3). It follows from Corollary 4.3 that $u < 12$ since $r_4(q) < 6$.

Since $r_4(q) = 2$, we may suppose by Lemma 4.6 that $q = [4, b, c, 2r, 2s, 2t]$ is a reduced form such that $b \neq 4$. If q is primitive, then b or c is odd. Since $b \neq 4$ and since b or c is odd, we obtain that $u \neq 6$ by Theorem 3.8(iii). Thus, (4.6) holds in this case.

If q is imprimitive, then since $b \neq 4$, we easily see by Corollary 3.11 that $u \neq 6$, and so (4.6) holds in all cases.

We now start to prove the first assertion (i).

($\Rightarrow$) By what was discussed above, we have that $q = [4, b, c, 2r, 2s, 2t]$, with $b \neq 4$, and $u = 2$, 4 or 8 (see (4.6)). Recall from (3.7) that $\mathrm{cont}(q) = 4$. Thus, we see by Corollary 3.15 that $u \neq 8$ since $b \neq 4$ and since q cannot be a diagonal form (by Theorem 2 of [**25**]). Thus, we are left with the cases $u = 2$ and 4 to analyze.

Suppose first that $u = 4$. By Theorem 3.16(ii), we have eight possible families of ternary quadratic forms. Since q satisfies the two properties that $\mathrm{cont}(q) = 4$ and that $\frac{1}{2}q$ is improperly primitive, three of these families are excluded. Indeed, the forms $q_{4,b,c}$, $[4, b, c, -b, 0, 0]$ and $[4, b, c, 2, 4, 4]$ cannot have these properties, and so we are left with five possible cases, which are q_1, q_2, q_3, q_4 and $[4, c, c, 2r, 0, 0]$ (see Remark 3.17). Also, we note that if $q = [4, c, c, 2r, 0, 0]$, then $r = -2$ because $q/2$ is improperly primitive and reduced (see (3.1)), and so $q = [4, c, c, -4, 0, 0] = q_{4,c}$ with $4 \mid c$ (see (3.10) for the definition of the notation $q_{4,c}$). By Lemma 3.13 and Remark 3.10, we know that all (remaining five) forms are reduced. We see from (4.4) that if the second coefficient of these forms q is not 4, then $r_4(q) = 2$. Hence, if $u = 4$, then we have that $q \in \{q_1, q_2, q_3, q_4, q_{4,c}\}$ provided that $q(0, 1, 0) \neq 0$, as stated.

We next assume that $u = 2$. By Theorem 3.16(i) (and by Remark 3.17), it follows that $q \in \{q_5, q_6, q_7\}$. We again have that if the second coefficient of q is not 4, then $r_4(q) = 2$, and so we obtained that $q \in \{q_5, q_6, q_7\}$ such that $q_i(0, 1, 0) \neq 4$, for $i = 5, 6, 7$. Thus, this part follows.

($\Leftarrow$) First, observe from Proposition 4.1 that there exists a genus 2 curve C such that $q \sim q_C$ for each case. Recall again all forms in the statement are reduced (by Lemma 3.13 and Remark 3.10). Moreover, since $b \neq 4$ in each case, we see that $r_4(q_i) = 2$, for $1 \leq i \leq 7$ and that $r_4(q_{4,c}) = 2$, where $c > 4$ by (4.4). Therefore, we obtain from Table 1 that $\mathrm{Aut}(C) \simeq C_2 \times C_2$, so this part follows.

We next prove the second assertion (ii). By (4.6), we have again that $u = 2, 4, 8$, and a reduced form $q = [4, b, c, 2r, 2s, 2t] \sim q_C$ as above. Since $\mathrm{Aut}(C) \simeq C_2 \times C_2$ we have by Table 1 that $a(q) = r_4(q) = 2$, and so $b \neq 4$ by (4.4).

Suppose first that $u = 8$. Since $b \neq 4$, we have by Corollary 3.15 that $q = q_{4,b,b}$, or $q = [4, b, b, 2, 4, 4]$, where $b \equiv 1 \pmod 4$, $b > 4$. But the first case is not possible since $q_{4,b,b}$ clearly represents an integer $n \equiv 2 \pmod 4$, which contradicts (3.8). So, the assertion follows in this case.

Suppose next that $u = 2$ or $u = 4$. By Theorem 3.16, we have that q is one of the forms listed in Theorem 3.16(i) and (ii) with $b \neq 4$. Thus, the second assertion follows. □

We are now ready to prove Theorem 1.1.

PROOF OF THEOREM 1.1. By what was mentioned in Remark 2.1, we have that $J_C \sim E \times E$, for some CM elliptic curve E. So, the first assertion follows.

We now verify the second assertion. We first prove this assertion in the primitive case, so suppose that q_C is a primitive ternary form. For the necessity part, assume that $|\operatorname{Aut}(C)| > 2$. Since q_C is primitive, we see that $u := |\operatorname{Aut}(C)| = 4$, 8 or 12 by Proposition 3.9 and equation (1.5). If $u = 12$, then this part follows by Proposition 4.7. If $u = 4$ or 8, this part follows respectively by Propositions 4.9(ii) and 4.8. The sufficiency part also follows from those results.

We henceforth suppose that q_C is imprimitive. We observe that if $|\operatorname{Aut}(C)| > 2$, then we have that $u := |\operatorname{Aut}(C)| = 4$, 8, 12, 24 or 48; see Table 1 and Remark 2.1. Thus the necessity and sufficiency parts follow from Propositions 4.4, 4.7, 4.8(i) and 4.9(i). The last assertion directly follows from Propositions 4.1 and 4.2. □

We now provide a proof of Theorem 1.2.

PROOF OF THEOREM 1.2. Recall from (1.1) that $q_C \to q' := [4,4,4]$ if and only if $\operatorname{Aut}(C)$ contains a subgroup isomorphic to the dihedral group D_6. Recall also from (3.9) that $q_{1,c}(x,y,z) = 4x^2+4y^2+cz^2+4yz+4xz+4xy$ and $q_{2,c}(x,y,z) = 4x^2+4y^2+cz^2-4xy$.

($\Leftarrow$) Obviously, q' primitively represents itself. Also, we have that $q_{1,c}(1,1,0) = q'$ and $q_{2,c}(1,-1,0) = q'$, for any c as in the hypothesis, and so this part follows.

($\Rightarrow$) It is clear that $\operatorname{rank}(q_C) = 2$ or 3 since $q_C \to q'$. If q_C is a binary form, then we are done because $q_C \sim q'$.

We henceforth assume that q_C is a ternary form. As in the proof of Lemma 3.12(i), we know that $r_4(q') = 6$, and thus $r_4(q_C) \geq 6$ since $q_C \to q'$. Thus, we have three cases for the value of $r_4(q_C)$ as 6, 8 and 12 by Table 1. If $r_4(q_C) = 12$, or equivalently, if $\operatorname{Aut}(C) \simeq \mathrm{GL}_2(3)$ by Table 1, then we have from Proposition 4.4 that $q_C \sim q_{1,4}$. In a similar way, if $r_4(q_C) = 8$, we get from the same result that $q_C \sim q_{2,4}$.

We henceforth suppose that $r_4(q_C) = 6$. By Proposition 4.7, there exists $c \equiv 0,1 \pmod 4, c > 4$ such that $q_C \sim q_{1,c}$ or $q_{2,c}$. Thus, the first assertion follows.

If $q \sim q_{1,c}$ or $q_{2,c}$, for some $c \equiv 0,1 \pmod 4, c > 4$, then the last assertion directly follows from Propositions 4.1 and 4.2. If $q \sim q' = [4,4,4]$, then the last assertion follows from Corollary 33 of **[20]** because q' is a binary form of type $(2,1,3)$ in the sense of that article. □

From the results above we can deduce the following useful table (Table 5) of the automorphism groups $\operatorname{Aut}(C)$ for *all possible* imprimitive reduced ternary forms q_C. It also provides the explicit information for the formulas $a(q_C)$ and $k(q_C) = |\operatorname{Aut}^+(q_C)|/a(q_C)$. The second column contains all possible imprimitive ternary forms q_C which were listed in Theorem 1.1.

Like Table 5, we have the following Table 6 for all primitive ternary forms q_C with $a(q_C) > 2$. Similarly, the second column contains *all possible* primitive reduced ternary forms q_C which were listed in Theorem 1.1. The third column follows from Theorems 3.8 and 3.16. Note that the table does not contain the forms q_C with $a(q_C) = 2$ even though those forms were discussed in Proposition 4.9(ii). The reason why it was not added to the table is that we do not have the converse of Proposition 4.9(ii). To give an explicit list for which the forms listed in Proposition 4.9(ii) are equivalent to a refined Humbert invariant seems hard.

TABLE 5. Imprimitive ternary forms q_C with $\mathrm{Aut}(C)$ when $a(q_C) \neq 1$.

$\mathrm{Aut}(C)$	$q_C \sim$	$\lvert\mathrm{Aut}^+(q_C)\rvert$	$a(q_C)$	$k(q_C)$
$C_2 \times C_2$	$q_5 = [4, b, c, 2r, 4, 4]$ with $b \neq c$	2	2	1
	$q_6 = [4, b, c, 2r, 0, -4]$ with $r < 0$	2	2	1
	$q_7 = [4, b, c, 2r, -4, 0]$ with $r < 0$	2	2	1
	$q_1 = [4, b, c, 0, 0, -4]$	4	2	2
	$q_2 = [4, b, b, 2r, 4, 4]$ with $r > 0$	4	2	2
	$q_3 = [4, b, c, 0, -4, 0]$ with $b \neq c$	4	2	2
	$q_4 = [4, b, c, -b, -4, 0]$ with $b \neq c$	4	2	2
	$q_{4,c} = [4, c, c, -4, 0, 0]$	4	2	2
D_4	$q_{6,c} = [4, 4, c, 0, -4, 0]$	4	4	1
	$q_{3,c} = [4, 4, c, -4, -4, 0]$	8	4	2
D_6	$q_{1,c} = [4, 4, c, 4, 4, 4]$	6	6	1
	$q_{2,c} = [4, 4, c, 0, 0 - 4]$	12	6	2
$C_3 \rtimes D_4$	$q_{2,4} = [4, 4, 4, 0, 0 - 4]$	12	12	1
$\mathrm{GL}_2(3)$	$q_{1,4} = [4, 4, 4, 4, 4, 4]$	24	24	1

Notation: Here we suppose that $|2r| \leq b$, $4 \mid b$, c, $2r$ and $4 < b \leq c$. In addition, we suppose that if $r < 0$, then $b \neq -2r$.

PROPOSITION 4.10. *Suppose that q_C is a ternary form. If $a(q_C) \neq 1$ and if q_C is not equivalent to a form $q := [4, b, b, 2, 4, 4]$, for some $b \equiv 1 \pmod 4$, then $k(q_C) \leq 2$. Moreover, if $q_C \sim q$, then $k(q_C) = 4$.*

PROOF. The first assertion follows from Table 5 when q_C is imprimitive (see Table 1 for all possible cases of $a(q_C)$). It also follows from Table 6 when q_C is primitive with $a(q_C) \neq 2$.

Assume next that q_C is a primitive ternary form with $a(q_C) = 2$ which is not equivalent to q. It follows from Table 1 that $\mathrm{Aut}(C) \simeq C_2 \times C_2$. Then we see that $u = 2$, 4 or 8 by (4.6). But by Proposition 4.9(ii) and Corollary 3.15, we see that if $u = 8$, then $q_C \sim q = [4, b, b, 2, 4, 4]$, for some $b \equiv 1 \pmod 4$, $b > 1$, which is not possible, and so $u \neq 8$. Since $u = 2$ or 4, it follows that $k(q_C) \leq 2$, which proves the first assertion.

TABLE 6. Primitive ternary forms q_C with $\mathrm{Aut}(C)$ when $a(q_C) > 2$.

$\mathrm{Aut}(C)$	$q_C \sim$	$\lvert\mathrm{Aut}^+(q_C)\rvert$	$a(q_C)$	$k(q_C)$
D_4	$q_{6,c} = [4, 4, c, 0, -4, 0]$	4	4	1
	$q_{4,4,c} = [4, 4, c, 0, 0, 0]$	8	4	2
	$q_{3,c'} = [4, 4, c', -4, -4, 0]$	8	4	2
D_6	$q_{1,c} = [4, 4, c, 4, 4, 4]$	6	6	1
	$q_{2,c} = [4, 4, c, 0, 0 - 4]$	12	6	2

Notation: Here we suppose that $4 < c, c'$, and $c \equiv 1 \pmod 4$ and $c' \equiv 1 \pmod 8$.

Suppose now that $q_C \sim q$. We have by Corollary 3.15 that $|\operatorname{Aut}^+(q)| = 8$ and by Corollary 35 of [**24**] that $a(q) = 2$, which implies that $k(q_C) = 4$, which proves the last assertion. □

REMARK 4.11. We can discuss the case that $a(q_C) = 1$, for a ternary form q_C. Since $\operatorname{Aut}(C) \not\simeq C_{10}$ by Remark 2.1, we know by Table 1 that $\operatorname{Aut}(C) \simeq C_2$ in this case. If q_C is primitive, then we may assume by Proposition 3.9 that q_C is an Eisenstein reduced form. If q_C is imprimitive, then we know by (3.7) that $\operatorname{cont}(q_C) = 4$, and so we may suppose that q_C is reduced by Theorem 3.1. Since $r_4(q_C) = 0$ by Table 1 and $r_1(q_C) = 0$ by (2.2), we have that $q_C(1,0,0) > 4$ (see Lemma 4.5). Conversely, if a ternary form q is equivalent to a refined Humbert invariant q_C, for some genus 2 curve C, then again we can assume that q is reduced. In addition if $q(1,0,0) \neq 4$, then $r_4(q) = 0$ by Lemma 4.5, which implies that $\operatorname{Aut}(C) \simeq C_2$ by Table 1.

We also observe that we may have that $k(q_C) > 4$ (by Proposition 4.10) in the case that $a(q_C) = 1$. For this, consider a reduced positive definite primitive form $q = [9, 16, 16, -16, 0, 0]$. Since $q \equiv 0, 1 \pmod 4$ and since $q(1,1,1) = 5^2$, which is relatively prime to the discriminant $\operatorname{disc}(q) = -6912$, we conclude from Theorem 1 of [**22**] that q is equivalent to a refined Humbert invariant. Moreover, since $\min(q) = 9$ (by Theorem 3.2), we see that q does not represent 1, and hence, it follows from (2.2) that $q \sim q_C$, for some genus 2 curve C. Since we see that $|\operatorname{Aut}^+(q_C)| = 12$ by Theorem 3.8(ii), it follows that $k(q_C) = 12$.

5. Applications to the intersections

Kani [**19**] introduced the concept of a *generalized Humbert set* $H(q)$ associated with a given quadratic form q to understand the algebraic nature of the set of isomorphism classes of the curves lying on abelian product surfaces. This set is defined in terms of the refined Humbert invariant. Given any integral positive definite quadratic form q in r variables, let

$$H(q) = \{\langle A, \theta\rangle \in \mathcal{A}_2(K) \mid q_{(A,\theta)} \text{ primitively represents } q\},$$

where $\mathcal{A}_2(K)$ is the set of isomorphism classes $\langle A, \theta\rangle$ of principally polarized abelian surfaces (A, θ). In other words, $\langle A, \theta\rangle \in H(q)$ if and only if there exists an injective homomorphism $f : \mathbb{Z}^r \hookrightarrow \operatorname{NS}(A, \theta)$ with $\operatorname{NS}(A, \theta)/f(\mathbb{Z}^r)$ torsion-free such that $q = q_{(A,\theta)} \circ f$. The set $H(q)$ is called a *generalized Humbert set.*

Notice that Dickson has a slightly different definition for the concept of primitively representness; see [**9**, p. 25]. He also uses the notion of *properly* instead of primitively. By the invariant factor theorem, we see that both definitions are equivalent, and so we can use the results from [**9**] related to this notion. We use the symbol $f \to q$ when f primitively (or properly) represents q.

Recall from [**19**, p. 25] that if $q(x) = Nx^2$, i.e., $r = 1$, then $H(q) = H_N$ is the *classical Humbert surface of invariant* N.

By using Tables 5 and 6, we can get interesting results related to the intersections of the *generalized Humbert sets* $H(q)$ and Humbert surfaces H_N.

It may be useful to mention that Milio and Robert [**32**][2] recently found a relation between the denominators of the Hilbert modular polynomials and the generalized Humbert sets $H(q)$, for a binary form q; see §5.4 of [**32**].

As in the ternary quadratic forms, we simply use the abbreviation $q = [a, b, c]$ to denote an integral binary quadratic form $q(x, y) = ax^2 + bxy + cy^2$.

REMARK 5.1. As was mentioned in the introduction, we have by Theorem 7.1 of [**23**] that if q represents a square, then $H(q)$ is a scheme, in which case it is called *a generalized Humbert scheme*. To illustrate, if q is a ternary form, then this is always the case by the main results of [**22**] and [**25**], i.e., $H(q)$ is always a scheme.

Moreover, we know by Theorem 7.1 of [**23**] that for a binary quadratic form q representing a square the nonempty $H(q)$ is a one-dimensional subscheme of $\mathcal{A}_2(K)$. We also know that if q_1 and q_2 are not equivalent binary quadratic forms such that $H(q_i) \neq \varnothing$, for $i = 1, 2$, then the intersection $H(q_1) \cap H(q_2)$ consists of finitely many *CM points* by Lemma 10.7 of [**23**]. Hence, we have that $q_{(A,\theta)}$ is a ternary form for any principally polarized abelian surface (A, θ) in this intersection; see Section 3. Note that it follows from Lemma 10.7 of [**23**] that $H(q)$ is always finite, for a ternary quadratic form q.

To obtain some useful results from Theorem 1.2 we need the following arithmetic fact, which is interesting and useful itself.

PROPOSITION 5.2. *Let q be a positive definite binary quadratic form such that q does not represent n. If a positive definite ternary quadratic form f primitively represents q and f represents n, then*

$$|\operatorname{disc}(f)| \;\leq\; n|\operatorname{disc}(q)|.$$

PROOF. Let us put $q = [a, t, b]$. Since f primitively represents q, there is $f' \sim f$ such that $f'(x, y, 0) = q(x, y)$; see [**9**, p. 29]. Thus, it follows that $f' = [a, b, c, r, s, t]$, for some integers c, r, s.

By completing the square (Hermite's method), we see that $4af'(x, y, z) = (2ax+ty+sz)^2+g(y, z)$, for some positive definite binary form $g(y, z)$. We explicitly find that

$$g(y, z) = \left(4ab - t^2\right) y^2 + (4ar - 2st)zy + \left(4ac - s^2\right) z^2.$$

We thus see that

$$\operatorname{disc}(g) \;=\; (4ar - 2st)^2 - 4\left(4ab - t^2\right)\left(4ac - s^2\right) \;=\; 16a(c\operatorname{disc}(q) + q(-r, s)). \tag{5.1}$$

Since f represents n, f' also represents n, but q does not represent this integer by the hypothesis, we see that $f'(x_0, y_0, z_0) = n$, for some $(x_0, y_0, z_0) \in \mathbb{Z}^3$ with $z_0 \neq 0$. Recall that $\operatorname{disc}(q)$ and $\operatorname{disc}(g)$ are negative. If we apply equation (2.4) of

[2] Note that this paper is actually published. However, the authors removed this section for length reasons, and they declared that "they should probably make a separate article about this"; see [**29**, p. 99].

[**7**] to $g(y_0, z_0)$, then we have

$$\begin{aligned} g(y_0, z_0) &= \frac{\left(2\left(4ab - t^2\right)y_0 + (4ar - 2st)z_0\right)^2 - \operatorname{disc}(g)z_0^2}{4\left(4ab - t^2\right)} \\ &= \frac{\left(2|\operatorname{disc}(q)|y_0 + (4ar - 2st)z_0\right)^2 + |\operatorname{disc}(g)|z_0^2}{4|\operatorname{disc}(q)|} \\ &\geq \frac{|\operatorname{disc}(g)|z_0^2}{4|\operatorname{disc}(q)|} \geq \frac{|\operatorname{disc}(g)|}{4|\operatorname{disc}(q)|}. \end{aligned}$$

Therefore, we obtain that

$$4an = (2ax_0 + ty_0 + sz_0)^2 + g(y_0, z_0) \geq \frac{|\operatorname{disc}(g)|}{4|\operatorname{disc}(q)|}.$$

So, it follows from (5.1) that $4an \geq -\frac{16a(c\operatorname{disc}(q)+q(-r,s))}{4|\operatorname{disc}(q)|} = \frac{16a(c|\operatorname{disc}(q)|-q(-r,s))}{4|\operatorname{disc}(q)|}$. Since $a > 0$, we have that

$$n \geq \frac{(c|\operatorname{disc}(q)| - q(-r,s))}{|\operatorname{disc}(q)|}, \quad \text{and so,} \quad n|\operatorname{disc}(q)| \geq c|\operatorname{disc}(q)| - q(-r,s).$$

The assertion follows since we see that $|\operatorname{disc}(f)| = |\operatorname{disc}(f')| = c|\operatorname{disc}(q)| - q(-r,s)$. □

Before discussing the applications of Proposition 5.2, we prove the following lemma, which is used frequently below.

LEMMA 5.3. *Let q be a binary quadratic form. Assume that q does not represent n. If $\langle A, \theta\rangle \in H(q) \cap H_n$, then $q_{(A,\theta)}$ is a ternary quadratic form.*

PROOF. Recall from [**31**, p. 202] that $\mathrm{NS}(A) \simeq \mathbb{Z}^\rho$, where $\rho \leq 4$, and so $\operatorname{rank}(q_{(A,\theta)}) = \rho - 1 \leq 3$ since its associated module is $\mathrm{NS}(A)/\mathbb{Z}\theta$ by definition. Note that $q_{(A,\theta)}$ cannot be a binary form because then we would have $q_{(A,\theta)} \sim q$, which is not possible since $q_{(A,\theta)} \to n$. It is clear that its rank cannot be 1 because $q_{(A,\theta)} \to q$. Thus, $q_{(A,\theta)}$ is a ternary quadratic form, which proves the assertion. □

Here is one of the interesting applications of this arithmetic fact by using the finiteness of $H(f)$, where f is ternary.

PROPOSITION 5.4. *Let q be a binary quadratic form. If q does not represent n, then $H(q) \cap H_n$ is finite. Moreover, we have that*

$$H(q) \cap H_n = \bigcup_{\substack{f \to q, \\ f \to n}} H(f), \tag{5.2}$$

where f runs over all ternary quadratic forms with $|\operatorname{disc}(f)| \leq n|\operatorname{disc}(q)|$.

PROOF. If the intersection $H(q) \cap H_n$ is empty, there is nothing to prove, and so suppose that it is not empty. Let us take $\langle A, \theta\rangle$ in the intersection. We know from Lemma 5.3 that $q_{(A,\theta)}$ is a ternary quadratic form. We now observe that $|\operatorname{disc}(q_{(A,\theta)})|$ has an upper bound. Indeed, since $\langle A, \theta\rangle \in H(q) \cap H_n$, we have that $q_{(A,\theta)} \to q$ and n by definition. Also q does not represent n by hypothesis, and so, it follows by Proposition 5.2 that $|\operatorname{disc}(q_{(A,\theta)})| \leq n|\operatorname{disc}(q)|$. This proves that the left-hand side of (5.2) is contained in the right-hand side because $\langle A, \theta\rangle \in H(q_{(A,\theta)})$

by definition. Since the right-hand side of (5.2) is contained in the left-hand side follows immediately by definition, (5.2) holds.

For the finiteness assertion, recall that there are finitely many positive definite ternary quadratic forms f up to equivalence such that $|\operatorname{disc}(f)| \leq N$, for a fixed N by Theorem 11 of [**40**]. Therefore, the intersection $H(q) \cap H_n$ is the union of $H(f)$'s, for finitely many ternary forms f by (5.2). Then, since we know that $H(f)$ is finite for any ternary form f by what was mentioned in Remark 5.1, the intersection is finite. □

Note that one can provide a completely different (and a geometric) proof for this corollary by using Theorem 10.5 of [**23**] in the case that the binary form q represents a square. In this case, $H(q)$ is a finite union of irreducible certain curves, as was studied explicitly in [**23**].

We know that via the *Torelli map* $\langle C\rangle \mapsto \langle J_C, \theta_C\rangle$ we can view the moduli space $\mathcal{M}_2$ as a subset of $\mathcal{A}_2$, and hence let us put $H^*(q) = H(q) \cap \mathcal{M}_2$ and $H_n^* = H_n \cap \mathcal{M}_2$. Lastly, let us define an index set for a given integer n,

$$\mathcal{S}_n := \{c \in \mathbb{Z} : 1 < c \leq n+1,\ c \equiv 0, 1 \pmod 4\}. \tag{5.3}$$

Then we have the following result.

COROLLARY 5.5. *Let $q = [4,4,4]$, and let $c > 1$ with $c \equiv 0,1 \pmod 4$ such that q does not represent c. Then*

$$H^*(q) \cap H_c^* = \bigcup_{\substack{c_1 \in \mathcal{S}_c \\ q_{1,c_1} \to c}} H^*(q_{1,c_1}) \cup \bigcup_{\substack{c_2 \in \mathcal{S}_c \\ q_{2,c_2} \to c}} H^*(q_{2,c_2}). \tag{5.4}$$

Therefore, all curves C of genus 2 lying on the Humbert surface H_c^*, where $\operatorname{Aut}(C)$ contains a subgroup isomorphic to D_6 lie on the unions of the $H^*(f)$'s, where $f \sim q_{i,c_i}$ is as in (5.4).

PROOF. Since it is clear that both $q_{1,c}$ and $q_{2,c}$ primitively represent q and both primitively represent c, for any $c \equiv 0, 1 \pmod 4$, $c > 1$, it follows that the left-hand side of equation (5.4) contains the right-hand side of that equation.

For the other inclusion, we first see that for any $\langle C\rangle \in H^*(q) \cap H_c^*$, the form q_C is a ternary form by Lemma 5.3. Since $q_C \to q$, it follows from (1.1) that $\operatorname{Aut}(C)$ contains D_6 up to isomorphism, and so by Theorem 1.2 we have that either $q_C \sim q_{1,c_1}$ or that $q_C \sim q_{2,c_2}$, where $c_i \equiv 0,1 \pmod 4$, with $c_i > 1$. Since $q_C \to c$, we see that $q_{i,c_i} \to c$, for $i = 1,2$.

We next claim that $c_i \leq c+1$, which then implies that $c_i \in \mathcal{S}_c$. To prove this, recall first from the second and third rows of Table 4 respectively that $|\operatorname{disc}(q_{1,c_1})| = 16(3c_1 - 4)$ and $|\operatorname{disc}(q_{2,c_2})| = 48c_2$. Since $|\operatorname{disc}(q)| = 48$, we obtain from Proposition 5.2 that if $q_C \sim q_{1,c_1}$ or if $q_C \sim q_{2,c_2}$, then $16(3c_1 - 4) \leq 48c$ and $48c_2 \leq 48c$ respectively, and so we obtain that $c_i \leq c+1$, for $i = 1,2$ as claimed. Thus, the other inclusion holds, which proves the first assertion. The second assertion follows from the first assertion and (1.1). □

As was done in Theorem 1.2, we determine all refined Humbert invariants $q_C \to [4,0,4]$.

THEOREM 5.6. *Let C/K be a curve of genus 2. Then we have that q_C primitively represents the binary quadratic form $q := [4,0,4]$ if and only if $q_C \sim q$, or q_C is equivalent to a form in the set Q_c, with $c \equiv 0,1 \pmod 4$, $c > 1$.*

In addition, if $q' \in Q_c$, for any $c \equiv 0, 1 \pmod 4$, $c > 1$, then there is a curve C/K such that q_C is equivalent to q'.

PROOF. We see that the second assertion directly follows from Propositions 4.1 and 4.2. We now prove the first assertion.

($\Leftarrow$) As in Theorem 1.2, this part could be shown by using the theory of ternary quadratic forms. But, we also prove this part by using (1.2) together with the above results. For this, we now observe that $\mathrm{Aut}(C)$ contains a subgroup isomorphic to the dihedral group D_4 in our situation.

Indeed, firstly, if $q_C \sim q$, then the observation directly follows from (1.2). Secondly, if $q_C \sim q_{i,4}$, for $i = 1, 2$, then we have from Proposition 4.4 that $\mathrm{Aut}(C) \simeq \mathrm{GL}_2(3)$ or $C_3 \rtimes D_4$. Lastly, if $q_C \sim q_{j,c}$, for $j = 3, 6$, or $q_C \sim q_{4,4,c}$, where c satisfies the conditions in the hypothesis, then we see from Proposition 4.8 that $\mathrm{Aut}(C) \simeq D_4$. Thus, the observation follows since $D_4 \leqslant \mathrm{Aut}(C)$ up to isomorphism in all cases by the group theory. Hence, this part follows from (1.2).

($\Rightarrow$) It is clear that $\mathrm{rank}(q_C) = 2$ or 3 since $q_C \to q$. If q_C is a binary form, then the assertion directly follows from (1.2), and so assume that it is a ternary form. It is easy to see that $r_1(x^2 + y^2) = 4$, and thus we have that $r_4(q) = 4$, and so $r_4(q_C) \geq 4$. Thus, we have four cases for the value of $r_4(q_C)$ as 4, 6, 8 and 12 by Table 1. As in the proof of Theorem 1.2, if $r_4(q_C) = 12$ or 8, then we have from Table 1 that $\mathrm{Aut}(C) \simeq \mathrm{GL}_2(3)$ or $C_3 \rtimes D_4$ respectively, and thus it follows from Proposition 4.4 that $q_C \sim q \in Q_4$.

We next observe that $r_4(q_C) = 6$ is not a possible case. If it were possible, then we would have from Table 1 that $\mathrm{Aut}(C) \simeq D_6$, which contradicts (1.1) since D_4 is not a subgroup of D_6.

We henceforth suppose that $r_4(q_C) = 4$, and it follows from Table 1 that $\mathrm{Aut}(C) \simeq D_4$. Then, the assertion directly follows from Proposition 4.8. □

In particular, it turns out that we have a simple expression of the intersection $H^*([4, 0, 4] \cap H_c^*$.

COROLLARY 5.7. *Let $q = [4, 0, 4]$, and let $c > 1$ with $c \equiv 0, 1 \pmod 4$ such that q does not represent c. Then*

$$H^*(q) \cap H_c^* = \bigcup_{f \to c} H(f), \tag{5.5}$$

where the union is over all quadratic forms f in the set $\cup_{4 \leq c' \leq c+1} Q_{c'}$ with $c' \equiv 0, 1 \pmod 4$ which primitively represent c. Therefore, all curves C of genus 2 lying on the Humbert surface H_c^*, where $\mathrm{Aut}(C)$ contains a subgroup isomorphic to D_4 lie in the union of the $H^*(f)$'s, for certain ternary quadratic forms f.

PROOF. As in the proof of Proposition 5.5, the left-hand side of the equation (5.5) contains the right-hand side of that equation by Theorem 5.6.

For the other inclusion, for any $\langle C \rangle \in H^*(q) \cap H_c^*$, we see that q_C is a ternary form by Lemma 5.3. Also, since $q_C \to q$, it follows from Theorem 5.6 that $q_C \in Q_{c'}$, up to equivalence, for some $c' > 1$ with $c' \equiv 0, 1 \pmod 4$.

We now show that $c' \leq c + 1$ by using a similar argument in the proof of Corollary 5.5. Recall first from Table 4 that $|\operatorname{disc}(q_{3,c'})| = 64(c' - 2)$, $|\operatorname{disc}(q_{6,c'})| = 64(c' - 1)$ and $|\operatorname{disc}(q_{4,4,c'})| = 64c'$. Since $|\operatorname{disc}(q)| = 64$, we obtain from Proposition 5.2 that $64(c' - 2) \leq 64c$, $64(c' - 1) \leq 64c$ and $64c' \leq 64c$ respectively, and so it follows that $c' \leq c + 1$ for the last two cases and that $c' \leq c + 2$ for the first situation.

Since $c, c' \equiv 0, 1 \pmod 4$, we also have that $c' \leq c+1$ in the last situation, and so the claim follows. This completes the proof of the assertion. $\square$

EXAMPLE 5.8. Here we want to discuss the genus 2 curves C with $\mathrm{Aut}(C) \not\simeq C_2 \times C_2$ in the intersection of three Humbert surfaces $H_4^* \cap H_5^* \cap H_8^*$. We show that $\langle C \rangle \in H^*(q_{6,5}) \cup H^*(q_{4,4,5})$, for any such curve C. Moreover, one can see from the formulas in our joint work with Kani that $H^*(q_{6,5})$ and $H^*(q_{4,4,5})$ are singletons, and so there are only two such curves.

By Proposition 5.2(d) and Remark 5.3 of [**23**], we can describe the intersection $H_5 \cap H_8$ in terms of $H(q)$'s for binary forms q such that $|\operatorname{disc}(q)| \leq 4 \cdot 5 \cdot 8 = 160$. By using a list of reduced forms q with $|\operatorname{disc}(q)| \leq 160$, one can find that

$$H_5 \cap H_8 = H([1,0,4]) \cup H([5,2,5]) \cup H([5,4,8]) \cup H([5,0,8]). \tag{5.6}$$

Thus, it follows that

$$\begin{aligned} H_4 \cap (H_5 \cap H_8) = (H_4 \cap H([1,0,4])) \cup (H_4 \cap H([5,2,5])) \\ \cup (H_4 \cap H([5,4,8])) \cup (H_4 \cap H([5,0,8])). \end{aligned}$$

By the irreducibility criterion (2.2), we have that $\mathcal{M}_2 \cap H([1,0,4]) = \varnothing$, and hence

$$\mathcal{H}_8 := H_4 \cap H_5 \cap H_8 \cap \mathcal{M}_2 = H_4^* \cap (H^*([5,2,5]) \cup H^*([5,4,8]) \cup H^*([5,0,8])).$$

Now, it is easy to see that $[5,2,5]$, $[5,4,8]$ and $[5,0,8]$ cannot represent 4, and hence the intersection $\mathcal{H}_8$ is finite by Proposition 5.4. Moreover, for any curve $\langle C \rangle \in \mathcal{H}_8$, we have that J_C is a CM product abelian surface since q_C is ternary by Lemma 5.3. By Proposition 5.4, one can find all refined Humbert invariants q_C corresponding to the curves C on $\mathcal{H}_8$, i.e., one can write $\mathcal{H}_8$ as a union of $H(f)$'s, for certain ternary forms f. This is possible by employing the relation between the representation numbers of the reciprocal form F_{q_C} and the represented binary forms by the ternary form q_C; see Theorem 40 of [**9**].

As mentioned above, we want to determine all q_C such that $\langle C \rangle \in \mathcal{H}_8$ with $\mathrm{Aut}(C) \not\simeq C_2 \times C_2$. Since $q_C \to 4$, we see that $a(q_C) \geq 2$. Moreover, if $\mathrm{Aut}(C) \not\simeq C_2 \times C_2$, then it follows from Table 1 that $a(q_C) > 2$. Since $q_C \to 5$, for any $\langle C \rangle \in \mathcal{H}_8 \subset H_5^*$, we have that q_C is primitive by (3.7), and so we obtain from Table 6 that $\mathrm{Aut}(C) \simeq D_4$ or D_6. Thus, $q_C \to [4,0,4]$ or $[4,4,4]$ respectively by (1.2) and (1.1), which implies that $\langle C \rangle \in H^*[4,0,4] \cap H_5^*$ or $\langle C \rangle \in H^*[4,4,4] \cap H_5^*$ respectively, and so we apply Corollary 5.7 and (5.4). Hence, we obtain the possible list: $q_C \sim q \in \{q_{4,4,5}, q_{6,5}, q_{1,5}, q_{2,5}\}$.

On the one hand, $q_{1,5}$ and $q_{2,5}$ cannot represent 8 primitively. (One can see this by using the expressions in (3.11) and (3.12) respectively). On the other hand, we see that $q_{6,5}(-1,-1,0) = 8$, and also that $q_{6,5} \to [5,4,8]$. Indeed, we find that $F_{q_{6,5}} = [5,4,4,0,4,0]$ (see Table 4). Then, by applying the proof of Theorem 38 of [**9**] to $q_{6,5}$ and $F_{q_{6,5}}(1,1,0) = 9$, we find a binary form $q_{6,5}T = [17,20,8]$, where $T = \left(\begin{smallmatrix} 1 & 1 & -1 \\ 0 & 1 & -1 \end{smallmatrix}\right)^t$. Here $q_{6,5}T = T^t M(q_{6,5}) T$, where $M(q_{6,5})$ is the coefficient matrix of $q_{6,5}$ as defined in the proof of Proposition 4.2. Since we see that $[17,20,8] \sim [5,4,8]$ by the transformation matrix $\left(\begin{smallmatrix} -1 & 0 \\ 1 & -1 \end{smallmatrix}\right)$, it follows that $q_{6,5} \to [5,4,8]$. Therefore, we obtain that $H^*(q_{6,5}) \subseteq \mathcal{H}_8$. We similarly observe that $q_{4,4,5} \to 8$ and $q_{4,4,5} \to [5,0,8]$. In fact, the first one is clear. To show the second one, we apply the same method/theorem to $q_{4,4,5}$ and $F_{q_{4,4,5}}(1,1,0) = 10$ (see Table 4), and we obtain that $q_{4,4,5} \to [13,36,28]$ by the transformation matrix $\left(\begin{smallmatrix} 1 & 1 & -1 \\ 0 & 2 & -1 \end{smallmatrix}\right)^t$. Since $[13,36,28] \sim [5,0,8]$ by the matrix $\left(\begin{smallmatrix} -1 & -2 \\ 1 & 1 \end{smallmatrix}\right)$, it follows that $H^*(q_{4,4,5}) \subseteq \mathcal{H}_8$.

As a conclusion, we find that all points $\langle C\rangle$ in $\mathcal{H}_8$ with $\mathrm{Aut}(C) \not\simeq C_2 \times C_2$ lie in $H^*(q_{6,5})$ and $H^*(q_{4,4,5})$.

6. Elliptic subcovers of a curve of genus 2

Let C/K be a curve of genus 2 over K. Recall from [**16**] or [**15**] that an *elliptic subcover* is a finite morphism $\phi : C \to E$ to an elliptic curve E/K which does not factor over a non-trivial isogeny (i.e., maximal covering) of E. Its degree is the degree of the morphism.

There is a considerable body of literature on elliptic subcovers dating back to the 19th century; for references on this topic, see for example [**1**]. In recent decades, this topic has attracted significant interest from various authors, including but not limited to Bruin and Doerksen [**4**], Frey and Kani [**10**], Kani [**15**], [**16**], [**21**], [**23**], Kuhn [**26**], and Shaska [**33**] Shaska and Völklein [**34**].

To present the relation of elliptic subcovers and our main ingredient, namely the refined Humbert invariant, let us recall from Theorem 4.5 of Kani [**15**] that for a curve C/K of genus 2 we have

$$C/K \text{ has an elliptic subcover of degree } m \;\Leftrightarrow\; q_C \to m^2. \tag{6.1}$$

For a given group G and a given number m, let

$$\mathcal{H}(G, m) \;:=\; \{\langle C\rangle : \mathrm{Aut}(C) \simeq G \text{ and } C \text{ has an elliptic subcover of degree } m\}$$

denote the set of isomorphism classes of genus 2 curves C/K with $\mathrm{Aut}(C) \simeq G$ such that C/K has an elliptic subcover of degree m. Kani [**23**] discussed this kind of set from the moduli point of view. Also, Shaska [**33**] explicitly described the sets $\mathcal{H}(D_4, 3)$ and $\mathcal{H}(D_6, 3)$. We use the expressions of the sets $\mathcal{H}(D_6, m)$ and $\mathcal{H}(D_4, m)$ with the intersections $H^*([4,4,4]) \cap H^*_{m^2}$ and $H^*([4,0,4]) \cap H^*_{m^2}$ respectively. These expressions are essentially derived from the results of [**20**] and [**24**] (see the proof of Proposition 6.1 below). Then we can go further to provide a more explicit description of these intersections by employing Propositions 5.5 and 5.7.

PROPOSITION 6.1. *For any odd integer m, we have that*

$$\mathcal{H}(D_6, m) \;=\; H^*([4,4,4]) \cap H^*_{m^2} \quad \textit{and} \quad \mathcal{H}(D_4, m) \;=\; H^*([4,0,4]) \cap H^*_{m^2}. \tag{6.2}$$

In particular,

$$|\mathcal{H}(D_6, m)| \;=\; |H^*([4,4,4]) \cap H^*_{m^2}| \quad \textit{and} \quad |\mathcal{H}(D_4, m)| \;=\; |H^*([4,0,4]) \cap H^*_{m^2}|. \tag{6.3}$$

PROOF. If $\langle C\rangle \in H^*([4,4,4]) \cap H^*_{m^2}$, then $q_C \to [4,4,4]$ and $q_C \to m^2$, and so it follows by (1.1) that $D_6 \leqslant \mathrm{Aut}(C)$ (up to isomorphism), and by (6.1) that C has an elliptic subcover of degree m. Since m is odd by the hypothesis, it follows from (3.7) that q_C is primitive. Hence, this means that $\mathrm{Aut}(C) \simeq D_6$ by Table 6, and thus $\langle C\rangle \in \mathcal{H}(D_6, m)$.

Conversely, if $\langle C\rangle \in \mathcal{H}(D_6, m)$, then $q_C \to m^2$ by (6.1), and also $q_C \to [4,4,4]$ by (1.1). This means that $\langle C\rangle \in H^*([4,4,4]) \cap H^*_{m^2}$ by definition, and so the first equality of (6.2) follows. The first equality of (6.3) also follows from this assertion.

The second equations of (6.2) and (6.3) similarly follow. Indeed, if $\langle C\rangle \in H^*([4,0,4]) \cap H^*_{m^2}$, then it follows by (1.2) that $D_4 \leqslant \mathrm{Aut}(C)$, and by (6.1) that C has an elliptic subcover of degree m. Since m is odd, q_C is primitive as was discussed above, and hence $\mathrm{Aut}(C) \simeq D_4$ by Table 6, and thus $\langle C\rangle \in \mathcal{H}(D_4, m)$. The converse is similar, and so the second equalities of both (6.2) and (6.3) follow. □

REMARK 6.2. By similar arguments in the proof of Proposition 6.1, $\mathcal{H}(D_6, m)$ can be formulated in terms of $H^*([4,4,4]) \cap H^*_{m^2}$, for all integers m, where m^2 is not primitively represented by $[4,4,4]$ (i.e., *not necessarily odd*). For this generalization, one only has to note that there are two non-isomorphic curves C of genus 2 when $\mathrm{Aut}(C) \simeq \mathrm{GL}_2(3)$ or $C_3 \rtimes D_4$, and in those cases, we have that $q_C \to [4,4,4]$ by Proposition 4.4. This means that if we in addition have that $q_C \to m^2$, for such a curve, then $\langle C \rangle \in (H^*([4,4,4]) \cap H^*_{m^2}) \setminus \mathcal{H}(D_6, m)$.

For example, when $m = 4$, we have that $[4,4,4]$ cannot primitively represent 4^2, but, $q_{2,4}(2,1,1) = 16$ (see (3.9)). This means that there exists a curve C of genus 2 with $\mathrm{Aut}(C) \simeq C_3 \rtimes D_4$ (by Proposition 4.4), and thus

$$\langle C \rangle \in (H^*([4,4,4]) \cap H^*_{16}) \setminus \mathcal{H}(D_6, 4).$$

We now discuss the cardinality of $\mathcal{H}(D_6, m)$, for odd integers m. For any odd n we have that

$$H^*([4,4,4]) \cap H^*_{n^2} = \bigcup_{\substack{c_1 \in \mathcal{S}^*_c \\ q_{1,c_1} \to n^2}} H^*(q_{1,c_1}) \cup \bigcup_{\substack{c_2 \in \mathcal{S}^*_c \\ q_{2,c_2} \to n^2}} H^*(q_{2,c_2}). \tag{6.4}$$

This equation can be deduced from (5.4). Indeed, since n is odd, it is clear that $[4,4,4]$ cannot represent n. Also, since $\mathrm{cont}(q_{i,c})$ is even for $i = 1,2$ (see (3.9)) when c is even, the forms $q_{i,c}$ cannot represent an odd number. Hence, $c \equiv 1 \pmod 4$ by (3.8). This condition implies that the exceptional cases $q_{1,4}$ and $q_{2,4}$ cannot occur (see (3.9) for the notation). Therefore, (6.4) follows from (5.4).

By definition, we see that the union $H(q_{1,c_1}) \cup H(q_{2,c_2})$ is disjoint for any c because q_{1,c_1} and q_{2,c_2} are not equivalent (by Remark 3.10). Also, it follows from Proposition 4.2 that $q_{i,c} \sim q_C$, for $c \in \mathcal{S}^*_n$, and so $|H^*(q_{i,c})| = |H(q_{i,c})|$, for $i = 1,2$. Therefore, it follows from (6.4) and 6.3 that

$$|\mathcal{H}(D_6, n)| = \sum_{c_i \in \mathcal{S}^*_n,\ q_{i,c_i} \to n^2} |H(q_{1,c_1})| + |H(q_{2,c_2})|. \tag{6.5}$$

EXAMPLE 6.3. We want to illustrate the case that $m = 3$ in Proposition 6.1 because the set $\mathcal{H}(D_6, 3)$ was explicitly discussed in the main results of [**33**]. For this, observe first that $q_{i,c}$ cannot represent 9 for an even number c, and for $i = 1,2$. Moreover, it is clear that $q_{i,9}(0,0,1) = 9$, for $i = 1,2$ and that $q_{1,5}(-1,-1,1) = 9$ and $q_{2,5}(1,1,1) = 9$. Thus, by Corollary 5.5, we have that

$$H^*([4,4,4]) \cap H^*_9 = H^*(q_{1,5}) \cup H^*(q_{1,9}) \cup H^*(q_{2,5}) \cup H^*(q_{2,9}), \tag{6.6}$$

and it follows by (6.5) that

$$|\mathcal{H}(D_6, 3)| = |H^*(q_{1,5})| + |H^*(q_{1,9})| + |H^*(q_{2,5})| + |H^*(q_{2,9})|.$$

By using the properties of the refined Humbert invariant, it is possible to reprove many parts of Shaska's main results; cf. Remark 6.4 and Example 6.5 below. Recall that the sets $\mathcal{H}(D_6, 3)$ and $\mathcal{H}(D_4, 3)$ are determined, and the equations/models for the curves in these sets defined over $\mathbb{Q}$ are given in Theorem 1, Corollary 1 and Table 1 of [**33**].

REMARK 6.4. By (6.5), if one finds a formula for the number $|H(q)|$, for any ternary form q, then it leads to a formula for the cardinality $|\mathcal{H}(D_6, n)|$. Recently, the formulas for $|H(q)|$, for any ternary form q, have been found by a joint work with Kani. By using this formula, we obtain that if $h(\Delta)$ denotes the class number of the discriminant Δ, then

$$|\mathcal{H}(D_6, n)| = \sum_{\substack{c\in\mathcal{S}_n^* \\ q_{1,c}\to n^2}} h(4-3c) + \frac{1}{2}\sum_{\substack{c\in\mathcal{S}_n^{'} \\ q_{2,c}\to n^2}} h(-3c) + \sum_{\substack{c\in\mathcal{S}_n^{''} \\ q_{2,c}\to n^2}} h(-3c).$$

Notice that it easily follows that $|\mathcal{H}(D_6, 3)| = 6$ by this formula, as in Theorem 1 of [**33**].

Observe that one similarly obtains a formula for the cardinality $|\mathcal{H}(D_4, n)|$ by what was discussed in Remark 6.4.

By exploiting the properties of q_C, one can say more about the curves $\langle C\rangle \in \mathcal{H}(D_6, n)$. The following examples outline some ideas of how we obtain the geometric properties of the curves C by using the arithmetic properties of the quadratic form q_C, and so we verify many results in [**33**] by a different method.

EXAMPLE 6.5. We here take a closer look at the curve C with $q_C \sim q_{1,5}$ (cf. Example 6 of [**33**]). By Remark 2.1 and equation (2.2), we know that $J_C \simeq E_1 \times E_2$ is a CM product surface. Recall that for two isogenous elliptic curves E, E', we have the *degree* form $q_{E,E'}$ on $\mathrm{Hom}(E, E')$ which is defined by $q_{E,E'}(h) = \deg(h)$, for $h \in \mathrm{Hom}(E, E')$, and we know that it is a positive definite binary quadratic form in the CM case. Thus, q_{E_1,E_2} is a binary form, and by Lemma 30 and equation (37) of [**21**], we see that $\mathrm{disc}(q_C)/16 = \mathrm{disc}(q_{E_1,E_2}) = -11$, and so it easily follows that $q_{E_1,E_2} \sim [1, 1, 3]$. Since it represents 1, it follows that $E_1 \simeq E_2$. Let us put $F = \mathrm{End}(E_1) \otimes \mathbb{Q}$. By Corollary 42 of [**17**] we find that $\mathrm{disc}(q_{E_1,E_2}) = \Delta_F$, where Δ_F is the discriminant of the quadratic number field F and that $\mathrm{End}(E_1)$ is isomorphic to the maximal order of F (since -11 is a fundamental discriminant).

Moreover, since the class number $h(-11) = 1$, we know that the isomorphism class of the elliptic curve E_1/K is uniquely determined by the endomorphism ring of E_1/K; see equation (55) of [**17**]. Moreover, we see from [**36**, p. 483] that the j-invariant $j(E_1) = -2^{15}$ and that E_1/K is given by the equation $y^2 + y = x^3 - x^2 - 7x + 10$.

We use Hayashida's formula [**12**], §7 (cf. the computer calculations of [**11**] derived from Hayashida's formula) in this special case to show that there is a unique curve on J_C. In fact, we are in the case that $J_C \simeq E_1 \times E_1$, where $\mathrm{End}(E_1)$ is the maximal order of $\mathbb{Q}(\sqrt{-11})$, and thus Hayashida's formula applies, and it implies that there is a unique curve C' of genus 2 on $E_1 \times E_1 \simeq J_C$, and so $C \simeq C'$. Thus, we conclude from Table 4 of [**11**] that C is given by the equation $y^2 = 2x^6 + 11x^3 - 22$. Observe that this curve is isomorphic to the curve C'' : $y^2 = 11x^6 + 11x^3 - 4$, where C'' is the curve which was given in the sixth row of Table 1 of [**33**] because both curves have the same *Igusa-Clebsch invariants* $(I_2, I_4, I_6, I_{10}) = (180576, 361304064, 31762962874368, 3877056611888608051 2)$ by SageMath [**8**].

EXAMPLE 6.6. We briefly study the properties of the curve C with $q_C \sim q_{2,5}$ (cf. Example 7 of [**33**]). As in Example 6.5, we know that $J_C \simeq E_1 \times E_2$ is a CM abelian product surface, and $-16\,\mathrm{disc}\,(q_{E_1,E_2}) = \mathrm{disc}\,(q_{2,5}) = -16.15$ (cf. Table 4). Again, as in this example, we obtain that $\mathrm{End}(E_1) \simeq \mathrm{End}(E_2)$ is the maximal order $\mathbb{Z}[(1 + \sqrt{-15})/2]$ of $\mathbb{Q}(\sqrt{-15})$.

If q_{E_1,E_2} represents 1, then $E_1 \simeq E_2$, but then by Hayashida and Nishi [**13**] (or Theorem 2 of [**18**]), $E_1 \times E_2$ cannot be a Jacobian of a genus 2 curve, which is a contradiction, and so q_{E_1,E_2} cannot represent 1. Thus it follows from [**5**, p. 19] that $q_{E_1,E_2} \sim [2,1,2]$. Then this implies that $J_C \not\simeq E \times E$, for any CM elliptic curve E because $[2,1,2]$ does not lie in the principal genus. Since $h(-15) = 2$, there are two elliptic curves E/K with $\mathrm{End}(E) = \mathcal{O}$ up to isomorphism over K. One can determine the j-invariants of these elliptic curves by finding the roots of the Hilbert class polynomial of discriminant -15; e.g., see the elliptic curves with LMFDB labels [**27**, Elliptic Curve 81.1.a1] and [**27**, Elliptic Curve 81.1.a3].

By using the formulas in [**24**], we see that there is a unique genus 2 curve on J_C up to isomorphism.

We observe by equation (3.12) that

$$R_9(q_{2,5}) = \{\pm(0,1,1), \pm(0,-1,1), \pm(1,0,1), \pm(-1,0,1), \pm(-1,-1,1), \pm(1,1,1)\},$$

and so there are 12 primitive triples (a,b,c) such that $q_{2,5}(a,b,c) = 9$, and thus there are 12 elliptic subcovers of degree 3 of C (up to isomorphism) by Theorem 4.5 of [**15**].

Acknowledgments

Part of this project grew out of the PhD thesis of the author. He thanks his supervisor, Prof. Ernst Kani for his extensive guidance with particular suggestions on the present article, and for sharing of his current (preprint) papers with the author. In addition, the author would like to express gratitude to Prof. John Voight for his valuable suggestions and insightful discussions, which provided diverse perspectives on the content of this article. The author also wants to thank Eda Kırımlı for proofreading the present article numerous times and giving many suggestions on it, and pointing out many errors and typos. Finally, the author would like to thank the referee for the valuable comments and lists of corrections provided by the referee.

References

[1] R. D. M. Accola and E. Previato, *Covers of tori: genus two*, Lett. Math. Phys. **76** (2006), no. 2-3, 135–161, `https://doi.org/10.1007/s11005-006-0067-5`. MR2235401

[2] H. Brandt, *Über das Mass positiver ternärer quadratischer Formen*, Math. Nachr. **6** (1952), 315–318, `https://doi.org/10.1002/mana.19520060507`. MR51268

[3] H. Brandt, *Zur Zahlentheorie der ternären quadratischen Formen*, Math. Ann. **124** (1952), 334–342, `https://doi.org/10.1007/BF01343574`. MR51269

[4] Nils Bruin and Kevin Doerksen, *The arithmetic of genus two curves with* $(4,4)$*-split Jacobians*, Canad. J. Math. **63** (2011), no. 5, 992–1024, DOI 10.4153/CJM-2011-039-3. MR2866068

[5] Duncan A. Buell, *Binary quadratic forms: Classical theory and modern computations*, Springer-Verlag, New York, 1989, DOI 10.1007/978-1-4612-4542-1. MR1012948

[6] Gabriel Cardona and Jordi Quer, *Curves of genus 2 with group of automorphisms isomorphic to* D_8 *or* D_{12}, Trans. Amer. Math. Soc. **359** (2007), no. 6, 2831–2849, DOI 10.1090/S0002-9947-07-04111-6. MR2286059

[7] David A. Cox, *Primes of the form* $x^2 + ny^2$*: Fermat, class field theory and complex multiplication*, A Wiley-Interscience Publication, John Wiley & Sons, Inc., New York, 1989. MR1028322

[8] The Sage Developers, *Sagemath, the sage mathematics software system (version 9.6)*, (2022). `https://www.sagemath.org`.

[9] L. Dickson, *Studies in the theory of numbers*, The University of Chicago Press, Chicago, 1930. Reprinted by Chelsea Publishing Co., New York, (1957).

[10] Gerhard Frey and Ernst Kani, *Curves of genus 2 covering elliptic curves and an arithmetical application*, Arithmetic algebraic geometry (Texel, 1989), Progr. Math., vol. 89, Birkhäuser Boston, Boston, MA, 1991, pp. 153–176, DOI 10.1007/978-1-4612-0457-2_7. MR1085258

[11] A. Gélin, E. Howe, and C. Ritzenthaler, *Principally polarized squares of elliptic curves with field of moduli equal to* $\mathbb{Q}$, Proc. 13th Algorithmic Number Theory Symposium, (2019), pp. 257–274, https://msp.org/obs/2019/2-1/p16.xhtml. MR3952016

[12] Tsuyoshi Hayashida, *A class number associated with the product of an elliptic curve with itself*, J. Math. Soc. Japan **20** (1968), 26–43, DOI 10.2969/jmsj/02010026. MR233804

[13] Tsuyoshi Hayashida and Mieo Nishi, *Existence of curves of genus two on a product of two elliptic curves*, J. Math. Soc. Japan **17** (1965), 1–16, DOI 10.2969/jmsj/01710001. MR201434

[14] Jun-ichi Igusa, *Arithmetic variety of moduli for genus two*, Ann. of Math. (2) **72** (1960), 612–649, DOI 10.2307/1970233. MR114819

[15] E. Kani, *Elliptic curves on abelian surfaces*, Manuscripta Math. **84** (1994), no. 2, 199–223, https://doi.org/10.1007/BF02567454. MR1285957

[16] E. Kani, *The number of curves of genus two with elliptic differentials*, J. Reine Angew. Math. **485** (1997), 93–121, https://doi.org/10.1515/crll.1997.485.93. MR1442190

[17] E. Kani, *Products of CM elliptic curves*, Collect. Math. **62** (2011), no. 3, 297–339, https://doi.org/10.1007/s13348-010-0029-1. MR2825715

[18] Ernst Kani, *Jacobians isomorphic to a product of two elliptic curves and ternary quadratic forms*, J. Number Theory **139** (2014), 138–174, DOI 10.1016/j.jnt.2013.12.006. MR3173190

[19] Ernst Kani, *The moduli spaces of Jacobians isomorphic to a product of two elliptic curves*, Collect. Math. **67** (2016), no. 1, 21–54, DOI 10.1007/s13348-015-0148-9. MR3439838

[20] Ernst Kani, *Elliptic subcovers of a curve of genus 2 II. The refined Humbert invariant*, J. Number Theory **193** (2018), 302–335, DOI 10.1016/j.jnt.2018.05.011. MR3846811

[21] Ernst Kani, *Elliptic subcovers of a curve of genus 2. I. The isogeny defect* (English, with English and French summaries), Ann. Math. Qué. **43** (2019), no. 2, 281–303, DOI 10.1007/s40316-018-0105-6. MR3996071

[22] E. Kani, *The refined Humbert invariant for abelian product surfaces with complex multiplication*, (2021). Preprint, 23 pages, https://mast.queensu.ca/~kani/.

[23] E. Kani, *Subcovers of curves and moduli spaces*, Geometry at the frontier, (2021), pp. 229–250, https://doi.org/10.1090/conm/766/15384. MR4248056

[24] E. Kani, *Curves of genus 2 on abelian surfaces*, (2023). Preprint, 39 pages. Available at https://mast.queensu.ca/~kani/papers/CurvesAS2.pdf.

[25] H. Kir, *The refined Humbert invariant for imprimitive ternary forms*, Preprint, arXiv:2211.04396, 2022.

[26] R. M. Kuhn, *Curves of genus 2 with split Jacobian*, Trans. Amer. Math. Soc. **307** (1988), no. 1, 41–49, https://doi.org/10.2307/2000749. MR936803

[27] The LMFDB Collaboration, *The L-functions and modular forms database*, 2023, https://www.lmfdb.org.

[28] Curtis T. McMullen, *Teichmüller curves in genus two: discriminant and spin*, Math. Ann. **333** (2005), no. 1, 87–130, DOI 10.1007/s00208-005-0666-y. MR2169830

[29] E. Milio and D. Robert, *Modular polynomials on Hilbert surfaces*, (2017). Preprint, 55 pages. Available at https://hal.science/hal-01520262v2.

[30] J. S. Milne, *Jacobian varieties*, Arithmetic geometry (Storrs, Conn., 1984), Springer, New York, 1986, pp. 167–212. MR861976

[31] David Mumford, *Abelian varieties*, Tata Institute of Fundamental Research Studies in Mathematics, vol. 5, Published for the Tata Institute of Fundamental Research, Bombay; by Oxford University Press, London, 1970. MR282985

[32] D. Robert, *Efficient algorithms for abelian varieties and their moduli spaces.*, Université de Bordeaux (UB) (2021). Preprint, 163 pages. Available at https://theses.hal.science/tel-03498268/.

[33] T. Shaska, *Genus 2 curves with* $(3,3)$*-split Jacobian and large automorphism group*, Algorithmic number theory (Sydney, 2002), 2002, pp. 205–218, https://doi.org/10.1007/3-540-45455-1_17. MR2041085

[34] Tanush Shaska and Helmut Völklein, *Elliptic subfields and automorphisms of genus 2 function fields*, Algebra, arithmetic and geometry with applications (West Lafayette, IN, 2000), Springer, Berlin, 2004, pp. 703–723. MR2037120

[35] Tetsuji Shioda and Naoki Mitani, *Singular abelian surfaces and binary quadratic forms*, Classification of algebraic varieties and compact complex manifolds, Lecture Notes in Math., Vol. 412, Springer, Berlin-New York, 1974, pp. 259–287. MR382289

[36] Joseph H. Silverman, *Advanced topics in the arithmetic of elliptic curves*, Graduate Texts in Mathematics, vol. 151, Springer-Verlag, New York, 1994, DOI 10.1007/978-1-4612-0851-8. MR1312368

[37] J. H. S. Smith, *On the orders and genera of ternary quadratic forms* (1867), In: Collect. Math. Papers vol. I, Oxford, pp. 455–509, (1894).

[38] K. Tahara, *On the finite subgroups of* $\mathrm{GL}(3,\mathbb{Z})$, Nagoya Math. J. **41** (1971), 169–209, `http://projecteuclid.org/euclid.nmj/1118798212`. MR272910

[39] Gerard van der Geer, *Hilbert modular surfaces*, Ergebnisse der Mathematik und ihrer Grenzgebiete (3) [Results in Mathematics and Related Areas (3)], vol. 16, Springer-Verlag, Berlin, 1988, DOI 10.1007/978-3-642-61553-5. MR930101

[40] G. L. Watson, *Integral quadratic forms*, Cambridge Tracts in Mathematics and Mathematical Physics, No. 51, Cambridge University Press, New York, 1960. MR118704

Department of Mathematics and Statistics, Jeffery Hall, 99 University Avenue, Queen's University, Kingston, Ontario, K7L 3N6, Canada

Email address: `19hk16@queensu.ca; harun.kir@ens-lyon.fr`

Contemporary Mathematics
Volume **832**, 2026
https://doi.org/10.1090/conm/832/16609

On the coefficients of the zeta-function's L-polynomial for algebraic function fields over finite fields

M. Koutchoukali

Abstract. We give an explicit formula of the coefficients of the Zeta-Function's L-polynomial for algebraic function fields over finite fields. Thus, we deduce an expression of the class number of algebraic function fields defined over finite fields. Moreover, we give an application of this formula in the case of the curves of defect 2 defined over $\mathbb{F}_2$.

1. Introduction

1.1. Context and motivation. It is well known that the arithmetic and geometry of algebraic function fields is partly reflected in the properties of their Zeta-functions (cf. [**7**] of Tsfasman and Vladut). For example, it is the case if we are interested in the existence of non-special divisors of degree $g-1$ in algebraic function fields of genus g defined over a finite field. Indeed, such an existence was proved in [**1**] or in [**2**] from the properties of the Zeta-function's L-polynomial for algebraic function fields over finite constant fields. In particular, the existence is guaranteed when the coefficients of the L-polynomial $L(t) = \sum_{i=0}^{2g} a_i t^i$ are all positive (which is the case for maximal function fields). More generally, this is the case when the quantity $S = a_g + 2\sum_{i=0}^{g-1} a_i$ is positive. So the main motivation behind this article is ultimately to show methods for obtaining information about the sign of the coefficients or the sign of the quantity S.

1.2. New results and organisation. This paper is organized as follow. In Section 2, we introduce notations and we recall a few results which we will use to prove the main result Theorem 3.1. In Section 3, we give a new explicit formula of the coefficients of the L-polynomial. So far, these coefficients were defined only by induction (see [**5**]). Moreover, we apply this formula to give a new expression of the class number h (see notations in Section 2). Finally, in Section 4 and for a defect 2 curves over $\mathbb{F}_2$, we use the explicit formula and a result on the reciprocals of the roots of L, given by J.P. Serre to establish a few properties of the coefficients concluding with their sign and variation.

2020 *Mathematics Subject Classification.* Primary 11G20; Secondary 11R58, 14G05, 11T71.
Key words and phrases. Function fields, finite fields, rational points, Zeta-Function, L-polynomial.

2. Preliminaries

2.1. Notation. Let us recall the usual notation (for the basic notions related to an algebraic function field $\mathbf{F}/\mathbb{F}_q$ see [**5**]). Let $\mathbf{F}/\mathbb{F}_q$ be a function field of genus g. For any integer $k \geq 1$ we denote by $\mathsf{P}_k(\mathbf{F}/\mathbb{F}_q)$ the set of places of degree k, by $\mathcal{B}_k(\mathbf{F}/\mathbb{F}_q)$ the cardinality of this set and by $\mathsf{P}(\mathbf{F}/\mathbb{F}_q) = \cup_k \mathsf{P}_k(\mathbf{F}/\mathbb{F}_q)$. The divisor group of $\mathbf{F}/\mathbb{F}_q$ is denoted by $\mathsf{D}(\mathbf{F}/\mathbb{F}_q)$. If a divisor $D \in \mathsf{D}(\mathbf{F}/\mathbb{F}_q)$ is such that

$$D = \sum_{P \in \mathsf{P}(\mathbf{F}/\mathbb{F}_q)} n_P P,$$

the support of D is the following finite set

$$\operatorname{supp}(D) = \{P \in \mathsf{P}(\mathbf{F}/\mathbb{F}_q) \mid n_P \neq 0\}$$

and its degree is

$$\deg(D) = \sum_{P \in \mathsf{P}(\mathbf{F}/\mathbb{F}_q)} n_P \deg(P).$$

We denote by $\mathsf{D}_n(\mathbf{F}/\mathbb{F}_q)$ the set of divisors of degree n. We say that the divisor D is *effective* if for each $P \in \operatorname{supp}(D)$ we have $n_P \geq 0$ and we denote $\mathsf{D}_n^+(\mathbf{F}/\mathbb{F}_q)$ the set of effective divisors of degree n and $A_n = \#\mathsf{D}_n^+(\mathbf{F}/\mathbb{F}_q)$.

Let $x \in \mathbf{F}/\mathbb{F}_q$, we denote by (x) the divisor associated to the rational function x, namely

$$(x) = \sum_{P \in \mathsf{P}(\mathbf{F}/\mathbb{F}_q)} v_P(x) P,$$

where v_P is the valuation at the place P. Such a divisor (x) is called a principal divisor, and the set of principal divisors is a subgroup of $\mathsf{D}_0(\mathbf{F}/\mathbb{F}_q)$ denoted by $\mathsf{Princ}(\mathbf{F}/\mathbb{F}_q)$. The factor group

$$\mathcal{C}(\mathbf{F}/\mathbb{F}_q) = \mathsf{D}(\mathbf{F}_q)/\mathsf{Princ}(\mathbf{F}/\mathbb{F}_q)$$

is called the divisor class group. If D_1 and D_2 are in the same class, namely if the divisor $D_1 - D_2$ is principal, we will write $D_1 \sim D_2$. We will denote by $[D]$ the class of the divisor D.

The *dimension* of a divisor D, denoted by $\dim(D)$, is the dimension of the vector space $\mathcal{L}(D)$ defined by the formula

$$\mathcal{L}(D) = \{x \in \mathbf{F} \mid (x) \geq -D\} \cup \{0\}.$$

By Riemann-Roch Theorem we know that the dimension of this vector space is related to the genus of $\mathbf{F}$ and the degree of D by

$$\dim(D) = \deg(D) - g + 1 + \dim(\kappa - D), \tag{1}$$

where κ denotes a canonical divisor of $\mathbf{F}/\mathbb{F}_q$ of degree $2g-2$. In this relation, the complementary term $i(D) = \dim(\kappa - D)$ is called the *index of speciality* and is not easy to compute in general. In particular, a divisor D is non-special when the index of speciality $i(D)$ is zero.

If $D_1 \sim D_2$, the following holds

$$\deg(D_1) = \deg(D_2), \quad \dim(D_1) = \dim(D_2),$$

so that we can define the degree $\deg([D])$ and the dimension $\dim([D])$ of a class. Since the degree of a principal divisor is 0, we can define the subgroup $\mathcal{C}(\mathbf{F}/\mathbb{F}_q)^0$ of classes of degree 0 divisors in $\mathcal{C}(\mathbf{F}/\mathbb{F}_q)$. It is a finite group and we denote by h its order, called the *class number* of $\mathbf{F}/\mathbb{F}_q$. Moreover if

$$L(t) = \sum_{i=0}^{2g} a_i t^i = \prod_{i=1}^{g} [(1-\alpha_i t)(1-\overline{\alpha_i} t)]$$

with $|\alpha_i| = \sqrt{q}$ is the numerator of the Zeta function of $\mathbf{F}/\mathbb{F}_q$, we have $h = L(1)$.

2.2. Elementary results. The following results will be applied to prove the main results.

We consider for $r \geq 1$ the number

$$N_r := N(\mathbf{F}_r) = \mathcal{B}_1(\mathbf{F}_r/\mathbb{F}_{q^r}) = |\{P \in \mathsf{P}(\mathbf{F}_r/\mathbb{F}_{q^r}); deg(P) = 1\}|$$

where $\mathbf{F}_r = \mathbf{F}\mathbb{F}_{q^r}$ is the constant field extension of $\mathbf{F}/\mathbb{F}_q$ of degree r. Let us recall the equation from [**5**, Corollary 5.1.16], for all $r \geq 1$,

$$N_r = q^r + 1 - \sum_{i=1}^{2g} \alpha_i^r \tag{2}$$

where $\alpha_1, \ldots, \alpha_{2g} \in \mathbb{C}$ are the reciprocals of the roots of $L(t)$. In particular, since $N_1 = N(\mathbf{F})$, we have

$$N(\mathbf{F}) = q + 1 - \sum_{i=1}^{2g} \alpha_i.$$

PROPOSITION 2.1 ([**5**, Corollary 5.1.17]). *Let $L(t) = \sum_{i=0}^{2g} a_i t^i$ be the L-polynomial of $\mathbf{F}/\mathbb{F}_q$, and $S_r := N_r - (q^r + 1)$. Then we have:*

(a) $L'(t)/L(t) = \sum_{r=1}^{\infty} S_r t^{r-1}$.

(b) $a_0 = 1$ *and*

$$ia_i = S_i a_0 + S_{i-1} a_1 + \cdots + S_1 a_{i-1} \tag{3}$$

for $i = 1, \ldots, g$.

Given $N_1, \ldots, N_g$ and using $a_{2g-i} = q^{g-i} a_i$, we can determine $L(t)$ from (3).

Let

$$\mathbf{S} = (S_1, S_2, S_3, \ldots)$$

be an infinite sequence of integers with S_i defined in Proposition 2.1, then the equality (3) can be defined as a linear infinite recurrence relation, for $n \geq 1$

$$a_n = \frac{1}{n} \sum_{i=1}^{\infty} S_i a_{n-i} \tag{4}$$

with $a_0 = 1$ and $a_{<0} = 0$.

T. Goy and R. Zatorsky [**3**] studied a similar infinite recurrence using the parapermanent of a triangular matrix defined below. The same tools will be used in the proof of our main result.

A triangular number table

$$B_n = \begin{pmatrix} b_{1,1} & & & \\ b_{2,1} & b_{2,2} & & \\ \vdots & \vdots & \ddots & \\ b_{n,1} & b_{n,2} & \cdots & b_{n,n} \end{pmatrix} \tag{5}$$

is a *triangular matrix* of order n. We note $B_n = (b_{i,j})_{1 \leq j \leq i \leq n}$.

To each element $b_{i,j}$ of matrix (5) we assign $(i-j+1)$ elements $b_{i,k}$, where $k=j, \ldots, i$, which are called the *derived elements* of the matrix determined by the *key element* $b_{i,j}$. The product of all derived elements determined by an element $b_{i,j}$ is called the *factorial product* of the element $b_{i,j}$ and denoted by

$$\{b_{i,j}\} = \prod_{k=j}^{i} b_{i,k}.$$

A tuple of key elements is called a *normal tuple* if their derived elements form a set of elements of cardinality n, no two of which belong to the same column of the matrix.

Let $\mathbb{P}(n)$ be the set of all ordered partitions of a positive integer n into positive integer summands, we said that $\mathbb{P}(n)$ is the set of compositions of n, each integer of a composition is called *a part.* We know, also, that $\#\mathbb{P}(n) = 2^{n-1}$. A one-to-one correspondence between elements of $\mathbb{P}(n)$ and normal tuples of key elements of matrix (5) was established by Tarakanov and Zatorsky [**6**], namely,

$$(n_1, n_2, \ldots, n_r) \leftrightarrow \begin{cases} \text{the normal tuple: } (b_{N_1,N_0+1}, b_{N_2,N_1+1}, \ldots, b_{N_r,N_{r-1}+1}), \\ \text{where } N_0 = 0 \text{ and } N_s = \sum_{i=1}^{s} n_i \text{ for } s = 1, 2, \ldots, r. \end{cases}$$

The *parapermanent* of the matrix (5) is defined as

$$pper(B_n) = \sum_{(m_1,\cdots,m_r)\in\mathbb{P}(n)} \prod_{s=1}^{r} \{b_{i(s),j(s)}\}$$

where $b_{i(s),j(s)}$ is the key element corresponding to m_s, the s-th component of the composition $(m_1, \ldots, m_r) \in \mathbb{P}(n)$. It can also be written as follow

$$(6) \qquad pper(B_n) = \sum_{r=1}^{n} \sum_{m_1+\cdots+m_r=n} \prod_{s=1}^{r} \{b_{m_1+\cdots+m_s, m_1+\cdots+m_{s-1}+1}\}.$$

and can be decomposed by elements of the last row as mentioned in [**8**], *i.e.*,

$$(7) \qquad pper(B_n) = \sum_{s=1}^{n} \{b_{n,s}\} pper(B_{s-1}),$$

where by definition, $pper(B_0) = 1$.

EXAMPLE 2.2. Let us determine $pper(B_3)$, we have for

$$r = 3 \to (1,1,1) \to \begin{cases} s = 1 \to \{b_{i(1),j(1)}\} = \{b_{1,1}\} = b_{1,1}, \\ s = 2 \to \{b_{i(2),j(2)}\} = \{b_{2,2}\} = b_{2,2}, \\ s = 3 \to \{b_{i(3),j(3)}\} = \{b_{3,3}\} = b_{3,3} \end{cases}$$

$$r = 2 \to \begin{cases} (1,2) \to \begin{cases} s = 1 \to \{b_{i(1),j(1)}\} = \{b_{1,1}\} = b_{1,1}, \\ s = 2 \to \{b_{i(2),j(2)}\} = \{b_{3,2}\} = b_{3,2} \cdot b_{3,3}, \end{cases} \\ (2,1) \to \begin{cases} s = 1 \to \{b_{i(1),j(1)}\} = \{b_{2,1}\} = b_{2,1} \cdot b_{2,2}, \\ s = 2 \to \{b_{i(2),j(2)}\} = \{b_{3,3}\} = b_{3,3} \end{cases} \end{cases}$$

$$r = 1 \to (3) \to \{b_{i(1),j(1)}\} = \{b_{3,1}\} = b_{3,1} \cdot b_{3,2} \cdot b_{3,3}$$

Thus,

$$pper(B_3) = b_{1,1} \cdot b_{2,2} \cdot b_{3,3} + b_{1,1} \cdot b_{3,2} \cdot b_{3,3} + b_{2,1} \cdot b_{2,2} \cdot b_{3,3} + b_{3,1} \cdot b_{3,2} \cdot b_{3,3}.$$

3. Main result

In this Section, we provide an explicit formula of the coefficients of the L-polynomial. From this formula, we deduce a new expression of the class number h.

THEOREM 3.1. *Let $L(t) = \sum_{i=0}^{2g} a_i t^i$ be the L-polynomial of $\mathbf{F}/\mathbb{F}_q$, and $S_r := N_r - (q^r + 1)$. For $i = 1, \ldots, g$, the following equations are equivalent:*

(a)

$$a_i = \frac{1}{i} \sum_{j=1}^{\infty} S_j a_{i-j}$$

where $a_0 = 1$ and $a_{<0} = 0$.

(b)

$$a_i = pper \begin{pmatrix} S_1 & & & & \\ \frac{S_2}{S_1} & \frac{S_1}{2} & & & \\ \frac{S_3}{S_2} & \frac{S_2}{S_1} & \frac{S_1}{3} & & \\ \vdots & \vdots & \vdots & \ddots & \\ \frac{S_i}{S_{i-1}} & \frac{S_{i-1}}{S_{i-2}} & \frac{S_{i-2}}{S_{i-3}} & \cdots & \frac{S_1}{i} \end{pmatrix}$$

(c)

$$a_i = \sum_{(m_1,\ldots,m_r)\in\mathbb{P}(i)} \prod_{s=1}^{r} \frac{S_{m_s}}{m_1 + \cdots + m_s} = \sum_{(m_1,\ldots,m_r)\in\mathbb{P}(i)} \prod_{s=1}^{r} \frac{N_{m_s} - (q^{m_s} + 1)}{m_1 + \cdots + m_s}$$

PROOF. Let

$$B_n = \begin{pmatrix} S_1 & & & & \\ \frac{S_2}{S_1} & \frac{S_1}{2} & & & \\ \frac{S_3}{S_2} & \frac{S_2}{S_1} & \frac{S_1}{3} & & \\ \vdots & \vdots & \vdots & \ddots & \\ \frac{S_n}{S_{n-1}} & \frac{S_{n-1}}{S_{n-2}} & \frac{S_{n-2}}{S_{n-3}} & \cdots & \frac{S_1}{n} \end{pmatrix}$$

First, we prove the equivalence between (a) and (b) by induction. By definition, $pper(B_0) = 1 = a_0$. Suppose that the equivalence is true for all $i < n$. by (7), we have

$$pper(B_n) = \sum_{j=1}^{n} \{b_{n,j}\} pper(B_{j-1})$$

by the induction hypothesis, we have

$$pper(B_{j-1}) = a_{j-1}$$

finally, by construction we have

$$b_{n,j} = \frac{S_{n+1-j}}{S_{n-j}} \Leftrightarrow \{b_{n,j}\} = \left\{ \frac{S_{n+1-j}}{S_{n-j}} \right\} = \frac{S_{n+1-j}}{n}.$$

Thus,

$$a_n = pper(B_n) = \sum_{j=1}^{n} \left\{ \frac{S_{n+1-j}}{S_{n-j}} \right\} a_{j-1} = \sum_{j=1}^{n} \frac{S_{n+1-j}}{n} a_{j-1} = \frac{1}{n} \sum_{j=0}^{n-1} S_{n-j} a_j.$$

Now, we prove the equivalence between (b) and (c). Since by (6)

$$pper(B_n) = \sum_{r=1}^{n} \sum_{m_1+\dots+m_r=n} \prod_{s=1}^{r} \{b_{m_1+\dots+m_s, m_1+\dots+m_{s-1}+1}\}$$
$$pper(B_n) = \sum_{(m_1,\dots,m_r)\in\mathbb{P}(n)} \prod_{s=1}^{r} \{b_{m_1+\dots+m_s, m_1+\dots+m_{s-1}+1}\}$$

it is sufficient to observe, as in the first part of the proof, that

$$b_{n,j} = \frac{S_{n+1-j}}{S_{n-j}} \Leftrightarrow \{b_{n,j}\} = \left\{\frac{S_{n+1-j}}{S_{n-j}}\right\} = \frac{S_{n+1-j}}{n}$$

thus

$$\{b_{m_1+\dots+m_s, m_1+\dots+m_{s-1}+1}\} = \frac{S_{m_1+\dots+m_s+1-(m_1+\dots+m_{s-1}+1)}}{m_1+\ldots+m_s}$$
$$\{b_{m_1+\dots+m_s, m_1+\dots+m_{s-1}+1}\} = \frac{S_{m_s}}{m_1+\dots+m_s}.$$

□

EXAMPLE 3.2. We want to determine a_4. We have

$$\mathbb{P}(4) = \{(4),(1,3),(3,1),(2,2),(1,1,2),(1,2,1),(2,1,1),(1,1,1,1)\}$$

$$r=1 \longrightarrow \{(4)\} \longrightarrow \sum_{4} \prod_{s=1}^{1} \{b_{m_1+\dots+m_s, m_1+\dots+m_{s-1}+1}\} = \{b_{4,1}\} = \frac{S_4}{4}$$

$$r=2 \longrightarrow \{(1,3),(3,1),(2,2)\} \longrightarrow \sum_{\substack{1+3\\3+1\\2+2}} \prod_{s=1}^{2} \{b_{m_1+\dots+m_s, m_1+\dots+m_{s-1}+1}\}$$
$$= \{b_{1,1}\}\cdot\{b_{4,2}\} + \{b_{3,1}\}\cdot\{b_{4,4}\} + \{b_{2,1}\}\cdot\{b_{4,3}\} = \frac{S_1}{1}\frac{S_3}{4} + \frac{S_3}{3}\frac{S_1}{4} + \frac{S_2}{2}\frac{S_2}{4}$$

$$r=3 \longrightarrow \{(1,1,2),(1,2,1),(2,1,1)\} \longrightarrow \sum_{\substack{1+1+2\\1+2+1\\2+1+1}} \prod_{s=1}^{3} \{b_{m_1+\dots+m_s, m_1+\dots+m_{s-1}+1}\}$$
$$= \{b_{1,1}\}\cdot\{b_{2,2}\}\cdot\{b_{4,3}\} + \{b_{1,1}\}\cdot\{b_{3,2}\}\cdot\{b_{4,4}\} + \{b_{2,1}\}\cdot\{b_{3,3}\}\cdot\{b_{4,4}\}$$
$$= \frac{S_1}{1}\frac{S_1}{2}\frac{S_2}{4} + \frac{S_1}{1}\frac{S_2}{3}\frac{S_1}{4} + \frac{S_2}{2}\frac{S_1}{3}\frac{S_1}{4}$$

$$r=4 \longrightarrow \{(1,1,1,1)\} \longrightarrow \sum_{1+1+1+1} \prod_{s=1}^{4} \{b_{m_1+\dots+m_s, m_1+\dots+m_{s-1}+1}\}$$
$$= \{b_{1,1}\}\cdot\{b_{2,2}\}\cdot\{b_{3,3}\}\cdot\{b_{4,4}\} = \frac{S_1}{1}\frac{S_1}{2}\frac{S_1}{3}\frac{S_1}{4}$$

finally

$$\begin{aligned} a_4 &= \frac{S_4}{4} + \frac{S_1}{1}\frac{S_3}{4} + \frac{S_3}{3}\frac{S_1}{4} + \frac{S_2}{2}\frac{S_2}{4} \\ &+ \frac{S_1}{1}\frac{S_1}{2}\frac{S_2}{4} + \frac{S_1}{1}\frac{S_2}{3}\frac{S_1}{4} + \frac{S_2}{2}\frac{S_1}{3}\frac{S_1}{4} + \frac{S_1}{1}\frac{S_1}{2}\frac{S_1}{3}\frac{S_1}{4} \\ &= \frac{S_4}{4} + \frac{S_1S_3}{3} + \frac{S_2^2}{8} + \frac{S_1^2S_2}{4} + \frac{S_1^4}{24}. \end{aligned}$$

With this, we conclude our example.

We can also give an explicit formula of the coefficients a_i depending on the reciprocals of the roots of L. But first, let's give an expression of N_m depending on these roots, see [**5**, Corollary 5.1.16]:

$$N_m = q^m + 1 - \sum_{i=1}^{2g} \alpha_i^m$$

which can also be written

$$N_m = q^m + 1 - \sum_{i=1}^{g} 2q^{m/2}\cos(\pi m\phi_i) \tag{8}$$

where $(\alpha_1, \ldots, \alpha_{2g}) = (q^{1/2}e^{i\pi\phi_1}, \ldots, q^{1/2}e^{i\pi\phi_g}, q^{1/2}e^{-i\pi\phi_1}, \ldots, q^{1/2}e^{-i\pi\phi_g})$ are the reciprocals of the roots of $L(t)$ with $\phi_i \in [0,1]$. The following results are the consequences of Theorem 3.1, (2) and (8).

PROPOSITION 3.3. *Let $L(t) = \sum_{i=0}^{2g} a_i t^i$ be the L-polynomial of $\mathbf{F}/\mathbb{F}_q$. We have*

$$\begin{aligned} a_i &= \sum_{(m_1,\ldots,m_r)\in\mathbb{P}(i)} \prod_{s=1}^{r} (-1)^r \frac{\sum_{j=1}^{g} 2q^{m_s/2}\cos(\pi m_s\phi_j)}{m_1 + \cdots + m_s} \\ &= \sum_{(m_1,\ldots,m_r)\in\mathbb{P}(i)} \prod_{s=1}^{r} (-1)^r \frac{\sum_{j=1}^{2g} \alpha_j^{m_s}}{m_1 + \cdots + m_s} \end{aligned}$$

for $i = 1, \ldots, g$ and $a_{2g-i} = q^{g-i}a_i$ with $a_0 = 1$.

PROOF. We use Theorem 3.1 (c) and

$$N_m = q^m + 1 - \sum_{i=1}^{g} 2q^{m/2}\cos(\pi m\phi_i) = q^m + 1 - \sum_{i=1}^{2g} \alpha_i^m. \qquad \square$$

We conclude by a new expression of the class number h depending on the previous parameters.

PROPOSITION 3.4. *Let $L(t)=\sum_{i=0}^{2g} a_i t^i$ be the L-polynomial of $\mathbf{F}/\mathbb{F}_q$ and h the class number. We have*

$$h = 1+q^g+\sum_{i=1}^{g-1}(1+q^{g-i})\sum_{(m_1,\ldots,m_r)\in\mathbb{P}(i)}\prod_{s=1}^{r}(-1)^r\frac{\sum_{j=1}^{g}2q^{m_s/2}\cos(\pi m_s\phi_j)}{m_1+\cdots+m_s}+$$
$$\sum_{(m_1,\ldots,m_r)\in\mathbb{P}(g)}\prod_{s=1}^{r}(-1)^r\frac{\sum_{j=1}^{g}2q^{m_s/2}\cos(\pi m_s\phi_j)}{m_1+\cdots+m_s}$$
$$=1+q^g+\sum_{i=1}^{g-1}(1+q^{g-i})\sum_{(m_1,\ldots,m_r)\in\mathbb{P}(i)}\prod_{s=1}^{r}(-1)^r\frac{\sum_{j=1}^{2g}\alpha_j^{m_s}}{m_1+\cdots+m_s}+$$
$$\sum_{(m_1,\ldots,m_r)\in\mathbb{P}(g)}\prod_{s=1}^{r}(-1)^r\frac{\sum_{j=1}^{2g}\alpha_j^{m_s}}{m_1+\cdots+m_s}$$
$$=1+q^g+\sum_{i=1}^{g-1}(1+q^{g-i})\sum_{(m_1,\ldots,m_r)\in\mathbb{P}(i)}\prod_{s=1}^{r}\frac{N_{m_s}-(q^{m_s}+1)}{m_1+\cdots+m_s}+$$
$$\sum_{(m_1,\ldots,m_r)\in\mathbb{P}(g)}\prod_{s=1}^{r}\frac{N_{m_s}-(q^{m_s}+1)}{m_1+\cdots+m_s}$$

PROOF. Since $h=L(1)=\sum_{i=0}^{2g} a_i$ with $a_0=1$ and $a_{2g}=q^g a_0$, this result is a consequence of Proposition 3.3. Note that the case $i=g$ has to be treated separately, otherwise a_g will be counted twice. □

4. Example of application

In this Section, we will exhibit an application for the explicit formula in Proposition 3.3. This application is not the simple calculation of the coefficients of the L-polynomial, since one can easily show that the complexity of the formula in Theorem 3.1 is exponential.

This application consists of giving some of these coefficients properties. These latter will allow us to determine the sign and the variation of the sequence $(a_n)_{0\le n\le g}$. We will focus on curves of defect 2 over the finite field $\mathbb{F}_2$ *i.e.*

$$|N_1(\mathbf{F}/\mathbb{F}_2)-(q+1)| = |N_1(\mathbf{F}/\mathbb{F}_2)-3| = g[2\sqrt{q}]-2 = g[2\sqrt{2}]-2 = 2g-2,$$

where $[x]$ denotes the largest integer $\le x$. The following Theorem gives the general case studied by J.P. Serre in [**4**, Theorem 2.5.1].

THEOREM 4.1. *For a curve over $\mathbb{F}_q$ such that $|N_1(\mathbf{F}/\mathbb{F}_q)-(q+1)| = g[2\sqrt{q}]-2$, there are six possibilities for $(\alpha_1+\overline{\alpha_1},\ldots,\alpha_g+\overline{\alpha_g})$:*

(a) $\pm([2\sqrt{q}],\ldots,[2\sqrt{q}],[2\sqrt{q}]-2)$ *for $g\ge 1$.*
(b) $\pm([2\sqrt{q}],\ldots,[2\sqrt{q}],[2\sqrt{q}]+\sqrt{2}-1,[2\sqrt{q}]-\sqrt{2}-1)$ *for $g=2$.*
(c) $\pm([2\sqrt{q}],\ldots,[2\sqrt{q}],[2\sqrt{q}]-1,[2\sqrt{q}]-1)$ *for $g=2$.*
(d) $\pm([2\sqrt{q}],\ldots,[2\sqrt{q}],[2\sqrt{q}]+\sqrt{3}-1,[2\sqrt{q}]-\sqrt{3}-1)$ *for $g\ge 2$.*
(e) $\pm([2\sqrt{q}],\ldots,[2\sqrt{q}],[2\sqrt{q}]+\frac{-1+\sqrt{5}}{2},[2\sqrt{q}]+\frac{-1+\sqrt{5}}{2},[2\sqrt{q}]+\frac{-1-\sqrt{5}}{2},[2\sqrt{q}]+\frac{-1-\sqrt{5}}{2})$ *for $g=4$.*
(f) $\pm([2\sqrt{q}],\ldots,[2\sqrt{q}],[2\sqrt{q}]+1-4\cos^2(\frac{\pi}{7}),[2\sqrt{q}]+1-4\cos^2(\frac{2\pi}{7}),[2\sqrt{q}]+1-4\cos^2(\frac{3\pi}{7}))$ *for $g=3$.*

We are interested here by the case (a) of Theorem 4.1 for $q = 2$. This case offers possibly infinitely many examples of curves.

Let $q = 2$, we can easily deduce that under the conditions of Theorem 4.1 case (a), we have $(\alpha_1+\overline{\alpha_1}, \dots, \alpha_g+\overline{\alpha_g}) = \pm([2\sqrt{q}], \dots, [2\sqrt{q}], [2\sqrt{q}]-2) = \pm(2, \dots, 2, 0)$. Thus, with equation (8) notations we have $\phi_i = 1/4$ or $3/4$ since $2\sqrt{2}\cos(\pi\cdot\phi_i) = \pm 2$ for $i \in \{1, \dots, g-1\}$ and $\phi_g = 1/2$ since $2\sqrt{2}\cos(\pi\cdot\phi_g) = 0$. These observations lead us to the following result:

PROPOSITION 4.2. *Let $L(t) = \sum_{n=0}^{2g} a_n t^n$ be the L-polynomial of $\mathbf{F}/\mathbb{F}_2$, under the conditions of case (a) we have $a_n = a_{n,\pi/4}$ or $a_n = a_{n,3\pi/4}$ where*

$$a_{n,\pi/4} := \sum_{(m_1,\dots,m_r)\in\mathbb{P}(n)} (-1)^r \cdot 2^r \cdot 2^{n/2} \prod_{s=1}^{r} \frac{(g-1)\cos(\pi m_s/4) + \cos(\pi m_s/2)}{m_1+\cdots+m_s}$$

and

$$a_{n,3\pi/4} := \sum_{(m_1,\dots,m_r)\in\mathbb{P}(n)} (-1)^r \cdot 2^r \cdot 2^{n/2} \prod_{s=1}^{r} \frac{(g-1)\cos(3\pi m_s/4) + \cos(\pi m_s/2)}{m_1+\cdots+m_s}$$

for $n = 1, \dots, g$ and $a_{2g-n} = q^{g-n} a_n$ with $a_0 = 1$.

PROOF. By Proposition 3.3 and the conditions of case (a), we have

$$\begin{aligned}
a_n &= \sum_{(m_1,\dots,m_r)\in\mathbb{P}(n)} \prod_{s=1}^{r} (-1)^r \frac{\sum_{j=1}^{g} 2q^{m_s/2}\cos(\pi m_s \phi_j)}{m_1+\cdots+m_s} \\
&= \sum_{(m_1,\dots,m_r)\in\mathbb{P}(n)} \prod_{s=1}^{r} (-1)^r \frac{\left(\sum_{j=1}^{g-1} 2q^{m_s/2}\cos(m_s\cdot\theta)\right) + 2q^{m_s/2}\cos(m_s\cdot\pi/2)}{m_1+\cdots+m_s} \\
&= \sum_{(m_1,\dots,m_r)\in\mathbb{P}(n)} (-1)^r \cdot 2^r \cdot 2^{n/2} \prod_{s=1}^{r} \frac{(g-1)\cos(m_s\cdot\theta) + \cos(m_s\cdot\pi/2)}{m_1+\cdots+m_s}
\end{aligned}$$

where $\theta = \pi/4$ or $3\pi/4$. □

For what comes next, we introduce the following notations:

$$C_{\pi/4}(n) := (g-1)\cos(\pi n/4) + \cos(\pi n/2),$$
$$C_{3\pi/4}(n) := (g-1)\cos(3\pi n/4) + \cos(\pi n/2),$$

and

$$\mathbb{P}^+_{\pi/4}(n) := \{(m_1, \dots, m_r) \in \mathbb{P}(n) \mid (-1)^r \cdot 2^r \cdot 2^{n/2} \prod_{s=1}^{r} \frac{C_{\pi/4}(m_s)}{m_1+\cdots+m_s} > 0\},$$

$$\mathbb{P}^-_{\pi/4}(n) := \{(m_1, \dots, m_r) \in \mathbb{P}(n) \mid (-1)^r.2^r.2^{n/2} \prod_{s=1}^{r} \frac{C_{\pi/4}(m_s)}{m_1+\cdots+m_s} < 0\},$$

$$\mathbb{P}^+_{3\pi/4}(n) := \{(m_1, \dots, m_r) \in \mathbb{P}(n) \mid (-1)^r \cdot 2^r \cdot 2^{n/2} \prod_{s=1}^{r} \frac{C_{3\pi/4}(m_s)}{m_1+\cdots+m_s} > 0\},$$

$$\mathbb{P}^-_{3\pi/4}(n) := \{(m_1, \dots, m_r) \in \mathbb{P}(n) \mid (-1)^r \cdot 2^r \cdot 2^{n/2} \prod_{s=1}^{r} \frac{C_{3\pi/4}(m_s)}{m_1+\cdots+m_s} < 0\},$$

and denote their cardinalities as follows:

$$P^+_{\pi/4}(n) := \#\mathbb{P}^+_{\pi/4}(n), \qquad P^-_{\pi/4}(n) := \#\mathbb{P}^-_{\pi/4}(n),$$
$$P^+_{3\pi/4}(n) := \#\mathbb{P}^+_{3\pi/4}(n), \qquad P^-_{3\pi/4}(n) := \#\mathbb{P}^-_{3\pi/4}(n).$$

Let $CR_{\pi/4} : \mathbb{P}(n) \to \mathbb{Q}$ be the map defined by:

$$CR_{\pi/4}((m_1, \ldots, m_r)) = (-1)^r \cdot 2^r \cdot 2^{n/2} \prod_{s=1}^{r} \frac{C_{\pi/4}(m_s)}{m_1 + \cdots + m_s}$$

and let $CR_{3\pi/4} : \mathbb{P}(n) \to \mathbb{Q}$ be the map defined by:

$$CR_{3\pi/4}((m_1, \ldots, m_r)) = (-1)^r \cdot 2^r \cdot 2^{n/2} \prod_{s=1}^{r} \frac{C_{3\pi/4}(m_s)}{m_1 + \cdots + m_s}.$$

PROPOSITION 4.3. *Let $L(t) = \sum_{n=0}^{2g} a_n t^n$ be the L-polynomial of $\mathbf{F}/\mathbb{F}_2$. Under the conditions of case (a) we have, for $n = 1, \ldots, g$ and for a fixed i such that $1 \le i \le n$*

$$a_n = \sum_{\substack{(m_1,\ldots,m_r)\in\mathbb{P}(n) \\ m_r \neq i}} (-1)^r \cdot 2^r \cdot 2^{n/2} \prod_{s=1}^{r} \frac{C_\theta(m_s)}{m_1 + \cdots + m_s} - 2^{\frac{i+2}{2}} \frac{C_\theta(i)}{n} a_{n-i}$$

and

$$a_n = \sum_{j=1}^{n} -2^{\frac{j+2}{2}} \cdot \frac{C_\theta(j)}{n} a_{n-j}$$

with $a_0 = 1$ and $\theta = \pi/4$ or $3\pi/4$. For $n > g$ we have $a_{2g-n} = q^{g-n} a_n$.

PROOF. The main idea is to observe that

$$\mathbb{P}(n) = \bigcup_{i=1}^{n} \{(m_1, \ldots, m_r, i) \text{ with } (m_1, \ldots, m_r) \in \mathbb{P}(n-i)\}$$

then, by Proposition 4.2 and for $1 \leq i \leq n$

$$\begin{aligned}
a_n = & \sum_{(m_1,\ldots,m_r)\in\mathbb{P}(n)} (-1)^r \cdot 2^r \cdot 2^{n/2} \prod_{s=1}^{r} \frac{C_\theta(m_s)}{m_1+\cdots+m_s} \\
= & \sum_{\substack{(m_1,\ldots,m_r)\in\mathbb{P}(n)\\ m_r\neq i}} (-1)^r \cdot 2^r \cdot 2^{n/2} \prod_{s=1}^{r} \frac{C_\theta(m_s)}{m_1+\cdots+m_s} + \\
& \sum_{(m_1,\ldots,m_{r-1},i)\in\mathbb{P}(n)} (-1)^r \cdot 2^r \cdot 2^{n/2} \prod_{s=1}^{r} \frac{C_\theta(m_s)}{m_1+\cdots+m_s} \\
= & \sum_{\substack{(m_1,\ldots,m_r)\in\mathbb{P}(n)\\ m_r\neq i}} (-1)^r \cdot 2^r \cdot 2^{n/2} \prod_{s=1}^{r} \frac{C_\theta(m_s)}{m_1+\cdots+m_s} + \\
& -2 \cdot 2^{i/2} \frac{C_\theta(i)}{n} \sum_{(m_1,\ldots,m_{r-1},i)\in\mathbb{P}(n)} (-1)^{r-1} \cdot 2^{r-1} \cdot 2^{(n-i)/2} \prod_{s=1}^{r-1} \frac{C_\theta(m_s)}{m_1+\cdots+m_s} \\
= & \sum_{\substack{(m_1,\ldots,m_r)\in\mathbb{P}(n)\\ m_r\neq i}} (-1)^r \cdot 2^r \cdot 2^{n/2} \prod_{s=1}^{r} \frac{C_\theta(m_s)}{m_1+\cdots+m_s} + \\
& -2 \cdot 2^{i/2} \frac{C_\theta(i)}{n} \sum_{(m_1,\ldots,m_{r-1})\in\mathbb{P}(n-i)} (-1)^{r-1} \cdot 2^{r-1} \cdot 2^{(n-i)/2} \prod_{s=1}^{r-1} \frac{C_\theta(m_s)}{m_1+\cdots+m_s} \\
= & \sum_{\substack{(m_1,\ldots,m_r)\in\mathbb{P}(n)\\ m_r\neq i}} (-1)^r \cdot 2^r \cdot 2^{n/2} \prod_{s=1}^{r} \frac{C_\theta(m_s)}{m_1+\cdots+m_s} - 2^{(i+2)/2} \frac{C_\theta(i)}{n} a_{n-i}
\end{aligned}$$

The second equality follows by taking the sum over i. □

REMARK 4.4.

- The second equality could also be proved applying Proposition 2.1.
- For each choice of i, in the first equation, we obtain a different formula that gives the same coefficient a_n (for a fixed n).

Observe that for $E_n = \{n + 8k \mid k \in \mathbb{N}\}$,

$$C_{\pi/4}(m_s) = \begin{cases} (g-1)\sqrt{2}/2 & \text{if } m_s \in E_1 \cup E_7 \\ -(g-1)\sqrt{2}/2 & \text{if } m_s \in E_3 \cup E_5 \\ -1 & \text{if } m_s \in E_2 \cup E_6 \\ -(g-2) & \text{if } m_s \in E_4 \\ g & \text{if } m_s \in E_8 \end{cases}$$

and

$$C_{3\pi/4}(m_s) = \begin{cases} (g-1)\sqrt{2}/2 & \text{if } m_s \in E_3 \cup E_5 \\ -(g-1)\sqrt{2}/2 & \text{if } m_s \in E_1 \cup E_7 \\ -1 & \text{if } m_s \in E_2 \cup E_6 \\ -(g-2) & \text{if } m_s \in E_4 \\ g & \text{if} m_s \in E_8 \end{cases}$$

it is clear (for $g > 2$) that $m \in \mathbb{P}^+_{\pi/4}(n)$, if and only if, the number of parts of m that are in $E_2 \cup E_3 \cup E_4 \cup E_5 \cup E_6$ has the same parity as r the number of all parts. $m \in \mathbb{P}^-_{\pi/4}(n)$, if and only if, this number and r have not the same parity.

Similarly, we can characterize $\mathbb{P}^+_{3\pi/4}(n)$ and $\mathbb{P}^-_{3\pi/4}(n)$ by replacing E_3 and E_5 with E_1 and E_7.

The following proposition, is a second characterization which is less obvious.

PROPOSITION 4.5. *Let $L(t) = \sum_{n=0}^{2g} a_n t^n$ be the L-polynomial of $\mathbf{F}/\mathbb{F}_2$, $m = (m_1, \ldots, m_r) \in \mathbb{P}(n)$ a composition of an integer $n \leq g$ with $g > 2$. Under the conditions of case (a) we have*

i) *The cardinality of $(E_1 \cup E_7 \cup E_8) \cap \{m_1, \ldots, m_r\}$ is even, if and only if, $m \in \mathbb{P}^+_{\pi/4}(n)$. It is odd, if and only if, $m \in \mathbb{P}^-_{\pi/4}(n)$.*
ii) *The cardinality of $(E_3 \cup E_5 \cup E_8) \cap \{m_1, \ldots, m_r\}$ is even, if and only if, $m \in \mathbb{P}^+_{3\pi/4}(n)$. It is odd, if and only if, $m \in \mathbb{P}^-_{3\pi/4}(n)$.*

PROOF.

i) By induction, we have for $n = 1$: $\mathbb{P}(1) = \{(1)\}$ then $CR_{\pi/4}((1)) = -1 \cdot 2 \cdot 2^{1/2} \cdot \frac{(g-1)\sqrt{2}/2}{1} = -2(g-1) < 0$. For $n = 2$: $\mathbb{P}(2) = \{(1,1),(2)\}$ then

$$CR_{\pi/4}((1,1)) = (-1)^2 \cdot 2^2 \cdot 2^{2/2} \cdot \frac{(g-1)\sqrt{2}/2}{1} \cdot \frac{(g-1)\sqrt{2}/2}{2} = 2(g-1)^2 > 0,$$

and

$$CR_{\pi/4}((2)) = -1 \cdot 2 \cdot 2^{2/2} \cdot \frac{-1}{2} = 2 > 0.$$

Suppose that the equivalence is true for $n' < n$, andlet $m = (m_1, \ldots, m_r) \in \mathbb{P}^+_{\pi/4}(n)$.

It follows that

- If $(E_1 \cup E_7 \cup E_8) \cap \{m_1, \ldots, m_r\} = \emptyset$, the cardinality of this set is even.
- If $m_i < n$ such that $\{m_i\} \subset (E_1 \cup E_7 \cup E_8) \cap \{m_1, \ldots, m_r\}$, we have

$$CR_{\pi/4}(m) = (-1)^r \cdot 2^r \cdot 2^{n/2} \prod_{s=1}^{r} \frac{C_{\pi/4}(m_s)}{m_1 + \cdots + m_s}$$

$$= (-1)^r \cdot 2^r \cdot 2^{n/2} \prod_{s=1}^{i-1} \frac{C_{\pi/4}(m_s)}{m_1 + \cdots + m_s} \cdot \frac{C_{\pi/4}(m_i)}{m_1 + \cdots + m_i} \cdot$$

$$\prod_{s=i+1}^{r} \frac{C_{\pi/4}(m_s)}{m_1 + \cdots + m_i + \cdots + m_s}$$

$$= (-1)^r \cdot 2^r \cdot 2^{n/2} \prod_{s=1}^{i-1} \frac{C_{\pi/4}(m_s)}{m_1 + \cdots + m_s} \cdot \frac{C_{\pi/4}(m_i)}{m_1 + \cdots + m_i} \cdot$$

$$\prod_{s=i+1}^{r} \frac{C_{\pi/4}(m_s)}{m_1 + \cdots + m_{i-1} + m_{i+1} + \cdots + m_s} \cdot$$

$$\prod_{s=i+1}^{r} \frac{m_1 + \cdots + m_{i-1} + m_{i+1} + \cdots + m_s}{m_1 + \cdots + m_{i-1} + m_i + m_{i+1} + \cdots + m_s}$$

$$= -2 \cdot 2^{m_i/2} C_{\pi/4}(m_i) \prod_{s=i+1}^{r} \frac{m_1 + \cdots + m_{i-1} + m_{i+1} + \cdots + m_s}{m_1 + \cdots + m_{i-1} + m_i + m_{i+1} + \cdots + m_s} \cdot$$

$$(-1)^{r-1} \cdot 2^{r-1} \cdot 2^{(n-m_i)/2} \prod_{s=1}^{r-1} \frac{C_{\pi/4}(m_s)}{m_1 + \cdots + m_s}$$

$$= -2 \cdot 2^{m_i/2} C_{\pi/4}(m_i) \prod_{s=i+1}^{r} \frac{m_1 + \cdots + m_{i-1} + m_{i+1} + \cdots + m_s}{m_1 + \cdots + m_i + \cdots + m_s} \cdot$$

$$CR_{\pi/4}((m_1, \cdots, m_{i-1}, m_{i+1}, \cdots, m_r))$$

Since $C_{\pi/4}(m_i) > 0$ and $CR_{\pi/4}(m) > 0$, we have

$$CR_{\pi/4}((m_1, \ldots, m_{i-1}, m_{i+1}, \ldots, m_r)) < 0.$$

By the induction hypothesis, the cardinality of

$$(E_1 \cup E_7 \cup E_8) \cap \{m_1, \ldots, m_{i-1}, m_{i+1}, \ldots, m_r\}$$

is odd, thus the cardinality of $(E_1 \cup E_7 \cup E_8) \cap \{m_1, \ldots, m_r\}$ is even.
ii) The second equivalence follows similarly by replacing E_3, E_5 with E_1, E_7. □

PROPOSITION 4.6. *Let $L(t) = \sum_{n=0}^{2g} a_n t^n$ be the L-polynomial of $\mathbf{F}/\mathbb{F}_2$. Under the conditions of case (a) and with Proposition 4.2 notations, we have*

$$a_{n,\pi/4} = \begin{cases} a_{n,3\pi/4} & \text{if } n \text{ is even} \\ -a_{n,3\pi/4} & \text{if } n \text{ is odd} \end{cases}$$

for $n = 1, \ldots, g$ and $a_{2g-n} = q^{g-n} a_n$ with $a_0 = 1$.

PROOF. Let $m = (m_1, \ldots, m_r) \in \mathbb{P}(n)$ a composition of an integer $n \leq g$. The main idea here, is to observe that if n is even then the cardinality of

$$(E_3 \cup E_5 \cup E_8) \cap \{m_1, \ldots, m_r\}$$

and

$$(E_1 \cup E_7 \cup E_8) \cap \{m_1, \ldots, m_r\}$$

are both even or both odd (1, 3, 5, 7 are the only odd numbers modulo 8). If n is odd, then one is even and the second is odd. Thus, by Proposition 4.5

$$CR_{\pi/4}(m) = \begin{cases} CR_{3\pi/4}(m) & \text{if } n \text{ is even} \\ -CR_{3\pi/4}(m) & \text{if } n \text{ is odd} \end{cases}$$

we conclude by observing that

$$a_{n,\pi/4} = \sum_{m \in \mathbb{P}(n)} CR_{\pi/4}(m) \text{ and } a_{n,3\pi/4} = \sum_{m \in \mathbb{P}(n)} CR_{3\pi/4}(m) \qquad \square$$

PROPOSITION 4.7. *Let $n \leq g$ be an integer with $g > 2$. then $P^+_{\pi/4}(n) > P^-_{\pi/4}(n)$ if n is even, $P^+_{\pi/4}(n) < P^-_{\pi/4}(n)$ if n is odd and $P^+_{3\pi/4}(n) > P^-_{3\pi/4}(n)$ for all n. More precisely*

(a) $P^+_{\pi/4}(2) - P^-_{\pi/4}(2) = 2$.
(b) $P^-_{\pi/4}(3) - P^+_{\pi/4}(3) = 2$.
(c) $P^+_{\pi/4}(4) - P^-_{\pi/4}(4) = 4$.
(d) $P^-_{\pi/4}(5) - P^+_{\pi/4}(5) = 4$.
(e) *$P^+_{\pi/4}(n) - P^-_{\pi/4}(n) > n$ for $n \geq 6$ if n is even.*
(f) *$P^-_{\pi/4}(n) - P^+_{\pi/4}(n) > n$ for $n \geq 6$ if n is odd.*
(g) *$P^\delta_{\pi/4}(n) := |P^+_{\pi/4}(n) - P^-_{\pi/4}(n)|$ is a strictly increasing sequence for $n \geq 6$.*
(h) *$P^+_{3\pi/4}(n) - P^-_{3\pi/4}(n) = 2$ for $n = 2$ and 3.*
(i) *$P^+_{3\pi/4}(n) - P^-_{3\pi/4}(n) = 4$ for $n = 4$ and 5.*
(j) *$P^+_{3\pi/4}(n) - P^-_{3\pi/4}(n) > n$ and $P^\delta_{3\pi/4}(n) := P^+_{3\pi/4}(n) - P^-_{3\pi/4}(n)$ is a strictly increasing sequence for $n \geq 6$.*

PROOF. The equalities (a), (b), (c), (d), (h) and (i) follow from a simple calculation. (e), (f) and (g) follow from (j) by observing that

$$CR_{\pi/4}(m) = \begin{cases} CR_{3\pi/4}(m) & \text{if } n \text{ is even} \\ -CR_{3\pi/4}(m) & \text{if } n \text{ is odd} \end{cases}$$

see the proof of Proposition 4.6.

Let's prove inequality (j). For $n = 6$, $P^+_{3\pi/4}(6) - P^-_{3\pi/4}(6) = 8 > 6$. For $n = 7$, $P^+_{3\pi/4}(7) - P^-_{3\pi/4}(7) = 10 > 7$. By induction, suppose that for $n' < n$ the property (j) is true. By Proposition 4.3

$$a_{n,3\pi/4} = \sum_{i=1}^{n} -2^{(i+2)/2} \cdot \frac{C_{3\pi/4}(i)}{n} a_{n-i}$$

thus, for $1 \leq i \leq n$, $P^+_{3\pi/4}(n-i) - P^-_{3\pi/4}(n-i) > n-i$ and $P^\delta_{3\pi/4}(n-i)$ is strictly increasing by the induction hypothesis, for $i \in E_1 \cup E_2 \cup E_4 \cup E_6 \cup E_7$, $-2^{(i+2)/2} \cdot \frac{C_{3\pi/4}(i)}{n} > 0$. The difference between the positive and the negative terms of $-2^{(i+2)/2} \cdot \frac{C_{3\pi/4}(i)}{n} a_{n-i}$ is equal to $P^\delta_{3\pi/4}(n-i)$. and for $i \in E_3 \cup E_5 \cup E_8$,

$-2^{(i+2)/2}.\frac{C_{3\pi/4}(i)}{n} < 0$. the positive and the negative terms of $-2^{(i+2)/2}.\frac{C_{3\pi/4}(i)}{n}a_{n-i}$ are reversed, the difference is equal to $-P^{\delta}_{3\pi/4}(n-i)$. We conclude that

$$P^{\delta}_{3\pi/4}(n) = \sum_{i\in E_1\cup E_2\cup E_4\cup E_6\cup E_7} P^{\delta}_{3\pi/4}(n-i) - \sum_{i\in E_3\cup E_5\cup E_8} P^{\delta}_{3\pi/4}(n-i).$$

Finally, if $i \in E_3\cup E_5\cup E_8$, we have $P^{\delta}_{3\pi/4}(n-i+1)-P^{\delta}_{3\pi/4}(n-i) \geq 1$ for $n-i \neq 2$ or 4, and $P^{\delta}_{3\pi/4}(n-i+1)-P^{\delta}_{3\pi/4}(n-i) = 0$ for $n-i = 2$ or 4. Since $P^{\delta}_{3\pi/4}(n-1) > n-1$, we conclude that $P^{\delta}_{3\pi/4}(n) > n$ and $P^{\delta}_{3\pi/4}(n) > P^{\delta}_{3\pi/4}(n-1)$. □

THEOREM 4.8. *Let $L(t) = \sum_{n=0}^{2g} a_n t^n$ be the L-polynomial of a function field $\mathbf{F}/\mathbb{F}_2$ with $1 \leq g \leq 6$. Under the conditions of case (a) and for $n \in \{0,\dots,g\}$ we have*

(a) *$a_{n,\pi/4} < 0$ if n is odd and $a_{n,\pi/4} > 0$ if n is even.*
(b) *$a_{n,3\pi/4} > 0$.*
(c) *The sequence $|a_{n,\theta}|$ with $\theta = \pi/4$ or $3\pi/4$ is increasing.*

PROOF. We will prove the theorem for $\theta = \pi/4$, the second part is obvious by Proposition 4.6.

By Proposition 4.2 we have

$$a_{n,\pi/4} = \sum_{(m_1,\dots,m_r)\in\mathbb{P}(n)} (-1)^r \cdot 2^{r+n/2} \prod_{s=1}^{r} \frac{(g-1)\cos(m_s\cdot\pi/4)+\cos(m_s\cdot\pi/2)}{m_1+\cdots+m_s}$$

Suppose that n is even. Let $m = (m_1,\dots,m_r) \in \mathbb{P}^{-}_{\pi/4}(n)$. Our aim here, is to associate to m another composition m' such that $P_{m'} \in \mathbb{P}^{+}_{\pi/4}(n)$ and $|CR_{\pi/4}(m)| \leq |CR_{\pi/4}(m')|$.

First, let's prove that $\exists j \in \{1,\dots,r\}$ such that $m_j \in E_3 \cup E_5$. Suppose the opposite, since $m \in \mathbb{P}^{-}_{\pi/4}(n)$ by Proposition 4.5, the number $\#[E_1 \cap \{m_1,\dots,m_r\}]$ is odd (there are no elements in $E_7\cup E_8$ because $n \leq g < 7$), but the other numbers in m are all even (2, 4, 6), then $m_1+\cdots+m_r = n$ is odd. This contradicts the supposition that n is even. We conclude that $\exists j \in \{1,\dots,r\}$ such that $m_j \in E_3\cup E_5$.

- If m contains $m_j \in E_5$, we associate to

$$m = (m_1,\dots,m_{j-1},m_j,m_{j+1},\dots,m_r)$$

m' such that

$$m' = (m_1,\dots,m_{j-1},m_{j_1},m_{j_2},m_{j+1},\dots,m_r)$$

with $m_{j_1} \in E_1$ and $m_{j_2} \in E_4$. It is clear, by Proposition 4.5, that $P_{m'} \in \mathbb{P}^{+}_{\pi/4}(n)$. Moreover, we have

$$CR_{\pi/4}(m) = (-1)^r\cdot 2^r\cdot 2^{n/2}\prod_{s=1}^{j-1}\frac{(g-1)\cos(m_s\cdot\pi/4)+\cos(m_s\cdot\pi/2)}{m_1+\cdots+m_s}$$
$$\cdot\frac{-(g-1)\cdot\sqrt{2}/2}{m_1+\cdots+m_j}\prod_{s=j+1}^{r}\frac{(g-1)\cos(m_s\cdot\pi/4)+\cos(m_s\cdot\pi/2)}{m_1+\cdots+m_s}$$

and

$$CR_{\pi/4}(m') = (-1)^{r+1} \cdot 2^{r+1} \cdot 2^{n/2} \prod_{s=1}^{j-1} \frac{(g-1)\cos(m_s \cdot \pi/4) + \cos(m_s \cdot \pi/2)}{m_1 + \cdots + m_s}$$
$$\cdot \frac{-(g-2)}{m_1 + \cdots + m_{j-1} + m_{j_1}}$$
$$\cdot \frac{(g-1) \cdot \sqrt{2}/2}{m_1 + \cdots + m_j} \prod_{s=j+1}^{r} \frac{(g-1)\cos(m_s \cdot \pi/4) + \cos(m_s \cdot \pi/2)}{m_1 + \cdots + m_s}$$

since

$$2 \cdot \frac{(g-2)}{m_1 + \cdots + m_{j-1} + m_{j_1}} \cdot \frac{(g-1) \cdot \sqrt{2}/2}{m_1 + \cdots + m_j} > \frac{(g-1) \cdot \sqrt{2}/2}{m_1 + \cdots + m_j} \tag{9}$$

($m_1 + \ldots + m_{j-1} + m_{j_1} \leq n-4$ and $n \leq g$) we conclude that $|CR_{\pi/4}(m)| < |CR_{\pi/4}(m')|$.

- If m contains $m_j \in E_3$, we associate to

$$m = (m_1, \ldots, m_{j-1}, m_j, m_{j+1}, \ldots, m_r)$$

m' such that

$$m' = (m_1, \ldots, m_{j-1}, m_{j_1}, m_{j_2}, m_{j_3}, m_{j+1}, \ldots, m_r)$$

with $m_{j_1}, m_{j_2}, m_{j_3} \in E_1$. It is clear, by Proposition 4.5, that $P_{m'} \in \mathbb{P}^+_{\pi/4}(n)$. Moreover, we have

$$CR_{\pi/4}(m) = (-1)^r \cdot 2^r \cdot 2^{n/2} \prod_{s=1}^{j-1} \frac{(g-1)\cos(m_s \cdot \pi/4) + \cos(m_s \cdot \pi/2)}{m_1 + \cdots + m_s} \cdot$$
$$\frac{-(g-1) \cdot \sqrt{2}/2}{m_1 + \cdots + m_j} \cdot \prod_{s=j+1}^{r} \frac{(g-1)\cos(m_s \cdot \pi/4) + \cos(m_s \cdot \pi/2)}{m_1 + \cdots + m_s}$$

and

$$CR_{\pi/4}(m') = (-1)^{r+2} \cdot 2^{r+2} \cdot 2^{n/2} \prod_{s=1}^{j-1} \frac{(g-1)\cos(m_s \cdot \pi/4) + \cos(m_s \cdot \pi/2)}{m_1 + \cdots + m_s} \cdot$$
$$\frac{(g-1) \cdot \sqrt{2}/2}{m_1 + \cdots + m_{j-1} + m_{j_1}} \cdot \frac{(g-1) \cdot \sqrt{2}/2}{m_1 + \cdots + m_{j_2}} \cdot \frac{(g-1) \cdot \sqrt{2}/2}{m_1 + \cdots + m_j} \cdot$$
$$\prod_{s=j+1}^{r} \frac{(g-1)\cos(m_s \cdot \pi/4) + \cos(m_s \cdot \pi/2)}{m_1 + \cdots + m_s}$$

since

$$4 \cdot \frac{(g-1) \cdot \sqrt{2}/2}{m_1 + \cdots + m_{j-1} + m_{j_1}} \cdot \frac{(g-1) \cdot \sqrt{2}/2}{m_1 + \cdots + m_{j_2}} \cdot \frac{(g-1) \cdot \sqrt{2}/2}{m_1 + \cdots + m_j} > \frac{(g-1) \cdot \sqrt{2}/2}{m_1 + \cdots + m_j} \tag{10}$$

we conclude that $|CR_{\pi/4}(m)| < |CR_{\pi/4}(m')|$.

Observe that in both inequalities (9) and (10), we can have the stronger conclusion: $2 \cdot |CR_{\pi/4}(m)| < |CR_{\pi/4}(m')|$ (because of the coefficients 2 and 4). That means that we can associate at least two negative compositions with one positive. That is why we have limited the theorem to $1 \leq g \leq 6$. There are not enough compositions to be required to associate more than two compositions with only one, except for $g = 6$ where we associate

$$(1,1,1,3 \text{ with } (1,1,1,1,1,1)$$
$$(1,1,3,1 \text{ with } (1,1,1,1,1,1)$$
$$(1,3,1,1 \text{ with } (1,1,4)$$
$$(1,5 \text{ with } (1,1,4)$$
$$(3,1,1,1 \text{ with } (1,4,1)$$
$$(5,1 \text{ with } (1,4,1)$$

we prove this easily as we shown above. We conclude that $a_{n,\pi/4} > 0$ if n is even. Similarly, we prove that $a_{n,\pi/4} < 0$ if n is odd, and the exception here is for $g = 5$:

$$(1,1,3 \text{ with } (1,1,1,1,1)$$
$$(1,3,1 \text{ with } (1,1,1,1,1)$$
$$(3,1,1 \text{ with } (1,4)$$
$$(5 \text{ with } (1,4)$$

Let's prove (c). Suppose n is even, let $m = (m_1, \ldots, m_r) \in \mathbb{P}(n-1)$,

- If $m = (m_1, \ldots, m_r) \in \mathbb{P}^+_{\pi/4}(n-1)$, let $m' = (m'_1, \ldots, m'_{r'}) \in \mathbb{P}^-_{\pi/4}(n-1)$ such that they are associated as we saw previously. There is M and M' such that $M = (m_1, \ldots, m_r, 1) \in \mathbb{P}^-_{\pi/4}(n)$ and $M' = (m'_1, \ldots, m'_{r'}, 1) \in \mathbb{P}^+_{\pi/4}(n)$ (if we add "1" to a composition, its sign changes by Proposition 4.5). It is clear that M and M' are associated then
$$|CR_{\pi/4}(m)| < |CR_{\pi/4}(m')| \ , \ |CR_{\pi/4}(M)| < |CR_{\pi/4}(M')|$$
and
$$|CR_{\pi/4}(m) + CR_{\pi/4}(m')| < |CR_{\pi/4}(M) + CR_{\pi/4}(M')|$$
Since $2(g-1)/n.|CR_{\pi/4}(m) + CR_{\pi/4}(m')| = |CR_{\pi/4}(M) + CR_{\pi/4}(M')|$ (see the first part of the proof).
- let $m' = (m'_1, \ldots, m'_{r'}) \in \mathbb{P}^-_{\pi/4}(n-1)$ that is not associated with any $\mathbb{P}^+_{\pi/4}(n-1)$ elements (it is possible by Proposition 4.7), then
$$M' = (m'_1, \ldots, m'_{r'}, 1) \text{ or } M' = (1, m'_1, \ldots, m'_{r'})$$
are in $\mathbb{P}^+_{\pi/4}(n)$ and one of them is associated with no element of $\mathbb{P}^-_{\pi/4}(n)$ (because $g \leq 6$, we can avoid the paths $1,1,1$ and $1,4$). Moreover, $|CR_{\pi/4}(m')| < |CR_{\pi/4}(M')|$.

We conclude that
$$|a_{n-1,\pi/4}| = |\sum_{m \in \mathbb{P}(n-1)} CR_{\pi/4}(m)| < |a_{n,\pi/4}| = |\sum_{m \in \mathbb{P}(n)} CR_{\pi/4}(m)|. \qquad \square$$

REMARK 4.9. To conclude this paper, we would like to highlight the following remarks, which provides additional insight into our results.

- In fact, Theorem 4.8 is true for $1 \leq g$ and the proof is still valid while we know how to deal with the exceptions, as we have seen for $g = 5$ and 6. The problem is that the number of the exceptions is quickly very large. We keep seeking for a stronger proof with no restriction for g, the idea is to organize $\mathbb{P}(n)$ following an ascending order of $CR_\theta(m)$ for analysing their distribution (Proposition 4.7 is a good beginning).

- We know that if $N_1 \geq g+1$, there exists a non-special divisor of degree $g-1$. Moreover we know all the curves which contain this kind of divisors over $\mathbb{F}_2$ for $g = 1$ and 2 (see [**1**]). Since for defect 2 curves over $\mathbb{F}_2$ with $g \geq 3$ one has
$$|N_1 - 3| = 2g - 2 \Rightarrow N_1 = 2g + 1 \Rightarrow N_1 \geq g + 1$$
the existence of these divisors is obvious. Thus, we do not need to use the coefficients a_n for this purpose. Nevertheless, for the defect 3 curves over $\mathbb{F}_2$ with $g \geq 3$, one has
$$|N_1 - 3| = 2g - 3 \Rightarrow N_1 = 2g \text{ or } N_1 = -2g + 6$$
our next goal, is to use the method of Section 4 to prove the existence of non-special divisor of degree $g-1$ for these curves (the case $N_1 = -2g+6$ with $g = 3$).

Acknowlegments

I would like to express my deepest appreciation and gratitude to my PhD supervisors Stéphane Ballet and Julia Pieltant for their priceless advice, guidance and continues support. Indeed, their involvement highly contributed to improve the paper. I would like to extend my sincere thanks to my supportive wife, who had always believed in my professional projects and my parents who helped me through different stages of my life.

References

[1] S. Ballet and D. Le Brigand, *On the existence of non-special divisors of degree g and $g-1$ in algebraic function fields over $\mathbb{F}_q$*, J. Number Theory **116** (2006), no. 2, 293–310, DOI 10.1016/j.jnt.2005.04.009. MR2195927

[2] S. Ballet, C. Ritzenthaler, and R. Rolland, *On the existence of dimension zero divisors in algebraic function fields defined over $\mathbb{F}_q$*, Acta Arith. **143** (2010), no. 4, 377–392, DOI 10.4064/aa143-4-4. MR2652586

[3] Taras Goy and Roman Zatorsky, *Infinite linear recurrence relations and superposition of linear recurrence equations*, J. Integer Seq. **20** (2017), no. 5, Art. 17.5.3, 14. MR3669600

[4] Jean-Pierre Serre, *Rational points on curves over finite fields*, Documents Mathématiques (Paris) [Mathematical Documents (Paris)], vol. 18, Société Mathématique de France, Paris, 2020. With contributions by Everett Howe, Joseph Oesterlé and Christophe Ritzenthaler; Edited by Alp Bassa, Elisa Lorenzo García, Christophe Ritzenthaler and René Schoof. MR4242817

[5] Henning Stichtenoth, *Algebraic function fields and codes*, 2nd ed., Graduate Texts in Mathematics, vol. 254, Springer-Verlag, Berlin, 2009. MR2464941

[6] V. E. Tarakanov and R. A. Zatorskiĭ, *On a relationship between determinants and permanents* (Russian, with Russian summary), Mat. Zametki **85** (2009), no. 2, 292–299, DOI 10.1134/S0001434609010301; English transl., Math. Notes **85** (2009), no. 1-2, 267–273. MR2548008

[7] M. A. Tsfasman and S. G. Vlăduţ, *Asymptotic properties of zeta-functions: Algebraic geometry, 7*, J. Math. Sci. (New York) **84** (1997), no. 5, 1445–1467, DOI 10.1007/BF02399198. MR1465522

[8] Roman A. Zatorsky, *Theory of paradeterminants and its applications*, Algebra Discrete Math. **1** (2007), 108–137. MR2367518

Aix Marseille Univ, CNRS, I2M, Marseille, France; and Institut de Mathématiques de Marseille, France

Email address: `koutchoukali.mahdi@hotmail.fr`; `mohamed-mahdi.koutchoukali@ac-aix-marseille.fr`

Contemporary Mathematics
Volume **832**, 2026
https://doi.org/10.1090/conm/832/16610

Generation of two "independent" points on an elliptic curve of j-invariant $\neq 0, 1728$

Dimitri Koshelev

ABSTRACT. This article is dedicated to a new generation method of two "independent" $\mathbb{F}_q$-points P_0, P_1 on almost any ordinary elliptic curve E over a finite field $\mathbb{F}_q$ of large characteristic. In particular, the method is relevant for all standardized and real-world elliptic curves of j-invariants different from 0, 1728. The points P_0, P_1 are characterized by the fact that nobody (even a generator) knows the discrete logarithm $\log_{P_0}(P_1)$ in the group $E(\mathbb{F}_q)$. Moreover, only one square root extraction in $\mathbb{F}_q$ (instead of two ones) is required in comparison with all previous generation methods.

1. Introduction

Despite ongoing calls for standardization and adaptation of post-quantum cryptography (PQC), elliptic curve cryptography (ECC) is now experiencing the all-time flourishing due to its active application when protecting most blockchain technologies, including cryptocurrencies. Their emphasis on ECC stems from concerns about efficiency of cryptographic protocols. Blockchains use multi-party computation (MPC) with plenty of parties, so the transition to PQC would lead to a catastrophic slowdown.

This article looks at an ECC optimization that is especially relevant to modern (blockchain) protocols. It is about the task of generating transparently several *"independent" $\mathbb{F}_q$-points* on an elliptic curve E over a finite field $\mathbb{F}_q$. "Independence" means that non-trivial linear relations between the points are unknown to anyone. We will focus on the case when the characteristic of $\mathbb{F}_q$ is large and $E\colon y^2 = f(x)$ is an ordinary (a.k.a. non-supersingular) curve. Today, other pairs $E/\mathbb{F}_q$ are considered to be suspicious and hence they are (almost) never used.

At the moment, in order to obtain one point in $E(\mathbb{F}_q)$ people frequently give preference to the naive try-and-increment method (represented, e.g., in [**22**, Section 8.2.1]) iterating the x-coordinate. Alternatively, one can resort to any constant-time map to $E(\mathbb{F}_q)$. In recent years, there are breakthroughs in constructing such maps [**30**, Section 3]. Nevertheless, it is worth stressing that constant-timeness is in fact superfluous in our context, because the seed of a generation process is not secret. On the contrary, it must be as public as possible so that everyone can make sure in

---

2020 *Mathematics Subject Classification.* Primary 14G50, 14H52, 14J27, 14Q20, 94A60.

Key words and phrases. Endomorphism rings, generation of "independent" points, isotrivial elliptic curves, Mordell–Weil lattices.

honesty of a generator. That is why, we can forget about tricky deterministic maps without loss of generality.

Given $x \in \mathbb{F}_q$, the condition $\sqrt{f(x)} \in \mathbb{F}_q$ can be checked without computing the square root, namely via the Legendre symbol $\left(\frac{f(x)}{q}\right)$. Generally speaking, this symbol (with whatever argument from $\mathbb{F}_q$) should be determined by Euclidean-type algorithms running in a bit time $O(\log^2(q))$. Moreover, they work well in practice [**26, 35**]. Meanwhile, two x-coordinates are on average enough to meet the desired condition. As is well known (e.g. from [**38**]), extracting $\sqrt{\cdot} \in \mathbb{F}_q$ costs $O(\log^3(q))$ bit operations, but never $O(\log^2(q))$ ones. So, the primitive $\sqrt{\cdot}$ is an order of magnitude more expensive than $\left(\frac{\cdot}{q}\right)$. Finally, when one needs more than one $\mathbb{F}_q$-point on E, nothing prevents from repeating multiple times the try-and-increment method with another seed.

A novel approach for solving the generation task is suggested in [**29**]. That article dwells on the extreme, but important case of j-invariant 0. The current article follows the same concept, but in the general case, that is, without significant restrictions on j-invariant. In contrast to [**29**], we will succeed in constructing just two "independent" points in $E(\mathbb{F}_q)$. Nonetheless, this already significantly affects performance, taking into account the cumulative effect. Indeed, it seemed earlier that $n \in \mathbb{N}$ "independent" points cannot be generated faster than by finding n roots (usually square) in the field $\mathbb{F}_q$. We will justify that $\lceil n/2 \rceil$ quadratic roots are sufficient for that purpose.

Throughout the text, $D < 0$ stands for the complex multiplication (CM) discriminant of E. As we will see, unlike $j(E) = 0$ (equivalently, $D = -3$), a new difficulty appears whenever D is large in absolute value. Fortunately, we will successfully circumvent this obstacle. It is worth noting that conservative cryptographers prefer and, at the same time, regulators standardize elliptic curves of huge CM discriminants, because they seem more secure as opposed to others. For instance, the source [**9**] recommends taking $D = D_0 c^2$ with $c \in \mathbb{N}$ and the square-free natural $-D_0 > 2^{100}$. The given condition is fulfilled for a random curve E with a high probability.

The problem of "independent" points has long history. For example, such points occur in the notorious *Dual Elliptic Curve Deterministic Random Bit Generator (Dual_EC_DRBG)* invented by NIST. It was heavily criticized, because NIST proposed its own points P, Q without any transparent explanation of their origin. Therefore, the cryptographic community was suspicious concerning a possible backdoor (see, e.g., [**10, 11**]). In addition, Dual_EC_DRBG had other disadvantages [**16, 39**], which raised even more doubts about the suitability of its use. Under public pressure, the last actual version of the standard [**8**] no longer contains this generator. Nevertheless, as shown in [**23**], it can be modified to be safe if the discrete logarithm between the points is truly unknown.

It is also worth mentioning a *Password-Authenticated Key Exchange (PAKE)* protocol under the name *SPAKE2* [**1, 2**]. It was one of the candidates (but not the winner) of the relatively recent public selection process [**20**] under the auspices of CFRG. SPAKE2 likewise requires two "independent" points denoted by M, N in the RFC [**32**]. Bearing in mind the bitter experience of Dual_EC_DRBG, the given RFC explicitly indicates where the points come from.

The third classical example is the so-called *Pedersen hash function* to the group $E(\mathbb{F}_q)$ (a.k.a. *Pedersen commitment*) [**34**, Section 3], which is also contained in

[**18**, Section 3]. The Pedersen hash is naturally generalized to an arbitrary number n of "independent" points (see, e.g., [**14**]). The larger the number n, the better the compression provided by the hash function, since it always returns one $\mathbb{F}_q$-point on E regardless of n. Thereby, it is tempting to take huge values of n (such as 2^{28}), which really happens in practice [**17**].

Entities of any cryptographic scheme operating with a large number of "independent" points face a difficult choice. They are obliged either to store/transmit the points or to regenerate them (or others) every time from a short seed. The first approach requires large amount of memory or a channel with good bandwidth, respectively. In turn, the second one forces entities to spend a part of their running time on repetitive point regeneration. Consequently, any noticeable acceleration of the given subroutine deserves attention to lean towards the second solution.

To be honest, little attention is paid to the problem of generating "independent" points, since it is not the bottleneck in contrast to other primitives on elliptic curves such as a *Multi-Scalar Multiplication (MSM)* [**13**]. By the way, MSM is essentially evaluating the Pedersen hash function. Nevertheless, the problem under consideration should be rigorously discussed by the scientific society. Otherwise, there is a risk that an ad hoc speed up proposed by a junior developer may come to a catastrophic security error not identified until the release of a software product.

The mathematics underlying the new generation method concerns elliptic curves over the rational function field $\mathbb{F}_q(u)$ and their *Mordell–Weil lattices* (see, e.g., [**40**], [**42**, Chapter III] as sources on the topic). The author hopes that Mordell–Weil lattices will sooner or later find other exciting applications in ECC or PQC, just as this was done in the past by pairings and isogenies, respectively. Such a prospect has a right to exist, because any $\mathbb{F}_q$-isogeny $E_0 \to E_1$ between elliptic $\mathbb{F}_q$-curves is realized as a point of E_1 over the function field $\mathbb{F}_q(E_0)$. In fact, $\mathbb{F}_q$-isogenies can be also imagined as points of a certain non-constant elliptic $\mathbb{F}_q(u)$-curve. This view enables to establish in Appendix an interesting relationship between isogeny-based cryptography (IBC) and yet another type of PQC.

2. Mathematical preliminaries

Consider $E\colon y^2 = f(x) := x^3 + ax + b$, an elliptic curve over a finite field $\mathbb{F}_q$ of characteristic 5 or greater. Since the curve E is ordinary by our assumption, its endomorphism ring is an order $\mathbb{Z}[\tau] = \mathbb{Z} \oplus \tau\mathbb{Z}$ in some imaginary quadratic field. Throughout the present paper, we will freely use basic facts from the theory of such quadratic orders (see, e.g., [**19**, Section 7]).

As above, denote by $D < 0$ the complex multiplication discriminant of E, that is, the discriminant of $\mathbb{Z}[\tau]$. As is customary, $\widehat{\tau}$ stands for the dual, i.e., the complex conjugate of τ. The integers

$$d := \deg(\tau) = \tau\widehat{\tau}, \qquad t := \operatorname{tr}(\tau) = \tau + \widehat{\tau}$$

are commonly called the degree (or norm) and trace of τ, respectively. Note that t is not in general the Frobenius trace of E, which is usually associated with this letter in the literature on elliptic curves. Clearly,

$$\tau = \frac{t + \sqrt{D}}{2}, \qquad \widehat{\tau} = \frac{t - \sqrt{D}}{2}$$

are the roots of the quadratic polynomial $m_\tau(x) := x^2 - tx + d$ with discriminant $D = t^2 - 4d$. Besides, it will be convenient to have before our eyes the elementary formula

$$\deg(n_0 + n_1\tau) = n_0^2 + n_0 n_1 t + n_1^2 d, \tag{2.1}$$

where $n_0, n_1 \in \mathbb{Z}$.

Let's extend the field $\mathbb{F}_q$ to the (infinite) function field $F := \mathbb{F}_q(u)$ in one variable u. Now, fix the quadratic twist $E^g\colon g(u)y^2 = f(x)$ by means of a function $g \in F^*$. Certainly, E^g is the trivial twist if $\sqrt{g} \in F$ because of the isomorphism

$$E^g \to E \qquad (x, y) \mapsto (x, \sqrt{g}\cdot y).$$

Evidently, $j(E^g) = j(E)$ and, conversely, every elliptic F-curve with the given j-invariant $\neq 0, 1728$ is isomorphic over F to the twist E^h for some $h \in F^*$. Such curves are said to be *isotrivial*. Thus, without loss of generality, we can deal with the form E^g rather than with the conventional Weierstrass form.

Hereafter, put $\mathcal{E} := E^g$ to simplify the notation. Recall that the (finitely generated) Mordell–Weil group $\mathcal{E}(F)$ is equipped with the so-called *naive height*

$$h\colon \mathcal{E}(F) \to \mathbb{N} \qquad h(P) := \begin{cases} \deg(x) & \text{if} \quad P = (x, y), \\ 0 & \text{if} \quad P = (0:1:0). \end{cases}$$

In particular, $h(P) = h(-P)$. Based on h and on [**42**, Theorem III.4.3], we get the *canonical height* and group *height pairing*

$$\begin{aligned} &\widehat{h}\colon \mathcal{E}(F) \to \mathbb{Q}_{\geqslant 0} \quad \widehat{h}(P) := \frac{1}{2}\lim_{n\to\infty}\frac{h(2^nP)}{4^n}, \\ &\langle\cdot,\cdot\rangle\colon \mathcal{E}(F)^2 \to \mathbb{Q} \quad \langle P, Q\rangle := \widehat{h}(P+Q) - \widehat{h}(P) - \widehat{h}(Q), \end{aligned} \tag{2.2}$$

respectively. It is useful to remember that $2\widehat{h}(P) = \langle P, P\rangle$ for each point P and $\widehat{h}(P) = 0$ if and only if the order of P is finite. The fact that $\widehat{h}$ always takes rational values will be at the core of proving Theorem 2.3. Curiously, as stressed in [**42**, Remark III.4.3.1], the analogue of $\widehat{h}$ for elliptic curves over number fields (probably) takes transcendental values at all non-torsion points.

Since $\mathcal{E}$ is a twist of E, the endomorphism rings of E and $\mathcal{E}$ are identical. Therefore, $\mathcal{E}(F)$ is not only a group, but also a $\mathbb{Z}[\tau]$-module.

LEMMA 2.1. *For any $\varphi \in \mathbb{Z}[\tau]$ and $P, Q \in \mathcal{E}(F)$ there are the identities*

$$\widehat{h}(\varphi(P)) = \deg(\varphi)\widehat{h}(P), \qquad \langle\varphi(P), \varphi(Q)\rangle = \deg(\varphi)\langle P, Q\rangle.$$

PROOF. The second identity immediately follows from the first one, hence we can concentrate on establishing the former. For compactness, we introduce the even function $\varphi_x := x \circ \varphi$ whose $\deg(\varphi_x) = 2\deg(\varphi)$. We lack the generalized naive height

$$h_{\varphi_x}(P) := \deg(\varphi_x(P)) = h(\varphi(P))$$

from [**43**, Section VIII.6]. In this notation, $h = h_x$. Note that

$$h_{\varphi_x}(2^nP) = h(\varphi(2^nP)) = h(2^n\varphi(P))$$

for each $n \in \mathbb{N}$. By virtue of [**43**, Proposition VIII.9.1], we have:

$$\widehat{h}(P) = \widehat{h}_{\varphi_x}(P) := \frac{1}{\deg(\varphi_x)}\lim_{n\to\infty}\frac{h_{\varphi_x}(2^nP)}{4^n} = \frac{\widehat{h}(\varphi(P))}{\deg(\varphi)}.$$

The lemma is proved. □

Suppose that we are given two short F-points P_i on the curve $\mathcal{E}$, where $i \in \mathbb{Z}/2$. "Short" means that their canonical heights $\widehat{h}_i := \widehat{h}(P_i)$ as well as $e := \langle P_0, P_1\rangle$ are small and hence they can be found. Also, we need the values $v_i := \langle P_i, \tau(P_{i+1})\rangle$. Provided that D is large, we cannot directly compute them. Fortunately, one value is expressed through the other.

LEMMA 2.2. *The linear relation $v_0 + v_1 = te$ holds.*

PROOF. We see that

$$dv_1 = \langle dP_1, \tau(P_0)\rangle = \langle \tau\widehat{\tau}(P_1), \tau(P_0)\rangle = d\langle \widehat{\tau}(P_1), P_0\rangle = d\langle P_0, \widehat{\tau}(P_1)\rangle.$$

The third equality is due to Lemma 2.1. So, $v_1 = \langle P_0, \widehat{\tau}(P_1)\rangle$ and ultimately

$$v_0 + v_1 = \langle P_0, (\tau + \widehat{\tau})(P_1)\rangle = \langle P_0, tP_1\rangle = te.$$

We obtain the statement of the lemma. □

Define the (symmetric) Gram matrix

$$M_2 := \begin{pmatrix} \langle P_0, P_0\rangle & \langle P_0, P_1\rangle \\ * & \langle P_1, P_1\rangle \end{pmatrix} = \begin{pmatrix} 2\widehat{h}_0 & e \\ * & 2\widehat{h}_1 \end{pmatrix}$$

with determinant $\det(M_2) = 4\widehat{h}_0\widehat{h}_1 - e^2$. Since the canonical height is positive definite on $\mathcal{E}(F)/\mathcal{E}(F)_{tor}$, the points P_0, P_1 are linearly independent over $\mathbb{Z}$ (in the strict sense) if and only if the matrix M_2 is non-degenerate (see [**42**, Lemma III.11.5]). We are able to say more by virtue of the following theorem.

THEOREM 2.3. *If $\Delta := -D\det(M_2)$ is not a square in $\mathbb{Q}$, then the points P_0, P_1 are linearly independent over $\mathbb{Z}[\tau]$.*

PROOF. Let's introduce yet another (symmetric) Gram matrix

$$M_4 := \begin{pmatrix} \langle P_0, P_0\rangle & \langle P_0, \tau(P_0)\rangle & \langle P_0, P_1\rangle & \langle P_0, \tau(P_1)\rangle \\ * & \langle \tau(P_0), \tau(P_0)\rangle & \langle \tau(P_0), P_1\rangle & \langle \tau(P_0), \tau(P_1)\rangle \\ * & * & \langle P_1, P_1\rangle & \langle P_1, \tau(P_1)\rangle \\ * & * & * & \langle \tau(P_1), \tau(P_1)\rangle \end{pmatrix}.$$

By applying Lemma 2.1 and the formula (2.1), it is easily verified that

$$M_4 = \begin{pmatrix} 2\widehat{h}_0 & t\widehat{h}_0 & e & v_0 \\ * & 2d\widehat{h}_0 & v_1 & de \\ * & * & 2\widehat{h}_1 & t\widehat{h}_1 \\ * & * & * & 2d\widehat{h}_1 \end{pmatrix}.$$

The points P_0, P_1 are independent over $\mathbb{Z}[\tau]$ if and only if P_0, $\tau(P_0)$, P_1, $\tau(P_1)$ are independent over $\mathbb{Z}$, that is, the matrix M_4 is non-degenerate. To avoid manual computation of its determinant the reader can resort to the code [**28**] written in Magma. It says that after the substitution $v_0 = te - v_1$ (cf. Lemma 2.2) we have:

$$\det(M_4) = \rho(v_1)^2, \qquad \text{where} \qquad \rho(v_1) := v_1^2 - tev_1 + (de^2 + D\widehat{h}_0\widehat{h}_1).$$

Further, the discriminant of the quadratic $\mathbb{Q}$-polynomial ρ equals

$$\operatorname{disc}(\rho) = (te)^2 - 4(de^2 + D\widehat{h}_0\widehat{h}_1) = D(e^2 - 4\widehat{h}_0\widehat{h}_1),$$

which is nothing but Δ in the statement of the theorem. If $\sqrt{\Delta} \notin \mathbb{Q}$, then the polynomial ρ does not have roots in the field $\mathbb{Q}$. Since v_1 is on the contrary a rational number, $\rho(v_1) \neq 0$ and thus the matrix M_4 is non-degenerate. □

LEMMA 2.4. *Assume that the initial curve E is not isogenous to a curve of j-invariant 1728 (i.e., of $D = -4$). If in addition $e = 0$ and $\widehat{h}_0 = \widehat{h}_1$, then the premise of Theorem 2.3 is fulfilled.*

PROOF. Under the lemma conditions, $\Delta = -4D\widehat{h}_1^2$. Consequently, $\sqrt{\Delta} \in \mathbb{Q}$ if and only if $\sqrt{-D} \in \mathbb{Q}$ or, equivalently, $D = -4c^2$ for some $c \in \mathbb{N}$. This implies that E is isogenous (necessarily over $\mathbb{F}_q$) to a curve of $j = 1728$ as stems, e.g., from [**25**, Theorems 25.1.2 and 25.4.6]. The lemma is proved. □

3. Generation of two "independent" points

First of all, recall for completeness (in Algorithm 1) the naive try-and-increment method of generating just one point in $E(\mathbb{F}_q)$.

Algorithm 1: Naive generation method of one point

Data: a seed $\in \{0,1\}^*$ and a cryptographic hash function $\eta\colon \{0,1\}^* \to \mathbb{F}_q$.
Result: a point in $E(\mathbb{F}_q)$.
begin
  $i := 0$;
  $x := \eta(seed||i)$;
  while $\left(\frac{f(x)}{q}\right) = -1$ **do**
    $i := i + 1$;
    $x := \eta(seed||i)$;
  end
  $y := \sqrt{f(x)}$;
  return (x, y).
end

In this section, we will deal solely with points $P_0, P_1 \in \mathcal{E}(F)$ independent over $\mathbb{Z}[\tau]$. In particular, they are of infinite order. It is necessary to give a strict definition of "independent" $\mathbb{F}_q$-points on E. It is reasonable to consider for this role the specializations $P_0(u)$, $P_1(u)$ at an element $u \in \mathbb{F}_q$. In addition, the reduction $\mathcal{E}(u)$ is assumed to be $\mathbb{F}_q$-isomorphic to E, that is, $\sqrt{g(u)} \in \mathbb{F}_q$. We are obliged to require independence of P_0, P_1 over $\mathbb{Z}[\tau]$, not only over $\mathbb{Z}$. The fact is that in the cryptographic context the group $E(\mathbb{F}_q)$ is cyclic. Thereby, the endomorphism τ acts on $E(\mathbb{F}_q)$ as the scalar multiplication $[\lambda]$, where λ is one of the two roots of the polynomial m_τ modulo the group order.

Evidently, there are at most a finite number of elements u at which P_0, P_1, or g is not correctly defined. Besides, it is worth emphasizing that $\log_{P_0(u)}(P_1(u))$ may be in principle a simple instance of the discrete logarithm problem (DLP) for specific values u. For a finite number of them, the equality $P_0(u) = P_1(u)$ even takes place. However, the DLP between $P_0(u)$, $P_1(u)$ seems intractable for general u. Otherwise, it would be surprising and perhaps would affect solving the DLP for two abstract points of $E(\mathbb{F}_q)$.

The new try-and-increment method of generating two "independent" points in $E(\mathbb{F}_q)$ is formalized in Algorithm 2. Up to an F-isomorphism of $\mathcal{E}$, it is enough to take $g \in \mathbb{F}_q[u]$. This permits to avoid the inversion operation in $\mathbb{F}_q$ during the evaluation $g(u)$. It is of paramount importance to pick a canonical seed and a cryptographically strong hash function η. Failing that, a dishonest entity can choose a value u (and then a preimage from $\eta^{-1}(u)$) for which the samples $P_0(u)$, $P_1(u)$ are weak from the viewpoint of the DLP. The same security requirement has to be respected in the case of Algorithm 1 executed twice.

Algorithm 2: New generation method of two "independent" points

Data: a seed $\in \{0,1\}^*$ and a cryptographic hash function $\eta\colon \{0,1\}^* \to \mathbb{F}_q$, a polynomial $g \in \mathbb{F}_q[u]$ and points $(x_0, y_0), (x_1, y_1) \in E^g(F)$ independent over $\mathbb{Z}[\tau]$.
Result: two "independent" points in $E(\mathbb{F}_q)$.
begin
  $i := 0$;
  $u := \eta(seed||i)$;
  while $\left(\frac{g(u)}{q}\right) = -1$ **do**
    $i := i + 1$;
    $u := \eta(seed||i)$;
  end
  $v := \sqrt{g(u)}$;
  return $(x_0(u), vy_0(u)),\ (x_1(u), vy_1(u))$.
end

3.1. Mestre–Kuwata–Wang's formulas. It remains to concretize the second data line of Algorithm 2. Given two elliptic $\mathbb{F}_q$-curves E_i of j-invariant $\neq 0, 1728$ (equivalently, the coefficients $a, b \neq 0$), Mestre [**33**] and Kuwata–Wang [**31**] separately found a function $g \in F$ and two non-torsion points $P_i \in E_i^g$. Whenever E_1 is a quadratic twist of E_0 (unique over $\mathbb{F}_q$), their result gives rise to a natural deterministic map $\mathbb{F}_q \to (E_0 \times E_1)(\mathbb{F}_q)$ or just $\mathbb{F}_q \to E_0(\mathbb{F}_q)$ if we do not need a point in $E_1(\mathbb{F}_q)$. The latter map is known in the cryptographic literature under the name *simplified SWU map* [**15**, Section 7], [**46**, Section 4.1]. By the way, its bottleneck likewise consists in extracting one square root in $\mathbb{F}_q$.

Here, we are on the contrary interested in the tweaked case $E_0 = E_1 =: E$. Constant-timeness is sacrificed for the second point on the same curve. In this case, Mestre–Kuwata–Wang's formulas have the form

$$x_0 := \frac{b(u^6 - 1)}{au^2(1 - u^4)}, \qquad x_1 := x_0 u^2, \qquad y := u^3.$$

These functions satisfy the equation $f(x_0)y^2 = f(x_1)$. Therefore, we possess the points

$$P_0 := (x_0, 1), \qquad P_1 := (x_1, y) \tag{3.1}$$

on the twist $\mathcal{E} = E^g$ with respect to

$$g := f(x_0) = \frac{f(x_1)}{y^2} = \frac{num}{den},$$

where

$$\begin{aligned} num &:= b\big(b^2u^{12} + 3b^2u^{10} + (a^3 + 6b^2)u^8 \\ &\quad + (2a^3 + 7b^2)u^6 + (a^3 + 6b^2)u^4 + 3b^2u^2 + b^2\big), \\ den &:= -\big(au^2(u^2 + 1)\big)^3. \end{aligned}$$

All the above formulas are verified in Magma [**28**].

Furthermore, this computer algebra system allows to compute the canonical height (pairing). Note that Magma's height $\widehat{h}$ is two times larger than in the definition (2.2), while the pairing $\langle \cdot, \cdot \rangle$ is consistent. It turns out that $\widehat{h}_0 = \widehat{h}_1 = 2$ and $e = 0$ for the current points P_0, P_1. Consequently, $\det(M_2) = 16 \neq 0$ and so P_0, P_1 are independent over $\mathbb{Z}$, confirming the fact already established in [**33**]. Thus, the *Mordell–Weil rank* r of $\mathcal{E}$, i.e., the rank of $\mathcal{E}(F)$ is no less than 2. In accordance with Theorem 2.3 and Lemma 2.4, the points P_0, P_1 are moreover independent over $\mathbb{Z}[\tau]$ (in particular, $r \geqslant 4$) unless E is isogenous to a $j = 1728$ curve. If so, this does not imply that there is a $\mathbb{Z}[\tau]$-dependency between P_0, P_1. We just do not know an answer concerning this question.

3.2. Elliptic curves for which $-D$ is a perfect square. Let E_4 denote an elliptic curve with j-invariant 1728. Even if $E \sim E_4$, another pair of short F-points on $\mathcal{E}$ probably exists for which the premise of Theorem 2.3 holds. Nonetheless, there is no essential advantage of such a curve E as compared with E_4 itself. First, a (multi-)scalar multiplication on E is slower than on E_4, because the latter enjoys the GLV decomposition technique [**25**, Section 11.3.3] with an order 4 automorphism as τ. And second, the curve E is not much secure than E_4, generally speaking. With rare exception, we are able to reduce the DLP from E to E_4 by evaluating an $\mathbb{F}_q$-isogeny $E \to E_4$ (of degree $c = \sqrt{-D}/2$) in a polylogarithmic time. These words are justified by the recent breakthrough [**36**] (cf. [**37**]) in evaluating isogenies of large prime degrees, not to mention those of smooth degrees (see, e.g., [**25**, Section 25.6]).

In fact, $\mathbb{F}_q$-curves E_4 are not so popular among implementers in contrast to $\mathbb{F}_q$-curves E_3 of j-invariant 0. This is due to the fact that $\mathrm{Aut}(E_3) \simeq \mathbb{Z}/6$ is greater than $\mathrm{Aut}(E_4) \simeq \mathbb{Z}/4$. As a consequence, curves E_3 have more symmetries than ones E_4, which impacts on performance of various cryptographic primitives such as pairings [**22**, Section 3.2.5]. To sum up, Algorithm 2 instantiated by the points (3.1) is always relevant in real-world cryptography unless $j(E) = 0$. In turn, the previous source [**29**] treats this special case by providing a generator up to four "independent" $\mathbb{F}_q$-points on E_3. If desired, a multi-point generator can be likewise constructed without any problems for curves E_4. The author is sure about that, because as well as for F-curves of $j = 0$, there are many F-curves of $j = 1728$ having moderate Mordell–Weil ranks (see again [**33**]).

Appendix. Relationship of algebraic surface cryptography and isogeny-based cryptography

Consider two elliptic curves $E_i\colon y_i^2 = f_i(x_i) := x_i^3 + a_i x_i + b_i$ (where $i \in \{0, 1\}$) over a finite field $\mathbb{F}_q$ of characteristic > 3. Let's introduce the *Kummer surface* $K := (E_0 \times E_1)/[-1]$ discussed, e.g., in [**40**, Example 11.6 and Section 11.2.1]. It has the form

$$K\colon f_0(x_0)y^2 = f_1(x_1) \quad \subset \quad \mathbb{A}^3_{(x_0,x_1,y)}, \qquad \text{where} \qquad y := \frac{y_1}{y_0}.$$

For the sake of compactness, put $F := \mathbb{F}_q(x_0)$. The F-curve $K \subset \mathbb{A}^2_{(x_1,y)}$ is the quadratic twist of E_1 with respect to f_0. In Section 3.1 we already encountered K in the case $E_0 = E_1$.

As said in [**41**, Proposition 3.1], there is the natural group isomorphism

$$\varPhi\colon K(F) \to \mathrm{Hom}(E_0, E_1) \oplus E_1(\mathbb{F}_q)[2] \qquad \big(X_1(x_0), Y(x_0)\big) \mapsto \big(X_1(x_0), Y(x_0)y_0\big),$$

where $\mathrm{Hom}(E_0, E_1)$ is the group of all $\mathbb{F}_q$-isogenies $E_0 \to E_1$. Moreover, taking into account the definition (2.2), we have:

$$2\widehat{h}(P) = h(P) = \deg(\Phi(P))$$

for each $P \in K(F)$. In other words, $\mathrm{Hom}(E_0, E_1)$ turns out to be a Mordell–Weil lattice (up to the multiplication by 2). Among other things, $K(F)_{tor} \simeq E_1(\mathbb{F}_q)[2]$ and $K(F)$ is of rank

$$(3.2) \quad r = \begin{cases} 0 & \text{if } E_0 \not\sim_{\mathbb{F}_q} E_1, \\ 2 & \text{if } E_0 \sim_{\mathbb{F}_q} E_1 \text{ and } \mathrm{End}(E_i) \text{ are quadratic orders,} \\ 4 & \text{if } E_0 \sim_{\mathbb{F}_q} E_1 \text{ and } \mathrm{End}(E_i) \text{ are (maximal) quaternion orders.} \end{cases}$$

Here, $\mathrm{End}(E_i) := \mathrm{Hom}(E_i, E_i)$ as usual. In contrast to the first two cases, the last one is possible only for the supersingular curves E_i. In this way, K (as a surface) is said to be supersingular too (see, e.g., [**40**, Section 12.4]).

The above view on isogenies allows to reformulate the famous isogeny-finding problem (IFP) for the two given elliptic curves E_i. In the new notation, it consists in searching for a non-torsion F-point on K. Note that Mestre–Kuwata–Wang's formulas give (if $a_i, b_i \neq 0$) a rational $\mathbb{F}_q$-curve on the Kummer surface K, but it is not the image of an F-point on K as a curve. The task under consideration is a specific instance of the so-called *section-finding problem (SFP)* on an irreducible (not necessarily elliptic) F-curve $C \subset \mathbb{A}^2$. Let's dive into a brief historical excursion on the topic.

Based on the SFP, Akiyama–Goto invented (in a series of works [**3–5**]) a type of PQC under the name *algebraic surface cryptography (ASC)*. To be more precise, they proposed several versions of some public-key encryption scheme. Afterwards, all their ciphers were fully broken in [**24, 27**]. The last noteworthy attempt to repair the ASC encryption scheme was timed to the NIST PQC competition. Akiyama–Goto in collaboration with many other Japanese cryptographers announced a cryptosystem dubbed *Giophantus* [**6, 7**]. However, the latter was also successfully attacked in [**12, 21**].

It is worth stressing that the original SFP itself remains resistant (even to a quantum computer) provided an appropriate choice of parameters. Generally speaking, the given problem boils down (in a natural way) to solving a large system of polynomial equations, which is intractable at the moment. In the situation $C = K$, the SFP in addition becomes equivalent to the IFP. This makes the SFP more reliable, because numerous isogenists fail to break the IFP. Besides, it happened historically that in ASC the coordinates of F-points on C have small degrees. Nevertheless, without any problems they can have smooth degrees to resemble even more chains of short isogenies.

Several years ago, Voloch [**45**] tried to revive ASC. Instead of public-key encryption, he discusses possibility of constructing a commitment scheme based on the SFP. This is relevant, because the Pedersen commitment is not quantum-safe. In order to prevent from obvious cheating in Voloch's commitment, the curve C has to have a unique F-point. This requirement seems quite restrictive and thereby affects negatively ease of generating the desired curve C.

One year later, Sterner [**44**] invented a similar commitment scheme in the language of supersingular isogenies, that is, for the supersingular $C = K$. Unfortunately, he missed Voloch's article. According to (3.2), there are a lot of F-points on K, hence this curve does not meet the uniqueness requirement. Sterner circumvented this trouble with the help of a trusted setup, whose goal is to return a random starting curve E_0 with unknown $\mathrm{End}(E_0)$. In this circumstance, the committer knows solely the initial F-point on K corresponding via Φ to a cyclic isogeny. The fact is that having another "cyclic" F-point amounts to obtaining a non-trivial endomorphism on E_0.

To summarize, the present appendix establishes an unexpected connection between two areas of PQC that earlier seemed too far from each other. Consequence of the new notice is twofold. On the one hand, this should animate research on ASC, which is in long-term decline in comparison with popular IBC. Perhaps, certain isogeny-based schemes may be carried over to the more general setting $C \neq K$. Who knows if this will lead to a faster or more compact cryptosystem for a carefully chosen curve C. On the other hand, attacks discovered in ASC may be potentially extended to some suspicious protocols on isogenies. The author hopes that the isomorphism Φ (despite its elementary form) will become the cornerstone to bring together representatives of the two PQC communities.

Acknowledgments

The author expresses his gratitude to Damien Stehlé and Michael Tsfasman for introducing him deeper to the theory of Euclidean lattices. Also, it is necessary to note the help with Mordell–Weil lattices, provided in different years on `https://www.mathoverflow.net` by Joseph H. Silverman, Noam D. Elkies, and Will Sawin.

References

[1] M. Abdalla and M. Barbosa, *Perfect forward security of SPAKE2*, 2019, https://eprint.iacr.org/2019/1194.

[2] M. Abdalla and D. Pointcheval, *Simple password-based encrypted key exchange protocols*, Topics in cryptology—CT-RSA 2005, Lecture Notes in Comput. Sci., vol. 3376, Springer, Berlin, 2005, pp. 191–208, DOI 10.1007/978-3-540-30574-3_14. MR2174378

[3] K. Akiyama and Y. Goto, *An algebraic surface public-key cryptosystem*, ISEC2004-80, vol. 104, 2004, pp. 13–20.

[4] K. Akiyama and Y. Goto, *A public-key cryptosystem using algebraic surfaces*, Post-Quantum Cryptography. PQCrypto 2006, 2006, pp. 119–138.

[5] K. Akiyama, Y. Goto, and H. Miyake, *An algebraic surface cryptosystem*, Public key cryptography—PKC 2009, Lecture Notes in Comput. Sci., vol. 5443, Springer, Berlin, 2009, pp. 425–442, DOI 10.1007/978-3-642-00468-1_24. MR2549997

[6] K. Akiyama, Y. Goto, S. Okumura, T. Takagi, K. Nuida, and G. Hanaoka, *A public-key encryption scheme based on non-linear indeterminate equations*, Selected areas in cryptography—SAC 2017, Lecture Notes in Comput. Sci., vol. 10719, Springer, Cham, 2018, pp. 215–234, DOI 10.1007/978-3-319-72565-9_11. MR3775586

[7] K. Akiyama, Y. Goto, S. Okumura, T. Takagi, K. Nuida, G. Hanaoka, H. Shimizu, and Y. Ikematsu, *A public-key encryption scheme based on non-linear indeterminate equations (Giophantus)*, 2017, https://eprint.iacr.org/2017/1241.

[8] E. Barker and J. Kelsey, *Recommendation for random number generation using deterministic random bit generators (NIST Special Publication 800-90A Revision 1)*, 2015, https://csrc.nist.gov/publications/detail/sp/800-90a/rev-1/final.

[9] D. J. Bernstein and T. Lange, *CM field discriminants*, 2013, https://safecurves.cr.yp.to/disc.html.

[10] D. J. Bernstein, T. Lange, and R. Niederhagen, *Dual EC DRBG*, 2013, https://projectbullrun.org/dual-ec.

[11] D. J. Bernstein, T. Lange, and R. Niederhagen, *Dual EC: a standardized back door*, The new codebreakers, Lecture Notes in Comput. Sci., vol. 9100, Springer, Berlin, 2016, pp. 256–281, DOI 10.1007/978-3-662-49301-4_17. MR3516335

[12] W. Beullens, W. Castryck, and F. Vercauteren, *IND-CPA attack on Giophantus*, 2018, `https://csrc.nist.gov/CSRC/media/Projects/Post-Quantum-Cryptography/documents/round-1/official-comments/Giophantus-official-comment.pdf`.

[13] G. Botrel and Y. El Housni, *Faster Montgomery multiplication and multi-scalar-multiplication for SNARKs*, IACR Transactions on Cryptographic Hardware and Embedded Systems (TCHES) **2023** (2023), no. 3, 504–521.

[14] P. Bottinelli, *Breaking Pedersen hashes in practice*, 2023, `https://research.nccgroup.com/2023/03/22/breaking-pedersen-hashes-in-practice`.

[15] E. Brier, J.-S. Coron, T. Icart, D. Madore, H. Randriam, and M. Tibouchi, *Efficient indifferentiable hashing into ordinary elliptic curves*, Advances in cryptology—CRYPTO 2010, Lecture Notes in Comput. Sci., vol. 6223, Springer, Berlin, 2010, pp. 237–254, DOI 10.1007/978-3-642-14623-7_13. MR2725597

[16] D. R. L. Brown and K. Gjøsteen, *A security analysis of the NIST SP 800-90 elliptic curve random number generator*, Advances in cryptology—CRYPTO 2007, Lecture Notes in Comput. Sci., vol. 4622, Springer, Berlin, 2007, pp. 466–481, DOI 10.1007/978-3-540-74143-5_26. MR2423865

[17] V. Buterin, *Open problem: ideal vector commitment*, 2020, `https://ethresear.ch/t/open-problem-ideal-vector-commitment/7421`.

[18] D. Chaum, E. van Heijst, and B. Pfitzmann, *Cryptographically strong undeniable signatures, unconditionally secure for the signer*, Advances in Cryptology—CRYPTO 1991 (Berlin, Heidelberg) (J. Feigenbaum, ed.), Lecture Notes in Computer Science, vol. 576, Springer, 1992, pp. 470–484.

[19] D. A. Cox, *Primes of the form $x^2 + ny^2$—Fermat, class field theory, and complex multiplication*, AMS Chelsea Publishing, Providence, RI, 2022. MR4502401

[20] Crypto Forum Research Group (CFRG), *PAKE selection process*, 2020, `https://github.com/cfrg/pake-selection`.

[21] J. Ding, J. Deaton, and K. Schmidt, *Giophantus distinguishing attack is a low dimensional learning with errors problem*, Adv. Math. Commun. **14** (2020), no. 4, 573–577, DOI 10.3934/amc.2020030. MR4159992

[22] N. El Mrabet and M. Joye (eds.), *Guide to pairing-based cryptography*, Cryptography and Network Security Series, Chapman and Hall/CRC, New York, 2017.

[23] R. R. Farashahi, B. Schoenmakers, and A. Sidorenko, *Efficient pseudorandom generators based on the DDH assumption*, Public key cryptography—PKC 2007, Lecture Notes in Comput. Sci., vol. 4450, Springer, Berlin, 2007, pp. 426–441, DOI 10.1007/978-3-540-71677-8_28. MR2424184

[24] J.-C. Faugère and P.-J. Spaenlehauer, *Algebraic cryptanalysis of the PKC'2009 algebraic surface cryptosystem*, Public key cryptography—PKC 2010, Lecture Notes in Comput. Sci., vol. 6056, Springer, Berlin, 2010, pp. 35–52, DOI 10.1007/978-3-642-13013-7_3. MR2660734

[25] S. D. Galbraith, *Mathematics of public key cryptography*, Cambridge University Press, Cambridge, 2012, DOI 10.1017/CBO9781139012843. MR2931758

[26] M. Hamburg, *Computing the Jacobi symbol using Bernstein–Yang*, 2021, `https://eprint.iacr.org/2021/1271`.

[27] P. Ivanov and J. F. Voloch, *Breaking the Akiyama-Goto cryptosystem*, Arithmetic, geometry, cryptography and coding theory, Contemp. Math., vol. 487, Amer. Math. Soc., Providence, RI, 2009, pp. 113–118, DOI 10.1090/conm/487/09528. MR2555990

[28] D. Koshelev, *Magma code*, 2023, `https://github.com/dishport/Generation-of-two-independent-points-on-an-elliptic-curve-of-j-invariant-not-0-1728`.

[29] D. Koshelev, *Generation of "independent" points on elliptic curves by means of Mordell–Weil lattices*, Mathematical Cryptology **4** (2024), no. 1, 11–22.

[30] D. Koshelev, *Some remarks on how to hash faster onto elliptic curves*, Journal of Computer Virology and Hacking Techniques **20** (2024), no. 4, 593–605.

[31] M. Kuwata and L. Wang, *Topology of rational points on isotrivial elliptic surfaces*, Internat. Math. Res. Notices **4** (1993), 113–123, DOI 10.1155/S107379289300011X. MR1214702

[32] W. Ladd, *SPAKE2, a password-authenticated key exchange (RFC 9382)*, 2023, `https://datatracker.ietf.org/doc/rfc9382`.

[33] J.-F. Mestre, *Rang de courbes elliptiques d'invariant donné* (French, with English summary), C. R. Acad. Sci. Paris Sér. I Math. **314** (1992), no. 12, 919–922. MR1168325

[34] T. P. Pedersen, *Noninteractive and information-theoretic secure verifiable secret sharing*, Advances in cryptology—CRYPTO '91 (Santa Barbara, CA, 1991), Lecture Notes in Comput. Sci., vol. 576, Springer, Berlin, 1992, pp. 129–140, DOI 10.1007/3-540-46766-1_9. MR1243648

[35] T. Pornin, *X25519 implementation for ARM Cortex-M0/M0+*, 2020, https://github.com/pornin/x25519-cm0.

[36] D. Robert, *Evaluating isogenies in polylogarithmic time*, 2022, https://eprint.iacr.org/2022/1068.

[37] C. Hazay and M. Stam (eds.), *Advances in cryptology—EUROCRYPT 2023. Part V*, Lecture Notes in Computer Science, vol. 14008, Springer, Cham, 2023, DOI 10.1007/978-3-031-30589-4. MR4590988

[38] P. Sarkar, *Computing square roots faster than the Tonelli-Shanks/Bernstein algorithm*, Adv. Math. Commun. **18** (2024), no. 1, 141–162, DOI 10.3934/amc.2022007. MR4671487

[39] B. Schoenmakers and A. Sidorenko, *Cryptanalysis of the dual elliptic curve pseudorandom generator*, 2006, https://eprint.iacr.org/2006/190.

[40] M. Schütt and T. Shioda, *Mordell–Weil lattices*, Ergebnisse der Mathematik und ihrer Grenzgebiete. 3. Folge. A Series of Modern Surveys in Mathematics, vol. 70, Springer, Singapore, 2019, DOI 10.1007/978-981-32-9301-4. MR3970314

[41] T. Shioda, *Correspondence of elliptic curves and Mordell–Weil lattices of certain elliptic K3's*, Algebraic cycles and motives. Vol. 2, London Math. Soc. Lecture Note Ser., vol. 344, Cambridge Univ. Press, Cambridge, 2007, pp. 319–339. MR2385296

[42] J. H. Silverman, *Advanced topics in the arithmetic of elliptic curves*, Graduate Texts in Mathematics, vol. 151, Springer-Verlag, New York, 1994, DOI 10.1007/978-1-4612-0851-8. MR1312368

[43] J. H. Silverman, *The arithmetic of elliptic curves*, 2nd ed., Graduate Texts in Mathematics, vol. 106, Springer, Dordrecht, 2009, DOI 10.1007/978-0-387-09494-6. MR2514094

[44] B. Sterner, *Commitment schemes from supersingular elliptic curve isogeny graphs*, Mathematical Cryptology **1** (2021), no. 2, 40–51.

[45] J. F. Voloch, *Commitment schemes and diophantine equations*, ANTS XIV—Proceedings of the Fourteenth Algorithmic Number Theory Symposium, Open Book Ser., vol. 4, Math. Sci. Publ., Berkeley, CA, 2020, pp. 1–5, DOI 10.2140/obs.2020.4.1. MR4235102

[46] R. S. Wahby and D. Boneh, *Fast and simple constant-time hashing to the BLS12-381 elliptic curve*, IACR Transactions on Cryptographic Hardware and Embedded Systems **2019** (2019), no. 4, 154–179.

Parallel Computation Laboratory, École Normale Supérieure de Lyon, France
Current address: Department of Mathematics, University of Lleida, Catalonia, Spain
Email address: dimitri.koshelev@gmail.com
URL: https://www.researchgate.net/profile/dimitri-koshelev

Contemporary Mathematics
Volume **832**, 2026
https://doi.org/10.1090/conm/832/16611

Principal polarization on products of abelian varieties over finite fields

Sergey Rybakov

ABSTRACT. We refine and generalize the results of K. E. Lauter and E. W. Howe on principal polarizations on products of abelian varieties over finite fields. Firstly, we study the reasons for the absence of an irreducible principal polarization in the isogeny class of the product of an ordinary and a supersingular abelian variety. Secondly, we provide a necessary condition for the existence of a principal polarization on an abelian variety in the isogeny class of the product of a geometrically simple abelian surface and an elliptic curve. As an application, we prove that this abelian threefold or its quadratic twist is a Jacobian.

1. Introduction

In this paper, we assume that the ground field $k = \mathbb{F}_q$ is a finite field of characteristic p and cardinality q. The Jacobian $J(C)$ of a smooth projective curve C over k can be non-simple as an abstract abelian variety without polarization. This is the case for many interesting curves, for example, for maximal or minimal curves over finite fields. This observation is used by K. Lauter in [**Lau02**], where she constructs genus 3 curves over finite fields that have the maximum or minimum number of points.

Her work relies on the following idea of J.-P. Serre [**Serre83**]. If a curve C over k has many points (in other words, the defect is small), then there is an isogeny

$$\varphi : A \times B \to J(C),$$

where A and B are abelian varieties of positive dimension. The pullback to $A \times B$ of the canonical polarization on $J(C)$ is a product of polarizations L_A and L_B on A and B respectively. Moreover,

$$\deg L_A = \deg L_B = \deg \varphi,$$

and there is an anti-isometry $\ker L_A \cong \ker L_B$. This construction can be inverted: we can *glue* two polarizations L_A and L_B on A and B, if there is an anti-isometry $\ker L_A \cong \ker L_B$ (see Section 3). In many interesting cases it is possible to prove that there is no curve over a given finite field with a given number of points using the fact that the Jacobian of such a curve has to be the gluing of polarizations on two abelian varieties of positive dimension.

2020 *Mathematics Subject Classification.* Primary 14K99, 14G05, 14G15.
Key words and phrases. abelian variety, principal polarization, Jacobian, finite field.

For example, in some situations there is no irreducible principal polarization in the isogeny class of the product of an ordinary and a supersingular abelian variety. In this paper, we study the reasons for this in more detail; in particular, we reprove and clarify the theorem of K. Lauter and E. Howe [**HL12**, Theorem 3.1].

The gluing construction is also used for a complete classification of zeta functions of curves of genus 2 obtained in [**HNR06**], where the results of [**Ru90**] and [**Howe95**] are used. In this paper, we extend this approach to the curves of genus 3. By [**Howe96**, Theorem 1.2], if an abelian variety of dimension 3 is simple, then it always has a principal polarization. We are interested in a less general case where an abelian variety is isogenous to a product of a geometrically simple abelian surface and an elliptic curve.

In this case, we encounter the phenomenon of exceptional prime numbers. In Section 2.4 we study them in full generality, but for geometrically simple abelian surfaces, they can be defined using Lemma 4.1.(1) as follows.

Let A be a geometrically simple abelian surface with the Weil polynomial f_A, and the real Weil polynomial h (see section 2 for the definition of Weil polynomials). The endomorphism algebra $K = \mathrm{End}^\circ(A)$ of A is a CM-field, and the real subfield K^+ of K is isomorphic to the quadratic extension $\mathbb{Q}[t]/h(t)\mathbb{Q}[t]$. We say that a prime $\ell \in \mathbb{Z}$ is *exceptional* (with respect to A) if

- $f_A(t) \equiv f(t)^2 \bmod \ell^2$, for some irreducible $f \in \mathbb{Z}_\ell[t]$;
- ℓ is inert in K^+.

Let B be an elliptic curve over k. According to Remark 4.2, if ℓ is exceptional, then for any isogeny $A' \to A$, and any polarization L on A'

$$\ker L \not\cong B[\ell].$$

However, by Proposition 4.3, in this situation the gluing of polarizations is still possible, at least if A is ordinary. That is, there exist isogeny $A' \to A$ and a polarization L on A' such that $\ker L \cong B[\ell^2]$.

If the Weil polynomial f_B of the elliptic curve B is irreducible, then $\mathrm{End}^\circ(B)$ is a quadratic extension of $\mathbb{Q}$. Denote by Δ_B the discriminant of $\mathrm{End}^\circ(B)$. We are now ready to formulate the main result of the paper.

THEOREM 1.1. *Let A be a geometrically simple abelian surface with real Weil polynomial h. Suppose that B is an elliptic curve with irreducible Weil polynomial $f_B(t) = t^2 - bt + q$. Assume that there is a prime ℓ that divides $h(b)$ and such that the following conditions hold:*

- *$\Delta_B \neq -\ell$;*
- *if $f_B(t) \equiv (t - t_1)^2 \bmod \ell$, then ℓ^2 divides $f_B(t_1)$;*
- *if ℓ is exceptional, then A is ordinary.*

Then $A \times B$ is isogenous to an abelian variety with irreducible principal polarization.

Vice versa, if $A \times B$ is isogenous to an abelian variety with irreducible principal polarization, then $h(b) \neq \pm 1$.

To an abelian threefold with irreducible principal polarization one can apply the following result due to Serre based on the Torelli Theorem and [**OU73**].

THEOREM 1.2 ([**Lau01**, Section 7]). *Let A be an abelian variety of dimension 3 over k with a principal polarization. Assume that the polarization is irreducible over an algebraic closure of k. Then A or its quadratic twist (corresponding to -1 automorphism) is the Jacobian of a smooth curve over k.*

From Theorem 1.1 and Theorem 1.2 we immediately get the following result.

THEOREM 1.3. *Let A be a geometrically simple abelian surface with a real Weil polynomial h. Suppose that B is an elliptic curve with irreducible Weil polynomial $f_B(t) = t^2 - bt + q$. Assume that there is a prime ℓ that divides $h(b)$ and such that the following conditions hold:*

- $\Delta_B \neq -\ell$;
- *if* $f_B(t) \equiv (t - t_1)^2 \bmod \ell$, *then* ℓ^2 *divides* $f_B(t_1)$;
- *if* ℓ *is exceptional, then* A *is ordinary.*

Then the isogeny class $A \times B$ or its quadratic twist (corresponding to -1 automorphism) contains the Jacobian of a smooth curve of genus 3. Vice versa, if $A \times B$ is isogenous to a Jacobian, then $h(b) \neq \pm 1$.

2. Preliminaries

2.1. Endomorphism algebras of abelian varieties. Let A and B be abelian varieties over k, and let $\mathrm{Hom}(A, B)$ be the group of homomorphisms from A to B over k. The group $\mathrm{Hom}(A, B)$ is finitely generated and torsion-free and $\mathrm{End}(A) = \mathrm{Hom}(A, A)$ has a ring structure with composition as multiplication. We will use the following notation:

$$\mathrm{Hom}^\circ(A, B) = \mathrm{Hom}(A, B) \otimes_{\mathbb{Z}} \mathbb{Q}, \text{ and } \mathrm{End}^\circ(A) = \mathrm{End}(A) \otimes_{\mathbb{Z}} \mathbb{Q}.$$

We will call an abelian variety A *simple* if it does not contain nontrivial abelian subvarieties. We will call an abelian variety B *isotypic* if there exists a simple abelian variety A such that B is isogenous to A^r for some r. Any abelian variety A over k is isogenous to a product of isotypic abelian varieties A_i. This decomposition corresponds to the decomposition of $\mathrm{End}^\circ(A)$ into a product of simple algebras $\mathrm{End}^\circ(A_i)$. In particular, $\mathrm{End}^\circ(A)$ is a semi-simple $\mathbb{Q}$-algebra [**Mum71**, IV.19. Corollaries 1 and 2].

An element $\varphi \in \mathrm{End}(A)$ is called an *isogeny* if φ is finite and surjective. The kernel $\ker \varphi$ of an isogeny φ is a finite group scheme; the *degree* $\deg \varphi$ of an isogeny is defined to be the order of its kernel. Note that since the kernel could be non-reduced, the order of $\ker L$ is not the same as the order of its group of points $\ker L(\bar{k})$. If $\varphi \in \mathrm{End}(A)$ is not an isogeny, we put $\deg \varphi = 0$. Then deg is a homogeneous polynomial function $\deg : \mathrm{End}^\circ(A) \to \mathbb{Q}$ of degree $2 \dim A$ [**Mum71**, IV.19.Theorem 2], i.e., for any $v_0, v_1 \in \mathrm{End}^\circ(A)$ the function $\deg(x_0 v_0 + x_1 v_1)$ is a polynomial in x_0 and x_1 of degree $2 \dim A$.

An important example of an isogeny is *the arithmetic Frobenius morphism* $F_A : A \to A$. By definition, F_A is trivial on schematic points and raises functions to their q-th powers. It is known that F_A is an isogeny of degree $q^{\dim A}$ [**Mum71**, Page 205].

Let $A[m]$ be the group subscheme of A annihilated by m. Fix a prime number $\ell \neq p$. The Frobenius morphism induces an action on the group $A[\ell^r](\bar{k})$, on the Tate module

$$T_\ell(A) = \varprojlim_r A[\ell^r](\bar{k}),$$

and on $V_\ell(A) = T_\ell(A) \otimes_{\mathbb{Z}_\ell} \mathbb{Q}_\ell$. The following theorem goes back to Weil.

THEOREM 2.1. *The module $T_\ell(A)$ is a free $\mathbb{Z}_\ell$-module of rank $2\dim A$. The arithmetic Frobenius morphism F_A induces on $V_\ell(A)$ a semisimple linear operator F. Its characteristic polynomial $f_A(t) = \det(t - F)$ is a monic polynomial of degree $2\dim A$ with integer coefficients. It does not depend on the choice of ℓ. For any $n \in \mathbb{Z}$ we have $f_A(n) = \deg(n - F)$. If π is a root of f_A, then $f_A(q/\pi) = 0$, and $|\pi| = \sqrt{q}$.* □

PROOF. Combine [**Mum71**, IV.19. Theorem 4], and [**Mum71**, IV.21. Theorem 4]. □

The polynomial f_A is called *the Weil polynomial of A*. The key result of the Tate–Honda theory states that abelian varieties A and B over k are isogenous if and only if $f_A = f_B$ [**Ta66**]. If A is simple, then f_A is irreducible; in particular, f_A is uniquely determined by any root π_A. Thus, the isogeny class of a simple abelian variety is determined by π_A.

The polynomial f_A uniquely determines the endomorphism algebra of the abelian variety. For example, the endomorphism algebra of a non-supersingular elliptic curve B is the imaginary quadratic extension $\mathbb{Q}[t]/f_B(t)\mathbb{Q}[t]$ of $\mathbb{Q}$ [**Wa69**]. If a simple abelian surface A is ordinary or mixed, then its endomorphism algebra is a CM-field and is isomorphic to $\mathbb{Q}(\pi) = \mathbb{Q}[t]/f_A(t)\mathbb{Q}[t]$ [**Ta66**].

The roots of f_A are called the *Weil numbers*. If π is a Weil number, then q/π is a Weil number by Theorem 2.1. Therefore,

$$f_A(t) = \prod_{i=1}^{\dim A} (t - \pi_i)(t - q/\pi_i)$$

for some Weil numbers $\pi_1, \ldots, \pi_{\dim A}$. Note that $\pi + q/\pi$ is a real algebraic number. We say that

$$h_A(t) = \prod_{i=1}^{\dim A} (t - \pi_i - q/\pi_i)$$

is *the real Weil polynomial of A*.

We need the following lemma, which allows us to construct new abelian varieties within a given isogeny class.

LEMMA 2.2 ([**Mil08**, IV.2.3]). *Let $\varphi : B \to A$ be an isogeny. Then $T_\ell(\varphi) : T_\ell(B) \to T_\ell(A)$ is a $\mathbb{Z}_\ell$–linear embedding commuting with the Frobenius action, whose image generates $V_\ell(A)$.*

Conversely, for each $\mathbb{Z}_\ell$–submodule $T \subset T_\ell(A)$, invariant under the Frobenius morphism and generating $V_\ell(A)$, there exist an abelian variety B and an isogeny $\varphi : B \to A$ over k such that $T_\ell(\varphi) : T_\ell(B) \to T$ is an isomorphism.

2.2. Finite subschemes of abelian varieties. In this section we recall some results on finite group schemes from [**Howe96**]. We fix an isogeny class $\mathcal{C}$ of abelian varieties of dimension g. We have a corresponding endomorphism algebra $\mathrm{End}^\circ(\mathcal{C})$ and an integer element $F \in \mathrm{End}^\circ(\mathcal{C})$ such that for every A in $\mathcal{C}$ there is an isomorphism $i_A : \mathrm{End}^\circ(\mathcal{C}) \cong \mathrm{End}^\circ(A)$ such that $i_A(F) = F_A$. Consider a subring $R = \mathbb{Z}[F, V]$ in $\mathrm{End}^\circ(\mathcal{C})$ generated over $\mathbb{Z}$ by F and a shift (Verschiebung) $V = q/F$. The algebra $K = R \otimes \mathbb{Q}$ is a product of real or CM-fields; therefore, the complex conjugation defines an involution $r \mapsto \bar{r}$ on K. Moreover, $\bar{F} = V$; therefore, R is invariant under this involution.

If $\ell \neq p$, the group scheme $A[\ell]$ is reduced and is uniquely determined by the Frobenius action on the vector space $A[\ell](\bar{k})$ of dimension $2\dim A$ over $\mathbb{F}_\ell$. By Theorem 2.1, the characteristic polynomial of this action on $A[\ell](\bar{k})$ is equal to $f_A(t) \bmod \ell$. Note that the Frobenius action is not always semi-simple on $A[\ell](\bar{k})$, and $A[\ell]$ is not always uniquely determined by the polynomial f_A. This issue can be avoided using Grothendieck groups.

Let $\mathcal{IK}$ be the full subcategory of the category of finite commutative group schemes Δ such that there exists a monomorphism from Δ to some abelian variety from $\mathcal{C}$. In other words, $\mathcal{IK}$ is the category of kernels of isogenies between varieties in $\mathcal{C}$. The category $\mathcal{IK}$ is a product of four subcategories $\mathcal{IK}_{rr}$, $\mathcal{IK}_{rl}$, $\mathcal{IK}_{lr}$, and $\mathcal{IK}_{ll}$, where the first index means that objects of these categories are respectively reduced or local, and the second means the same for the Cartier dual of each object.

For a finite $\mathbb{Z}$-algebra S we denote by mod_S the category of finite S-modules. There is a natural functor $\mathcal{P}$ from $\mathcal{IK}_{rr} \oplus \mathcal{IK}_{rl}$ to mod_R given by $X \mapsto X(\bar{k})$, where the R-module structure on $X(\bar{k})$ is given by the action of Frobenius and Verschiebung on X. The functor $\mathcal{P}$ can be extended to $\mathcal{IK}_{lr}$ as follows. If M is a finite R-module, then

$$\widehat{M} = \mathrm{Hom}_{\mathbb{Z}}(M, \mathbb{Q}/\mathbb{Z})$$

is naturally an R-module: for every $r \in R$ and $\psi \in \widehat{M}$ we set

$$(r\psi)(m) = \psi(\bar{r}m)$$

for all $m \in M$. For a local group scheme X with reduced Cartier dual scheme $D(X)$ we define $\mathcal{P}(X)$ by the formula:

$$\mathcal{P}(X) = \widehat{D(X)(\bar{k})}.$$

The Grothendieck group $G(\mathcal{IK})$ of $\mathcal{IK}$ is defined as the quotient of the free abelian group generated by isomorphism classes of objects in $\mathcal{IK}$ by the subgroup generated by the expressions $Y - X - Z$ for any exact sequence

$$0 \to X \to Y \to Z \to 0$$

in $\mathcal{IK}$. Denote by $[X]$ an element of $G(\mathcal{IK})$ corresponding to X. For an R-module M we denote by $[M]_R$ the corresponding element of the Grothendieck group $G(\mathrm{mod}_R)$ of the category of finite R-modules.

There is a natural homomorphism of Grothendieck groups

$$\varepsilon : G(\mathcal{IK}) \to G(\mathrm{mod}_R).$$

On $\mathcal{IK}_{rr} \oplus \mathcal{IK}_{rl} \oplus \mathcal{IK}_{lr}$ the morphism ε is induced by $\mathcal{P}$. In the remaining case, when X and its Cartier dual $D(X)$ are local, we define ε as follows. The category K_{ll} (if it is not empty) is generated by a single simple object α_p. This is the group scheme $\mathrm{Spec}\, k[t]/(t^p)$ with co-multiplication $t \mapsto t \otimes 1 + 1 \otimes t$. Therefore, it is enough to define $\varepsilon([\alpha_p])$. The corresponding R-module is the abelian group $\mathbb{Z}/p\mathbb{Z}$ equipped with the trivial action of F and V.

THEOREM 2.3 ([**Howe96**, Theorem 3.1]). *The morphism* $\varepsilon : G(\mathcal{IK}) \to G(\mathrm{mod}_R)$ *is a well-defined isomorphism.* □

Let $\alpha = a/b \in K = R \otimes \mathbb{Q}$ be an invertible element. *The principal element* corresponding to α is

$$\mathrm{Pr}_R(\alpha) = [R/aR]_R - [R/bR]_R \in G(\mathrm{mod}_R).$$

THEOREM 2.4 ([**Howe96**, Theorem 3.5]). *If $\alpha \in K$ is an isogeny, then*

$$\varepsilon([\ker \alpha]) = \mathrm{Pr}_R(\alpha). \quad \square$$

2.3. Kernels of polarizations on abelian varieties over finite fields. This section is based on the main results of the paper [**Howe96**] about principal polarizations on abelian varieties over finite fields.

Let $R^+ = \mathbb{Z}[F+V] \subset R$, and let $K^+ = R^+ \otimes \mathbb{Q} \subset K = R \otimes \mathbb{Q}$. Denote by $\mathcal{O}$ the ring of integers of K, and let $\mathcal{O}^+ = K^+ \cap \mathcal{O}$ be the ring of integers of K^+. Choose an order $R \subset S$ in K stable under involution, and put $S^+ = S \cap \mathcal{O}^+$.

Define an involution $P \mapsto \bar{P}$ on $G(Mod_S)$ by the formula:

$$\overline{[M]_S} = [\widehat{M}]_S.$$

Since any S-module is naturally an S^+-module, the formula $N_{S/S^+}([M]_S) = [M]_{S^+}$ defines a norm $N_{S/S^+} : G(Mod_S) \to G(Mod_{S^+})$. Let $Z(S)$ be the subgroup of symmetric elements in the kernel of

$$G(Mod_S) \to G(Mod_{S^+}) \otimes \mathbb{Z}/2\mathbb{Z},$$

induced by the norm. It is known that $\mathrm{Pr}(Tp(K)) \subset Z(R)$ [**Howe96**, Section 5].

An ample line bundle L on A is called a polarization. It defines an isogeny to the dual abelian variety

$$\varphi_L : A \to \check{A}.$$

The kernel of this isogeny $\ker L$ is called *the kernel of polarization.* A polarization is *principal* if the kernel is trivial. The group scheme $\ker L$ is endowed with the Weil pairing $e_L : \ker L \times \ker L \to G_m$ [**Mum71**, §23].

A finite group scheme X from $\mathcal{IK}$ is called *attainable in the isogeny class* $\mathcal{C}$ if there exists an abelian variety A in $\mathcal{C}$ and a polarization L on A such that $X = \ker L$. An element P of $G(\mathrm{mod}_R)$ is called *attainable* if it is *effective*, that is, $P = [M]_R$ for some R-module M, and there exists a group scheme X attainable in $\mathcal{C}$ such that $\varepsilon([X]_{\mathcal{C}}) = P$. We say that an ideal $\mathfrak{p} \subset R$ is *attainable*, if $[R/\mathfrak{p}]_R$ is attainable.

Let $B(S) = \{P + \bar{P} : P \in G(\mathrm{mod}_S)\}$, and let $\mathcal{B}(S)$ be $Z(S)/(\mathrm{Pr}(Tp(K))B(S))$.

THEOREM 2.5 ([**Howe96**, Theorem 1.3]). *Let $\mathcal{C}$ be an isogeny class of abelian varieties over k. Then there exists an element $I \in \mathcal{B}(R)$ such that attainable elements in $G(\mathrm{mod}_R)$ are exactly the effective elements of $Z(R)$ that belong to the class I.*

In particular, $\mathcal{C}$ contains a principally polarized abelian variety if and only if $I = 0$. $\square$

We are going to study attainable primes. The group $Z(R)/B(R)$ is a vector space over $\mathbb{F}_2$, with basis formed by the classes of the form $[R/\mathfrak{p}]_R$, where $\mathfrak{p}$ is the maximal ideal of R stable with respect to involution and such that the degree extension of $R/\mathfrak{p}$ over $R^+/(R^+ \cap \mathfrak{p})$ is two. We call such ideals $\mathfrak{p}$ *generating*, and denote the ideal $R^+ \cap \mathfrak{p}$ by $\mathfrak{p}^+$.

Let M be a finite R-module. We denote by $M_{\mathfrak{p}}$ the localization at a prime ideal $\mathfrak{p} \subset R$. It follows from the Chinese Remainder Theorem [**E95**, Theorem 2.13] that

$$M \cong \oplus_{\mathfrak{p}} M_{\mathfrak{p}},$$

where the sum is taken over all maximal ideals R. The set of prime ideals $\mathfrak{p}$ such that $M_{\mathfrak{p}} \neq 0$ is called *the support* of M.

REMARK 2.6. Since the localization functor is exact, if M and N are equivalent in $G(\mathrm{mod}_R)$, then $M_{\mathfrak{p}}$ and $N_{\mathfrak{p}}$ are also equivalent. Let $[M]_R = \varepsilon([\Delta])$, where Δ is a finite group scheme. Then $[M_{\mathfrak{p}}]_R = \varepsilon([\Delta_{\mathfrak{p}}])$, where $\Delta = \oplus_{\mathfrak{p}} \Delta_{\mathfrak{p}}$.

We now prove that if $\mathfrak{p}$ is not a generating ideal, because $R/\mathfrak{p} \cong R^+/\mathfrak{p}^+$, then it is not attainable.

LEMMA 2.7. *Let Δ be a kernel of polarization on A, and let $\mathfrak{p}$ be a symmetric ideal such that $R/\mathfrak{p} \cong R^+/\mathfrak{p}^+$. Then $\epsilon([\Delta_{\mathfrak{p}}]) \neq [R/\mathfrak{p}]_R$.*

PROOF. If the symmetric ideal $\mathfrak{p}$ has the property $R/\mathfrak{p} \cong R^+/\mathfrak{p}^+$, then the class $[R/\mathfrak{p}]_R$ does not belong to $Z(R)$. Therefore, we have to prove that $\epsilon([\Delta_{\mathfrak{p}}])$ belongs to $Z(R)$.

Let M be an R-module such that $[M]_R = \epsilon([\Delta])$; in particular, $\epsilon([\Delta_{\mathfrak{p}}]) = [M_{\mathfrak{p}}]_R$. According to Theorem 2.5, $N_{R/R^+}(M)$ is equivalent to $N \oplus N$ for some R^+-module N. Since $\mathfrak{p}$ is symmetric, $N_{R/R^+}(M_{\mathfrak{p}}) \cong N_{R/R^+}(M)_{\mathfrak{p}^+}$. By Remark 2.6, $N_{R/R^+}(M)_{\mathfrak{p}^+}$ is equivalent to $N_{\mathfrak{p}^+} \oplus N_{\mathfrak{p}^+}$; therefore, $[M_{\mathfrak{p}}]_R \in Z(R)$. □

We say that an isogeny class $\mathcal{C}$ is *exact*, if the following sequence is exact

$$(*) \qquad Z(\mathcal{O})/B(\mathcal{O}) \xrightarrow{N} Z(R)/B(R) \to \mathcal{B}(R) \to 0,$$

where N is induced by the norm $N_{\mathcal{O}/R} : G(\mathrm{mod}_{\mathcal{O}}) \to G(\mathrm{mod}_R)$. In particular, an element $x \in Z(R)/B(R)$ vanishes in $\mathcal{B}(R)$ if and only if it lies in the image of N.

REMARK 2.8. If the isogeny class is exact, and R is maximal at a generating ideal $\mathfrak{p}$, then $\mathfrak{p}$ is attainable.

THEOREM 2.9. *An isogeny class $\mathcal{C}$ is exact if and only if there exists a principally polarized abelian variety in $\mathcal{C}$.*

An isogeny class $\mathcal{C}$ is exact in the following cases:

(1) $\dim \mathcal{C}$ *is odd;*
(2) $\mathcal{C}$ *is the class of a geometrically simple abelian surface.* □

PROOF. According to [**Howe95**, Proposition 6.4], there is a push-out diagram:

$$\begin{array}{ccc} Z(\mathcal{O})/B(\mathcal{O}) & \longrightarrow & \mathcal{B}(\mathcal{O}) \\ \big\downarrow {\scriptstyle N} & & \big\downarrow {\scriptstyle i^*} \\ Z(R)/B(R) & \longrightarrow & \mathcal{B}(R) \end{array}$$

It follows that the isogeny class $\mathcal{C}$ is exact if and only if i^* is the zero map. According to [**Howe95**, Proposition 7.1], and Theorem 2.5, if i^* is the zero map, then there exists a principally polarized abelian variety in $\mathcal{C}$. On the other hand, by [**Howe95**, Proposition 6.2], $\mathcal{B}(\mathcal{O})$ is either trivial or isomorphic to $\mathbb{Z}/2\mathbb{Z}$. Therefore, if the obstruction element I is trivial, then i^* is the zero map.

The second part of the theorem follows from [**Howe95**, Theorem 1.2], and [**MN02**, Theorem 4.3]. □

2.4. Exceptional primes. In this section, we examine the case where a generating prime is not attainable.

Let $\mathfrak{p} \subset R$ be a maximal ideal over $\ell \in \mathbb{Z}$. Let L be an unramified extension of $\mathbb{Q}_\ell$ with the residue field $R/\mathfrak{p}$. Denote by $\Lambda_{\mathfrak{p}}$ the ring of integers of L. According to [**Ry14**, Proposition 2.5], the localization $R_\ell \cong R \otimes_{\mathbb{Z}} \mathbb{Z}_\ell$ is a $\Lambda_{\mathfrak{p}}$–algebra. In the same way we can start with $\mathfrak{p}^+ = \mathfrak{p} \cap R^+$, and define $\Lambda_{\mathfrak{p}}^+$ as a ring of integers in an unramified extension L^+ of $\mathbb{Q}_\ell$ with residue field $R^+/\mathfrak{p}^+$. Note that if $\mathfrak{p}$ is generating, then $[L : L^+] = 2$. In applications we need only the case $\Lambda_{\mathfrak{p}}^+ = \mathbb{Z}_\ell$.

We say that ℓ is *an exceptional prime* if there exists a prime ideal $\mathfrak{p}_1 \subset \mathcal{O}$ over ℓ such that

(1) $\mathfrak{p} = \mathfrak{p}_1 \cap R$ is a generating ideal;
(2) if $\mathfrak{p}_1^+ = \mathfrak{p}_1 \cap \mathcal{O}^+$, then $\dim(\mathcal{O}^+/\mathfrak{p}_1^+) > \dim(R^+/\mathfrak{p}^+)$;

In this situation, we say that $\mathfrak{p}_1$ is *an exceptional prime ideal.*

THEOREM 2.10. *Let $\mathcal{C}$ be an exact isogeny class, and let $\mathfrak{p}_1 \subset \mathcal{O}$ be a maximal ideal such that $\mathfrak{p} = \mathfrak{p}_1 \cap R$ is generating.*

(1) *If $\dim(\mathcal{O}^+/\mathfrak{p}_1^+ \cap \mathcal{O}^+) = \dim(R^+/\mathfrak{p}^+)$, then the class $[R/\mathfrak{p}]_R$ is attainable.*
(2) *If $\mathfrak{p}_1$ is an exceptional prime ideal, then $\mathfrak{p}_2 = \overline{\mathfrak{p}_1}$ is an exceptional prime ideal such that $\mathfrak{p} = \mathfrak{p}_2 \cap R$. Moreover, $\mathfrak{p}_1 \neq \mathfrak{p}_2$, and $\mathfrak{p}$ is* not *attainable.*

PROOF. Since $\mathfrak{p}$ is symmetric, $\mathfrak{p} \subset \mathfrak{p}_2$. The ideal $\mathfrak{p} \subset R$ is maximal; therefore, $\mathfrak{p} = \mathfrak{p}_2 \cap R$. Since $\mathfrak{p}$ is generating, the image Λ of a natural monomorphism from $\Lambda_{\mathfrak{p}}$ to the localization K_ℓ is not contained in R^+; therefore,

$$\mathcal{O}_{\mathfrak{p}_1} = \mathcal{O}^+_{\mathfrak{p}_1^+} \cdot \Lambda \cong \mathcal{O}^+_{\mathfrak{p}_1^+} \otimes_{\Lambda_{\mathfrak{p}}^+} \Lambda_{\mathfrak{p}}.$$

Assume first that $\dim(\mathcal{O}^+/\mathfrak{p}_1^+) = \dim(R^+/\mathfrak{p}^+)$. Then the maximal unramified subring of $\mathcal{O}^+_{\mathfrak{p}_1^+}$ is equal to the image of $\Lambda_{\mathfrak{p}}^+$. It follows that

$$\mathcal{O}_{\mathfrak{p}_1} \cong \mathcal{O}^+_{\mathfrak{p}_1^+} \otimes_{\Lambda_{\mathfrak{p}}^+} \Lambda_{\mathfrak{p}}$$

is a domain and $\mathfrak{p}_1 = \mathfrak{p}_2 = \mathfrak{p}_1^+\mathcal{O}$ is symmetric.

Since $\mathfrak{p}$ is generating,

$$\dim(R/\mathfrak{p}) = 2\dim(R^+/\mathfrak{p}^+).$$

On the other hand,

$$\dim(\mathcal{O}/\mathfrak{p}_1) \leq 2\dim(\mathcal{O}^+/\mathfrak{p}_1^+) = 2\dim(R^+/\mathfrak{p}^+),$$

and therefore the embedding $R/\mathfrak{p} \to \mathcal{O}/\mathfrak{p}_1\mathcal{O}$ is an isomorphism, that is, $[R/\mathfrak{p}]_R$ lies in the image of the norm, and $\mathfrak{p}$ is attainable.

Assume now that $\dim(\mathcal{O}^+/\mathfrak{p}_1^+) > \dim(R^+/\mathfrak{p}^+)$. Then the maximal unramified subring of $\mathcal{O}^+_{\mathfrak{p}_1^+}$ is greater than the image of $\Lambda_{\mathfrak{p}}^+$; since $[L : L^+] = 2$ we have

$$\mathcal{O}^+_{\mathfrak{p}_1^+} \otimes_{\Lambda_{\mathfrak{p}}^+} \Lambda_{\mathfrak{p}} \cong \mathcal{O}^+_{\mathfrak{p}_1^+} \oplus \mathcal{O}^+_{\mathfrak{p}_1^+}.$$

In this case we get that $\mathfrak{p}_1^+\mathcal{O} = \mathfrak{p}_1\mathfrak{p}_2$, and $\mathfrak{p}_1 \neq \mathfrak{p}_2$.

Clearly, $[\mathcal{O}/\mathfrak{p}_1]_{\mathcal{O}} \notin Z(\mathcal{O})$, and $[R/\mathfrak{p}]_R \notin B(R)$, because $R/\mathfrak{p}$ is a simple R-module. The support of $[R/\mathfrak{p}]_R$ is equal to $\mathfrak{p}$; therefore, if $[R/\mathfrak{p}]_R$ is in the image of the norm, then it is in the image of the submodule of $Z(\mathcal{O})$ generated by $[\mathcal{O}/\mathfrak{p}_1]_{\mathcal{O}} + [\mathcal{O}/\mathfrak{p}_2]_{\mathcal{O}}$. This is nonsense, because the natural morphism $R/\mathfrak{p} \to \mathcal{O}/\mathfrak{p}_1$ is injective, and $\dim(R/\mathfrak{p}) < \dim(\mathcal{O}/\mathfrak{p}_1) + \dim(\mathcal{O}/\mathfrak{p}_2)$. The theorem is proved. □

3. Polarizations on the product of two abelian varieties

Recall the Serre construction of the principal polarization on a product of two polarized abelian varieties [**Lau02**]. Let A_1 and A_2 be abelian varieties with polarizations L_1 and L_2 of degree N^2. Assume that we are given an isomorphism $\psi : \operatorname{Ker} L_1 \to \operatorname{Ker} L_2$ with the following property:

$$e_{L_2}(\psi(x), \psi(y)) = e_{L_1}(x, y)^{-1},$$

in other words, ψ is an anti-isometry. Denote by Δ the image of the mapping

$$Id \times \psi : \operatorname{Ker} L_1 \to \operatorname{Ker} L_1 \times \operatorname{Ker} L_2.$$

On $J = (A_1 \times A_2)/\Delta$ one get a principal polarization as follows. The kernel of the product polarization on $A_1 \times A_2$ is $\operatorname{Ker} L_1 \times \operatorname{Ker} L_2$. The restriction of the product form $e_{L_1} \times e_{L_2}$ to Δ is trivial. According to [**Mum71**, §23,Theorem 2], the polarization on $A_1 \times A_2$ descends to a principal polarization M on J. This construction is called *the gluing of polarizations.*

We say that the principal polarization L on J is *reducible* if there exist abelian subvarieties J_1 and J_2 with polarizations L_1 and L_2 such that $J \cong J_1 \times J_2$ and L is the product of polarizations L_1 and L_2. Otherwise, we call the polarization *irreducible.*

Let E be the set of integers e such that for any finite group scheme Δ over k that can be embedded in a variety isogenous to A_1 and in a variety isogenous to A_2 we have $e\Delta = 0$. By definition, *the gluing exponent* $e(A_1, A_2)$ is the greatest common divisor of the set E [**HL12**]. Note that $e(A_1, A_2)$ is finite if and only if there is no abelian variety with a nonzero morphism to both A_1 and A_2.

LEMMA 3.1. *Let A_1 and A_2 be simple abelian varieties, and let there exist polarizations L_1 on A_1 and L_2 on A_2 of degree $N^2 > 1$ and an anti-isometry $\operatorname{Ker} L_1 \to \operatorname{Ker} L_2$. If $e(A_1, A_2)$ is finite, then the gluing of L_1 and L_2 is irreducible.*

PROOF. Assume that the gluing of the polarizations $\alpha : A_1 \times A_2 \to J$ is reducible: $J \cong J_1 \times J_2$. Let the composition $\alpha : A_1 \times A_2 \to J_1 \times J_2$ be given by the matrix $\alpha_{ij} : A_i \to J_j$. Since A_1 and A_2 are simple and not isogenous, α splits into the product of two isogenies, say $\alpha_{11} : A_1 \to J_1$, and $\alpha_{22} : A_2 \to J_2$. Therefore, the kernel of α is the direct sum of the kernels of α_{11} and α_{22}, but this contradicts the fact that α induces embeddings $A_1 \to J$, and $A_2 \to J$. □

We will compute $e(A_1, A_2)$ in some cases. For this, we need a lemma.

LEMMA 3.2. *Let $f(t)$ be a separable Weil polynomial with a real Weil polynomial $h(t)$, and let Λ be a ring. For any $r \in \Lambda$ there exists a polynomial $\lambda(t) \in \Lambda[t]$ such that*

$$f(t) = \lambda(t)(t^2 - rt + q) + t^{\deg h} h(r).$$

PROOF. The lemma follows from the equality:

$$\begin{aligned} f(t) &= \prod_x (t - x)(t - q/x) = \prod_x (t^2 - b_x t + q) = \prod_x ((t^2 - rt + q) + (r - b_x)t) \\ &= \lambda(t)(t^2 - rt + q) + t^{\deg h} \prod_x (r - b_x) = \lambda(t)(t^2 - rt + q) + t^{\deg h} h(r), \end{aligned}$$

where $b_x = x + q/x$, and $h(t) = \prod_x (t - b_x)$. □

PROPOSITION 3.3. *Let A be a simple abelian variety with real Weil polynomial $h(t)$, and let B be an elliptic curve with real Weil polynomial $h_B(t) = t-b$. Suppose that both f_A and f_B are separable. Then there exists a natural number r such that $e(A,B) = |h(b)|/p^r$. Moreover, if A is ordinary and B is supersingular, then $e(A,B) = |h(b)|$.*

PROOF. There exists a polynomial $a(t) \in \mathbb{Z}[t]$ such that

$$h(t) = a(t)(t-b) + h(b).$$

Therefore, the reduced resultant of $h(t)$ and $h_B(t)$ is equal to $h(b)$ [**HL12**, Lemma 2.7]. According to [**HL12**, Proposition 2.8], $e(A,B)$ divides $h(b)$. Hence, it is enough to prove that for every prime $\ell \neq p$, if ℓ^n divides $h(b)$, then there exists an ℓ-primary group scheme Δ such that $\ell^{n-1}\Delta \neq 0$, and there are homomorphisms $\Delta \to A'$ and $\Delta \to B'$, where A' is isogenous to A and B' is isogenous to B.

By the Lemma 2.2, there exists an elliptic curve B' such that

$$T_\ell(B') \cong \mathbb{Z}_\ell[t]/f_B(t)\mathbb{Z}_\ell[t],$$

where F_B acts on the right as multiplication by t. Put $\Delta = B'[\ell^n]$. To prove the proposition we now have to construct a submodule $T \subset V_\ell(A)$ such that $\mathcal{P}(\Delta)$ is a submodule of $T/\ell^n T$, and T generates $V_\ell(A)$; then, according to Lemma 2.2, there exists an abelian variety A' such that $T_\ell(A') \cong T$, and Δ is isomorphic to a submodule of $A'[\ell^n]$.

Define a sequence of polynomials

$$f_r(t) = t^r - \alpha_r t + \beta_r \in \mathbb{Z}[t]$$

for $r \geq 2$ as follows: $f_2 = f_B$, and

$$f_{r+1}(t) = tf_r(t) + \alpha_r f_B(t).$$

Clearly, $f_r(F_B) = 0$ for all $r \geq 2$. Since f_A is separable, there exists a cyclic vector $v \in V_\ell(A)$, i.e., $v, Fv, \ldots, F^{d-1}v$ is a basis, where $d = \deg f_A$. Let $v_0 = v, v_1 = Fv$, and

$$v_r = \frac{1}{\ell^n} f_r(F)v$$

for $d > r \geq 2$. According to Lemma 3.2 we have:

$$f_A(t) \equiv \lambda(t) f_B(t) \bmod \ell^n$$

for some $\lambda(t) \in \mathbb{Z}[t]$; using this relation it is straightforward to check that $v_0, \ldots, v_{d-1}$ is a basis of an F-invariant submodule $T \subset V_\ell(A)$. Moreover, by induction we get that

$$f_r(F)v \equiv 0 \bmod \ell^n.$$

In other words, v and Fv generate a submodule of $T/\ell^n T$ isomorphic to $\mathcal{P}(\Delta)$.

If A is ordinary, then $h(0)$ is a p-adic unit. Moreover, if B is supersingular, then p divides b; therefore, $h(b) \equiv h(0) \bmod p$, and $v_p(h(b)) = v_p(e(A,B)) = 0$. It follows that $e(A,B) = |h(b)|$. □

There is a simple way to prove that polarizations on two abelian varieties A and B cannot be glued together. Let ℓ be a prime divisor of $e(A,B)$. If B is supersingular, and all the roots of f_B are real, then $B[\ell]$ is semi-simple. If, on the other hand, any kernel of a polarization on A annihilated by $e(A,B)$ is not semi-simple, then there is no gluing of polarizations on A and B. The following proposition allows one to prove that the kernel of a polarization is not semi-simple.

PROPOSITION 3.4. *Let A be an abelian variety with polarization L, and let Δ be the kernel of L. Assume that the Weil polynomial f_A of A is separable. Let $\mathfrak{p}$ be a symmetric maximal ideal of R_A such that*

(1) $p \notin \mathfrak{p}$;
(2) R_A *is maximal at* $\mathfrak{p}$;
(3) $\mathfrak{p}^+$ *is ramified in the extension* K/K^+.

If the support of Δ contains $\mathfrak{p}$, then $\mathcal{P}(\Delta_\mathfrak{p}) \cong R_A/\mathfrak{p}^a$, where $a = \mathrm{length}(\Delta_\mathfrak{p}) > 1$. In particular, $\Delta_\mathfrak{p}$ is not semi-simple and $(\mathfrak{p}^+)^{\lceil a/2 \rceil}\Delta_\mathfrak{p} = 0$.

PROOF. According to Lemma 3.5, $\mathcal{P}(\Delta_\mathfrak{p}) \cong R_A/\mathfrak{p}^a$, where $a = \mathrm{length}(\Delta_\mathfrak{p})$. Since $\mathfrak{p}^2 = \mathfrak{p}^+\mathcal{O}$, we have

$$(\mathfrak{p}^+)^{\lceil a/2 \rceil}\Delta_\mathfrak{p} \subset \mathfrak{p}^a\Delta_\mathfrak{p} = 0.$$

Since $\mathfrak{p}^+$ is ramified, $R/\mathfrak{p} \cong R^+/\mathfrak{p}^+$. It follows from Lemma 2.7 that $\mathcal{P}(\Delta_\mathfrak{p}) \not\cong R_A/\mathfrak{p}$. Therefore, if the support of Δ contains $\mathfrak{p}$, then the multiplicity of $[R_A/\mathfrak{p}]_{R_A}$ in $\varepsilon([\Delta])$ is at least two, that is, $a > 1$. The proposition follows. □

LEMMA 3.5. *Let A be an abelian variety with a finite subgroup scheme Δ. Assume that the Weil polynomial f_A of A is separable. Let $\mathfrak{p}$ be a maximal ideal of R_A such that $p \notin \mathfrak{p}$ and the order R_A is maximal at $\mathfrak{p}$. If the support of Δ contains $\mathfrak{p}$, then $\mathcal{P}(\Delta_\mathfrak{p}) \cong R_A/\mathfrak{p}^a$, where $a = \mathrm{length}(\Delta_\mathfrak{p})$.*

PROOF. The group scheme Δ_ℓ corresponds to a quotient of the Tate module $M = T_\ell(A)/T_M$. Since the Weil polynomial f_A of A is square-free, the module $V_\ell(A)$ is free of rank 1 over $R_A \otimes \mathbb{Q}$. Since the order of R_A is maximal at $\mathfrak{p}$, we find that $T_\ell(A)_\mathfrak{p}$ is free of rank 1 over $R_\mathfrak{p} = (R_A)_\mathfrak{p}$. Therefore, the module $M_\mathfrak{p}$ is isomorphic to the quotient of $R_\mathfrak{p}$ by some power of the maximal ideal: $M_\mathfrak{p} \cong R_\mathfrak{p}/\mathfrak{p}^a$. □

As an application, we reprove a result E.Howe and K.Lauter [**HL12**]. We use the following particular case of the Dedekind criterion.

THEOREM 3.6 (Dedekind). *Let $f \in \mathbb{Z}_\ell[t]$ be a polynomial such that*

$$f(t) = g(t)^r + \ell\alpha(t),$$

where $g(t) \in \mathbb{Z}_\ell[t]$ is irreducible modulo ℓ, and $\alpha \in \mathbb{Z}_\ell[t]$. Then the ring $\Lambda = \mathbb{Z}_\ell[t]/f(t)\mathbb{Z}_\ell[t]$ is regular if and only if either $r = 1$, or $r > 1$ and the polynomials $\alpha(t)$ and $g(t)$ are coprime modulo ℓ.

PROOF. Follows from [**C93**, Theorem 6.1.2]. □

THEOREM 3.7 ([**HL12**, Theorem 3.1]). *Suppose that q is a square. Let A be an ordinary abelian variety with a real Weil polynomial $h(t)$, and let B be a power of a supersingular elliptic curve with a real Weil polynomial $(t - 2s)^n$, where $s^2 = q$, and $n > 0$. If the number $h(2s)$ is squarefree, then there is no variety in the isogeny class of $A \times B$ with irreducible principal polarization. In particular, there is no curve over k with the real Weil polynomial equal to $h(t)(t - 2s)^n$.*

PROOF. Assume that there is an irreducible principal polarization on a variety J in the isogeny class of $A \times B$. According to [**HL12**, Lemma 2.3] there exist abelian varieties A' and B' in the isogeny class of A and B respectively, and a non-trivial group scheme Δ with monomorphisms to A' and B'. Moreover, Δ is isomorphic to the kernel of a polarization on A'.

Since Δ is a subgroup scheme of B', the Frobenius action on Δ is constant. We are going to apply Proposition 3.4 to A' and show that the Frobenius action on Δ is non-constant. This contradiction proves the theorem.

Let ℓ be a prime divisor of $h(2s)$. Since A is ordinary, $\ell \neq p$, and since $h(2s)$ is squarefree, f_A is separable. Denote by $\mathfrak{p} \subset R_A$ the ideal generated by $F - s, V - s$, and ℓ. Clearly, $R_A/\mathfrak{p} \cong \mathbb{F}_\ell$; therefore, $\mathfrak{p}$ is prime and symmetric. The localization $R_\mathfrak{p}$ of R_A at $\mathfrak{p}$ is isomorphic to the quotient $\mathbb{Z}_\ell[t]/f_\mathfrak{p}$, where

$$f_\mathfrak{p} - (t-s)^{2r} = \ell\alpha(t)$$

for some $\alpha(t) \in \mathbb{Z}_\ell[t]$. According to the Dedekind criterion, $R_\mathfrak{p}$ is maximal if and only if $(t-s)$ is coprime to $\alpha(t)$ modulo ℓ. This is equivalent to

$$\alpha(s) \not\equiv 0 \bmod \ell, \text{ and } f_\mathfrak{p}(s) \not\equiv 0 \bmod \ell^2.$$

Apply Lemma 3.2 with $r = 2s$. We get that $f_A(s) = s^{\dim A} h(2s)$. Since $h(2s)$ is squarefree, $f_\mathfrak{p}(s) \not\equiv 0 \bmod \ell^2$, and $R_\mathfrak{p}$ is maximal. It is now clear, that $\mathfrak{p}^+$ is ramified. According to Proposition 3.4, the Frobenius action on $\Delta(\bar{k})$ is non-constant. The theorem is proved. □

REMARK 3.8. The separability of f_B is crucial for Proposition 3.3. In fact, from the proof of Theorem 3.7 it follows that $e(A, B) \neq |h(b)|$, where $b = 2s$. One can use [**Ry14**] and prove that if f_A is separable, and $f_B = (t-s)^2$, where $s^2 = q$, then for any prime $\ell \neq p$ we have: ℓ^n divides $e(A, B)$ if and only if ℓ^{2n} divides $f(s)$, and ℓ^n divides $f'(s)$, where f' is the derivative of f.

We now apply the same technique to a product of a supersingular and a non-supersingular surface. Let $\ell \neq 2, p$ be a prime, and let $s = \pm\sqrt{q}$. The ring $\Lambda_\ell = \mathbb{Z}_\ell[t]/(t^2 - q)\mathbb{Z}_\ell[t]$ is either $\mathbb{Z}_\ell^2$ or the ring of integers in an unramified extension of $\mathbb{Q}_\ell$. We say that ℓ divides $h(2s)$ if ℓ divides $h(2s)$ in Λ_ℓ.

THEOREM 3.9. *Suppose that q is not a square. Let A be an ordinary abelian surface with the real Weil polynomial $h(t) = t^2 + a_1 t + a_2 - 2q$, and let B be a supersingular simple abelian surface with the Weil polynomial $(t^2 - q)^2$. Assume that if ℓ is any prime divisor of $h(2s)$, then ℓ is even, and ℓ^2 does not divide $h(2s)$. Then there is no variety in the isogeny class of $A \times B$ with irreducible principal polarization. In particular, there is no curve over k with the Weil polynomial equal to $f_A(t)(t^2 - q)^2$.*

PROOF. Assume that there exists a variety in the isogeny class of $A \times B$ with irreducible principal polarization. As in the proof of Theorem 3.7, there exist A' and B' in the isogeny class of A and B, respectively, and a kernel Δ of a polarization on A' such that there exists a monomorphism $\Delta \to B'$. Clearly, Δ is semi-simple, because, the polynomial $t^2 - q$ is separable modulo ℓ. We now deduce from Proposition 3.4 that Δ is not semi-simple.

Let ℓ be an odd prime divisor of $h(2s)$. Since A is ordinary, $\ell \neq p$, and since $h(2s)$ is squarefree, f_A is separable. Let $\pi_1, \pi_2, q/\pi_1, q/\pi_2$ be the roots of f_A, and let $b_i = \pi_i + q/\pi_i$, where $i = 1$, or $i = 2$. Since ℓ divides $h(2s)$ we have

$$b_1 b_2 = a_2 - 2q \equiv -4q \bmod \ell,$$

and ℓ divides $b_1 + b_2 = -a_1$. It follows that

$$f(t) = (t^2 - b_1 t + q)(t^2 - b_2 t + q) = (t^2 + q)^2 + a_1 t(t^2 + q) + b_1 b_2 t^2 \equiv (t^2 - q)^2 \bmod \ell.$$

Assume first that the polynomial $t^2 - q$ is irreducible modulo ℓ. In this case K^+ is inert in ℓ, and $\mathfrak{p}^+ = \ell\mathcal{O}^+$ is ramified in the extension K/K^+, i.e., $\ell\mathcal{O} = \mathfrak{p}^2$. We are going to prove that the order R_A is maximal at ℓ. According to the Dedekind criterion, we have to show that $(t^2 - q)$ is coprime modulo ℓ to

$$\frac{1}{\ell}(f_A(t) - (t^2 - q)^2) = \frac{t}{\ell}(a_1 t^2 + (a_2 + 2q)t + qa_1).$$

This easily follows from the assumption that ℓ^2 does not divide $h(2s)$.

Assume now that $s \in \mathbb{Z}_\ell$, and $t^2 - q \equiv (t - s)(t + s) \bmod \ell$. Then $f(t) = f_+(t)f_-(t)$ is a product of monic polynomials such that $f_\pm(t) \equiv (t \pm s)^2 \bmod \ell$. According to the Dedekind criterion, R_A is maximal if and only if $f_\pm(s) \not\equiv 0 \bmod \ell^2$; and the last assertion is true, because ℓ^2 does not divide $h(2s)$. □

4. Principal polarizations on abelian threefolds

In this section we assume that A is a geometrically simple abelian surface; since supersingular abelian surfaces are not geometrically simple, A is either an ordinary surface or a mixed one, so $K = \mathrm{End}^\circ(A)$ is a CM-field. According to Proposition 2.9, the isogeny class of A is exact, and we can assume that there is a principal polarization L on A. Let B be an elliptic curve with commutative endomorphism algebra $K_B = \mathrm{End}^\circ(B)$. Denote by Δ_B the discriminant of K_B. Let $R = R_A$ and R_B be the orders generated by the Frobenius and Verschiebung endomorphisms in $\mathrm{End}^\circ(A)$ and $\mathrm{End}^\circ(B)$ respectively. It is natural to claim that if there exists a divisor ℓ of $e(A, B)$, then there is a gluing of polarizations on A and B. Thanks to Proposition 3.3, we can use $h(b)$ instead of $e(A, B)$. Firstly, we examine the case of an exceptional divisor ℓ of $h(b)$.

LEMMA 4.1.

(1) *A prime ℓ is exceptional if and only if*
  - $f_A(t) \equiv f(t)^2 \bmod \ell^2$, *where $f \in \mathbb{Z}_\ell[t]$ is irreducible modulo ℓ;*
  - *ℓ is inert in K^+.*

(2) *If $f_A(t) \equiv f(t)^2 \bmod \ell$, and $f(t)$ is irreducible modulo ℓ, then $[A[\ell]]_R = 2[X]_R$, where $X(\bar{k})$ is a two-dimensional vector space over $\mathbb{F}_\ell$, and F acts on $X(\bar{k})$ with characteristic polynomial f. Moreover, either $[X]_R$ is attainable, or ℓ is exceptional.*

(3) *If ℓ is exceptional, then ℓ^2 divides $a_1^2 - 4a_2 + 8q$.*

PROOF. It is straightforward to check that if

$$f_A(t) \equiv f(t)^2 \bmod \ell,$$

then ℓ divides the discriminant of the real Weil polynomial $a_1^2 - 4a_2 + 8q$; therefore, $\dim(R^+/\mathfrak{p}^+) = 1$. If ℓ is split or ramified in K^+, then Theorem 2.10.(1) shows that $[X]_R$ is attainable. If ℓ is inert in K^+, then

$$\dim(\mathcal{O}^+/\mathfrak{p}_1^+) = 2 > \dim(R^+/\mathfrak{p}^+),$$

i.e., ℓ is exceptional. This proves (2), and the "if" part of (1).

If ℓ is exceptional, then, by definition,

$$\dim(\mathcal{O}^+/\mathfrak{p}_1^+) > \dim(R^+/\mathfrak{p}^+),$$

i.e., $R_\ell^+ \neq \mathcal{O}_\ell^+$; therefore, ℓ divides the discriminant of the real Weil polynomial $a_1^2 - 4a_2 + 8q$. On the other hand, the same inequality shows that $\dim(\mathcal{O}^+/\mathfrak{p}_1^+) = 2$, i.e., ℓ is inert in K^+. It follows that ℓ^2 divides $a_1^2 - 4a_2 + 8q$, and there exists $s \in \mathbb{Z}_\ell$ such that

$$h(t) \equiv (t - s)^2 \bmod \ell^2.$$

In particular, $a_1 \equiv -2s \bmod \ell^2$. Put $f(t) = t^2 - st + q$. We have

$$\begin{aligned} f_A(t) &= (t-\pi_1)(t-q/\pi_1)(t-\pi_2)(t-q/\pi_2) \\ &= (t^2 - b_1 t + q)(t^2 - b_2 t + q) \\ &= (f(t) + (s-b_1)t)(f(t) + (s-b_2)t) \\ &= f(t)^2 + tf(t)(2s + a_1) + t^2 h(s) \equiv f(t)^2 \bmod \ell^2. \end{aligned}$$

Part (1) is proved. □

REMARK 4.2. If ℓ is exceptional, then, according to Theorem 2.10, the group scheme X from the previous lemma is not attainable. Therefore, for any isogeny $A' \to A$ and any polarization L on A'

$$\ker L \not\cong B[\ell].$$

This observation does not give an obstruction to gluing of polarizations. Indeed, assume that ℓ divides $h_A(b)$. According to Lemma 4.1, ℓ^2 divides the discriminant of the real Weil polynomial $a_1^2 - 4a_2 + 8q$, and we have

$$4h(b) = 4b^2 + 4a_1 b + 4a_2 - 8q \equiv 4b^2 + 4a_1 b + a_1^2 \equiv (2b + a_1)^2 \bmod 4\ell^2.$$

Therefore, ℓ^2 divides $h(b)$, and $f_A(t) \equiv f_B(t)^2 \bmod \ell^2$. Moreover, the proposition below shows that $2[X]_R$ is attainable.

PROPOSITION 4.3. *Let A be a geometrically simple ordinary abelian surface with a principal polarization, and let ℓ be an exceptional prime. Then*

$$f_A(t) \equiv f(t)^2 \bmod \ell^2$$

for some $f \in \mathbb{Z}_\ell[t]$, and there is an isogeny $A' \to A$ of degree ℓ, and a polarization L on A' such that

$$\mathcal{P}(\ker L) \cong \Lambda/\ell^2\Lambda,$$

where $\Lambda \cong \mathbb{Z}_\ell[t]/f\mathbb{Z}_\ell[t]$.

PROOF. There exists a morphism $\sigma : T_\ell(A) \to K_\ell$ such that the Weil pairing on $T_\ell(A)$ is given by

$$e_A(x,y) = \mathrm{tr}_{K_\ell/\mathbb{Q}_\ell}(\xi'\sigma(x)\overline{\sigma(y)})$$

for some $\xi' \in K$ such that $\overline{\xi'} = -\xi'$ [**Howe95**, Section 6].

Since ℓ is exceptional, $R_\ell^+ \neq \ell\mathcal{O}_\ell^+$; thus, the image Λ of a natural monomorphism from $\Lambda_\mathfrak{p}$ to the localization K_ℓ is not contained in R_ℓ^+. Moreover, the local rings $\mathcal{O}_\ell^+$ and Λ are isomorphic; therefore,

$$\mathcal{O}_\ell \cong \Lambda \otimes \Lambda \cong \Lambda \oplus \Lambda$$

as Λ-modules, and conjugation acts componentwise:

$$\overline{(x_1, x_2)} = (\bar{x}_1, \bar{x}_2) \in \Lambda \oplus \Lambda.$$

Let $\Lambda \cong \mathbb{Z}_\ell[z]$, where $z^2 \in \mathbb{Z}_\ell$, and $\bar{z} = -z$. Then $\xi'/z \in \mathcal{O}_\ell^+$, and the image of ξ' in $\Lambda \oplus \Lambda$ is $(z\xi, z\bar{\xi})$ for some $\xi \in \Lambda$.

It follows that the Weil pairing corresponds to the pairing

$$(*) \qquad e((x_1,x_2),(y_1,y_2)) = \mathrm{tr}_{L/\mathbb{Q}_\ell}(z\xi x_1 \bar{y}_1) + \mathrm{tr}_{L/\mathbb{Q}_\ell}(z\bar{\xi} x_2 \bar{y}_2)$$

on $\Lambda \oplus \Lambda$, where without loss of generality we can assume that $\xi \in \Lambda$ is a unit, and e is perfect on $\mathcal{O}_\ell$.

The induced pairing on the Λ-submodule $T = \sigma(T_\ell(A))$ of K_ℓ is perfect. Assume that $T \subset \mathcal{O}_\ell$. We claim that in this case $T = \mathcal{O}_\ell$. Indeed, if $T \subset \ell\mathcal{O}_\ell$, then clearly the pairing on T is not perfect; therefore, the Λ-module $M = \mathcal{O}_\ell/T$ is cyclic. Assume that $M \neq 0$. Let $v \in \mathcal{O}_\ell$ generate M; hence, there exists a minimal natural $r > 0$ such that $\ell^r v \in T$. Since the pairing on T is perfect, there exists $u \in T \subset \mathcal{O}_\ell$ such that $e(\ell^r v, u) \in \mathbb{Z}_\ell^*$. It follows that $e(v, u) \notin \mathbb{Z}_\ell$. Nonsense.

The inequality $\dim(\mathcal{O}^+/\mathfrak{p}_1^+) > \dim(R^+/\mathfrak{p}^+)$ shows that $R_\ell \neq \mathcal{O}_\ell$, and thus there exists a natural m such that

$$R_\ell = \Lambda + \ell^m \mathcal{O}_\ell.$$

It follows that $T' = \Lambda \oplus \ell\Lambda \subset \mathcal{O}_\ell$ is an R-submodule of T. Clearly, the kernel of e on T' is isomorphic to $\Lambda/\ell^2\Lambda$.

Assume now that $T \not\subset \mathcal{O}_\ell$. The Λ-module $M = T/(T \cap \mathcal{O}_\ell)$ is cyclic, because otherwise the pairing on T is not integral. Let $v \in T$ generate M over Λ, and let $u \in T \cap \mathcal{O}_\ell$ be a second basis element such that $e(v, u) \in \mathbb{Z}_\ell^*$. It follows that $T' = \ell\Lambda v \oplus \Lambda u$ is an R-submodule of T such that the kernel of the restriction of e to T' is isomorphic to $\Lambda/\ell^2\Lambda$.

According to Lemma 2.2, there exists an abelian surface A' and an isogeny $A' \to A$ such that the kernel of the induced polarization L on A' is isomorphic to $\Lambda/\ell^2\Lambda$. Now, Lemma 4.1 completes the proof. □

REMARK 4.4. The proof of Proposition 4.3 is based on the Deligne equivalence theorem, because it is used in Section 6 of [**Howe95**]. I think the proposition has to be true for any simple abelian surface with commutative endomorphism algebra.

LEMMA 4.5 ([**HNR06**, Lemma 4.2]). *Let ℓ be a prime number and K_1 be an imaginary quadratic field whose discriminant is not equal to $-\ell$. Then there are infinitely many primes r' that split in K_1 and are not squares modulo ℓ^n, where $n > 1$, if $\ell = 2$.*

We need the following generalization of [**HNR06**, Lemma 4.3].

LEMMA 4.6. *Let B be an elliptic curve. Suppose that there exists a surface A with a polarization L, and an isomorphism $\psi : \ker L \to B[\ell^n]$, where $n > 1$. If $\Delta_B \neq -\ell$, then there exists a curve B' in the isogeny class of B, and an anti-isometry $B'[\ell^n] \to \ker L$.*

PROOF. Let e_B and e_X be Weil pairings on $B[\ell^n]$ and $X = \ker L$, respectively. Then there exists r such that the following diagram is commutative:

$$\begin{array}{ccc} X \times X & \longrightarrow & B[\ell^n] \times B[\ell^n] \\ \downarrow{\scriptstyle e_X} & & \downarrow{\scriptstyle e_B} \\ G_m & \xrightarrow[r]{} & G_m \end{array}$$

Take a composition of ψ with an isogeny $\varphi : B \to B'$ of degree r', then the number r is multiplied by r'. If $-r$ is a square modulo ℓ^n, then it suffices to multiply ψ by r' such that $r(r')^2 \equiv -1 \bmod \ell$. Then $r'\psi$ is an anti-isometry.

Suppose that $-r$ is not a square modulo ℓ^n. If $\Delta_B \neq -\ell$, then, by Lemma 4.5, one can find r', which is not a square modulo ℓ^n and splits in K_1, hence $B[r'] \cong X_1 \oplus X_2$ splits, and there exists an isogeny $\varphi : B \to B' = B/X_1$ of degree r'. There are infinitely many such m by the Dirichlet theorem on primes in arithmetic progressions; as before, for such r' the number $-rr'$ is a square modulo ℓ. The lemma is proven. □

PROOF OF THEOREM 1.1. Suppose that $\ell \neq p$ divides $h(b)$. According to Lemma 3.2, $f_B(t)$ divides $f_A(t)$ modulo ℓ^n if and only if ℓ^n divides $h(b)$. We are going to prove that under conditions of the theorem there exists a gluing of polarizations on A and B.

If $f_B(t)$ is irreducible modulo ℓ, then $[A[\ell]]_R = [X]_R + [X']_R$, where $X \cong B[\ell]$. If $f_A(t) \not\equiv f_B(t)^2 \bmod \ell$, then $f_A \equiv f_B f' \bmod \ell$ is a product of two coprime monic polynomials modulo ℓ; therefore, $[X]_R \neq [X']_R$, and, by Hensel's lemma, $f_A = f_1 f_2$, where $f_1, f_2 \in \mathbb{Z}_\ell[t]$ are coprime and monic, and $f_1 \equiv f_B \bmod \ell$. According to the Chinese Remainder Theorem,

$$R \cong \mathbb{Z}_\ell[t]/f_1\mathbb{Z}_\ell[t] \oplus \mathbb{Z}_\ell[t]/f_2\mathbb{Z}_\ell[t],$$

and, by Theorem 3.6, the order $R_1 = \mathbb{Z}_\ell[t]/f_1\mathbb{Z}_\ell[t]$ is maximal. By Remark 2.8, $[X]_R$ is attainable. According to Lemma 4.1, if $f_A(t) \equiv f_B(t)^2 \bmod \ell$, then $[X]_R$ is either attainable, or ℓ is exceptional.

Assume that ℓ is exceptional. By Remark 4.2, ℓ^2 divides $h(b)$, and $f_A(t) \equiv f_B(t)^2 \bmod \ell^2$. According to Proposition 4.3, A is isogenous to an abelian surface A' with a polarization L such that $\ker L \cong B[\ell^2]$.

Let us now consider the case when the polynomial $f_B(t) \equiv (t-t_1)(t-t_2) \bmod \ell$ is reducible modulo ℓ. Then

$$[B[\ell]]_{R_B} = [Y_1]_{R_B} + [Y_2]_{R_B},$$

where both $Y_1(\bar{k}) \cong Y_2(\bar{k}) \cong \mathbb{F}_\ell$ and F acts on Y_i as multiplication by t_i. Therefore,

$$[A[\ell]]_R = [Y_1]_R + [Y_2]_R + [X']_R.$$

Clearly, $[Y_1]_R = \overline{[Y_2]_R}$, and from Theorem 2.5 it follows that A is isogenous to a surface A' with a polarization L, such that $[\ker L]_R = [Y_1]_R + [Y_2]_R$. If $t_1 \not\equiv t_2 \bmod \ell$, then

$$\ker L \cong Y_1 \oplus Y_2 = B[\ell].$$

Suppose that $t_1 \equiv t_2 \bmod \ell$. The Frobenius action on $\ker L(\bar{k})$ is either a multiplication by t_1, or is given by a non-trivial Jordan cell with eigenvalue t_1. Let S_ℓ be the localization of R_B at ℓ. By assumption, ℓ^2 divides $f_B(t_1)$, and, according to the Dedekind criterion, the order S_ℓ is not maximal. Let $\mathcal{O}_\ell$ be the maximal order in $S_\ell \otimes \mathbb{Q}_\ell$; then the Frobenius action on $\mathcal{O}_\ell/\ell\mathcal{O}_\ell$ is the multiplication by t_1, and on $S_\ell/\ell S_\ell$ is given by a non-trivial Jordan cell with eigenvalue t_1. By Lemma 2.2, there exist elliptic curves B_1 and B_2 such that $T_\ell(B_1) \cong S_\ell$, and $T_\ell(B_2) \cong \mathcal{O}_\ell$, as R_B-modules. Therefore, either $\ker L \cong B_1[\ell]$, or $\ker L \cong B_2[\ell]$.

Consider the case when p divides $h(b)$. Suppose that B is ordinary. Then $B[p] \cong X_r \oplus X_l$, where $X_r \otimes \bar{k} \cong \mathbb{Z}/p\mathbb{Z}$, and $X_l \otimes \bar{k} \cong \mu_p$, since B is not supersingular [**Wa69**]. The Weil pairing is non-degenerate; therefore, we have an isomorphism of X_r with its Cartier dual $D(X_l)$. We get the equality $[X_r]_{R_B} = \overline{[X_l]_{R_B}}$. By Theorem 2.5, the element $[X_r]_R + [X_l]_R$ is an attainable group subscheme of A. Suppose now, that B is supersingular. Since p divides both b and $h(b)$, the surface A is mixed. Thus, there exists a monomorphism $B[p] \to A$. Moreover, $[B[p]] = 2[\alpha_p]$ is an attainable group subscheme of A.

We have proved that there is an abelian surface A' with a polarization L and an elliptic curve B' such that $\ker L \cong B'[\ell^n]$, where $n = 1$ or $n = 2$. By Lemma 4.6, one can choose B' in its isogeny class in such a way that there is an anti-isometry $B'[\ell^n] \cong \ker L$. Let L_B be the polarization on B with kernel $B[\ell^n]$. Now, by Lemma 3.1, the gluing of L and L_B gives an irreducible principal polarization on a variety in the isogeny class of $A \times B$. The theorem is proved. □

Acknowledgments

The author thanks the anonymous reviewer for many useful remarks on the paper. The author is grateful to the Weizmann Institute Emergency Program to Host Visiting Scientists Affected by the War in Ukraine.

References

[C93] Henri Cohen, *A course in computational algebraic number theory*, Graduate Texts in Mathematics, vol. 138, Springer-Verlag, Berlin, 1993, DOI 10.1007/978-3-662-02945-9. MR1228206

[E95] David Eisenbud, *Commutative algebra: With a view toward algebraic geometry*, Graduate Texts in Mathematics, vol. 150, Springer-Verlag, New York, 1995, DOI 10.1007/978-1-4612-5350-1. MR1322960

[Howe95] Everett W. Howe, *Principally polarized ordinary abelian varieties over finite fields*, Trans. Amer. Math. Soc. **347** (1995), no. 7, 2361–2401, DOI 10.2307/2154828. MR1297531

[Howe96] Everett W. Howe, *Kernels of polarizations of abelian varieties over finite fields*, J. Algebraic Geom. **5** (1996), no. 3, 583–608. MR1382738

[HL12] Everett W. Howe and Kristin E. Lauter, *New methods for bounding the number of points on curves over finite fields*, Geometry and arithmetic, EMS Ser. Congr. Rep., Eur. Math. Soc., Zürich, 2012, pp. 173–212, DOI 10.4171/119-1/12. MR2987661

[HNR06] Everett W. Howe, Enric Nart, and Christophe Ritzenthaler, *Jacobians in isogeny classes of abelian surfaces over finite fields* (English, with English and French summaries), Ann. Inst. Fourier (Grenoble) **59** (2009), no. 1, 239–289, DOI 10.5802/aif.2430. MR2514865

[Lau01] Kristin Lauter, *Geometric methods for improving the upper bounds on the number of rational points on algebraic curves over finite fields*, J. Algebraic Geom. **10** (2001), no. 1, 19–36. With an appendix in French by J.-P. Serre. MR1795548

[Lau02] Kristin Lauter, *The maximum or minimum number of rational points on genus three curves over finite fields*, Compositio Math. **134** (2002), no. 1, 87–111, DOI 10.1023/A:1020246226326. With an appendix by Jean-Pierre Serre. MR1931964

[MN02] Daniel Maisner and Enric Nart, *Abelian surfaces over finite fields as Jacobians*, Experiment. Math. **11** (2002), no. 3, 321–337. With an appendix by Everett W. Howe. MR1959745

[Mil08] J. Milne. *Abelian varieties.* 2008. `http://www.jmilne.org/math/CourseNotes/av.html`

[Mum71] David Mumford, *Abelian varieties*, Tata Institute of Fundamental Research Studies in Mathematics, vol. 5, Published for the Tata Institute of Fundamental Research, Bombay; by Oxford University Press, London, 1970. MR282985

[OU73] Frans Oort and Kenji Ueno, *Principally polarized abelian varieties of dimension two or three are Jacobian varieties*, J. Fac. Sci. Univ. Tokyo Sect. IA Math. **20** (1973), 377–381. MR364265

[Ru90] Hans-Georg Rück, *Abelian surfaces and Jacobian varieties over finite fields*, Compositio Math. **76** (1990), no. 3, 351–366. MR1080007

[Ry14] Sergey Rybakov, *Finite group subschemes of abelian varieties over finite fields*, Finite Fields Appl. **29** (2014), 132–150, DOI 10.1016/j.ffa.2014.04.001. MR3225384

[Serre83] Jean-Pierre Serre, *Sur le nombre des points rationnels d'une courbe algébrique sur un corps fini* (French, with English summary), C. R. Acad. Sci. Paris Sér. I Math. **296** (1983), no. 9, 397–402. MR703906

[Ta66] John Tate, *Endomorphisms of abelian varieties over finite fields*, Invent. Math. **2** (1966), 134–144, DOI 10.1007/BF01404549. MR206004

[Wa69] William C. Waterhouse, *Abelian varieties over finite fields*, Ann. Sci. École Norm. Sup. (4) **2** (1969), 521–560. MR265369

Weizmann Institute of Science, Rehovot, Israel
Email address: `rybakov.sergey@gmail.com`

Contemporary Mathematics
Volume **832**, 2026
https://doi.org/10.1090/conm/832/16612

Superelliptic Jacobians and central simple representations

Yuri G. Zarhin

ABSTRACT. Let $f(x)$ be a polynomial of degree at least 5 with complex coefficients and without repeated roots. Suppose that all the coefficients of $f(x)$ lie in a subfield K of $\mathbb{C}$ such that:

- K contains a primitive p-th root of unity;
- $f(x)$ is irreducible over K;
- the Galois group $\mathrm{Gal}(f)$ of $f(x)$ acts doubly transitively on the set of roots of $f(x)$;
- the index of every maximal subgroup of $\mathrm{Gal}(f)$ does *not* divide $\deg(f)-1$.

Then the endomorphism ring of the Jacobian of the superelliptic curve

$$y^p = f(x)$$

is isomorphic to the pth cyclotomic ring for all primes $p > \deg(f)$.

1. Introduction

The aim of this paper is to explain how to compute the endomorphism algebra of Jacobians of smooth projective models of superelliptic curves $y^q = f(x)$ where $q = p^r$ is a prime power and $f(x)$ a polynomial of degree $n \geq 5$ with complex coefficients that is in "general position". Here "general position" means that there is a (sub)field K such that all the coefficients of $f(x)$ lie in K and the Galois group of $f(x)$ acts doubly transitively on the set of its roots (in particular, $f(x)$ is irreducible over K). It turns out that for a broad class of the doubly transitive Galois groups (and under certain mild restrictions on q) the corresponding endomorphism algebra is "as small as possible", i.e., is canonically isomorphic to a product of cyclotomic fields $\mathbb{Q}(\zeta_{p^i})$ $(1 \leq i \leq r)$.

In order to state explicitly our results, let us start with the notation and some basic facts related to cyclotomic fields and cyclotomic polynomials. As usual, $\mathbb{Z}, \mathbb{Q}, \mathbb{C}$ denote the ring of integers, the field of rational numbers and the field of complex numbers respectively.

2020 *Mathematics Subject Classification.* Primary 14H40, 14K05, 11G10.

Key words and phrases. Superelliptic curves, jacobians, endomorphisms of abelian varieties.

Partially supported by the Simons Foundation Collaboration grant # 585711 and the Travel Support for Mathematicians Grant MPS-TSM-00007756 from the Simons Foundation. Part of this work was done in December 2023 during my stay at the Max-Planck Institut für Mathematik (Bonn, Germany), whose hospitality and support are gratefully acknowledged.

Let p be an odd prime and $\mathbb{F}_p$ the corresponding (finite) prime field of characteristic p. We write $\mathbb{Z}_p$ and $\mathbb{Q}_p$ for the ring of p-adic integers and the field $\mathbb{Q}_p$ of p-adic numbers respectively. Let r be a positive integer and $q = p^r$. Let

$$\zeta_q \in \mathbb{C}.$$

be a primitive qth root of unity. We write $\mathbb{Q}(\zeta_q)$ be the qth cyclotomic field and

$$\mathbb{Z}[\zeta_q] = \sum_{i=0}^{\phi(q)-1} \mathbb{Z} \cdot \zeta_q^i$$

for its ring of integers. (Hereafter $\phi(q) := (p-1)p^{r-1}$ is the Euler function.)

Let us consider the polynomial

$$\mathcal{P}_q(t) := \sum_{j=0}^{q-1} t^j = \prod_{i=1}^{r} \Phi_{p^i}(t) \in \mathbb{Z}[t]$$

where

$$\Phi_{p^i}(t) = \sum_{j=0}^{p-1} t^{ip^{r-1}} \in \mathbb{Z}[t]$$

is the p^ith *cyclotomic* polynomial.

Let $f(x) \in \mathbb{C}[x]$ be a polynomial of degree $n \geq 4$ without repeated roots. In what follows we always assume that *either p does not divide n or q divides n.*

Let $C_{f,q}$ be a smooth projective model of the smooth affine curve

$$y^q = f(x).$$

It is well known ([**16**], pp. 401–402, [**30**], Prop. 1 on p. 3359, [**21**], p. 148) that the genus $g(C_{f,q})$ of $C_{f,q}$ is $(q-1)(n-1)/2$ if p does not divide n and $(q-1)(n-2)/2$ if q divides n. The map

$$(x, y) \mapsto (x, \zeta_q y)$$

gives rise to a non-trivial biregular automorphism

$$\delta_q : C_{f,q} \to C_{f,q}$$

of period q.

Let $J(C_{f,q})$ be the Jacobian of $C_{f,q}$, which is an abelian variety of dimension $g(C_{f,q})$. We write $\operatorname{End}(J(C_{f,q}))$ for its ring of endomorphisms and denote by $\operatorname{End}^0(J(C_{f,q}))$ its endomorphism algebra $\operatorname{End}(J(C_{f,q})) \otimes \mathbb{Q}$. By functoriality, δ_q induces an automorphism of $J(C_{f,q})$, which we still denote by δ_q. It is known ([**21**, p. 149], [**23**, p. 448], [**34**, Lemma 4.8]) that

$$\mathcal{P}_q(\delta_q) = 0 \tag{1}$$

in $\operatorname{End}(J(C_{f,q}))$. Then (1) gives rise to the ring homomorphism,

$$\mathbf{i}_{q,f} : \mathbb{Z}[t]/\mathcal{P}_q(t)\mathbb{Z}[t] \hookrightarrow \mathbb{Z}[\delta_q] \subset \operatorname{End}(J(C_{f,q})),\ t + \mathcal{P}_q(t)\mathbb{Z}[t] \mapsto \delta_q, \tag{2}$$

which is a *ring embedding* ([**21**, p. 149], [**23**, p. 448], [**34**, Lemma 4.8]). (The first map in (2) is actually a ring isomorphism.) This implies that the subring $\mathbb{Z}[\delta_q]$ of $\operatorname{End}(J(C_{f,q}))$ generated by δ_q is isomorphic to $\mathbb{Z}[t]/\mathcal{P}_q(t)\mathbb{Z}[t]$. It follows that the $\mathbb{Q}$-subalgebra

$$\mathbb{Q}[\delta_q] \subset \operatorname{End}^0(J(C_{f,q})) \tag{3}$$

generated by δ_q has $\mathbb{Q}$-dimension $q-1$, is isomorphic to

$$\mathbb{Q}[t]/\mathcal{P}_q(t)\mathbb{Q}[t] \cong \prod_{i=1}^{r} \mathbb{Q}(\zeta_{p^i})$$

and therefore has dimension $q-1$.

We will need the following elementary observation.

REMARK 1.1.

(i) Suppose that a prime p is greater than n. Then p does *not* divide $n!$. Since every subgroup H of $\mathrm{Gal}(f)$ is isomorphic to a subgroup of $\mathbf{S}_n$, its order $|H|$ divides $n!$ and therefore is *not* divisible by p. Hence, if $p>n$, then $|H|$ is *not* divisible by p.

(ii) Suppose that H is a transitive subgroup of $\mathrm{Gal}(f)$ with respect to the action on the roots of $f(x)$. Then its order $|H|$ is divisible by n.

Let us formulate our main results.

First, we start with the case $q=p$. Then $\mathcal{P}_p(t)$ coincides with $\Phi_p(t)$ and there is a natural ring isomorphism

$$\mathbb{Z}[t]/\mathcal{P}_p(t)\mathbb{Z}[t] \cong \mathbb{Z}[\zeta_p]$$

that sends (the coset of) t to ζ_p. This gives us the the *ring embedding*

$$\mathbf{i}_{p,f}:\mathbb{Z}[\zeta_p] \hookrightarrow \mathbb{Z}[\delta_p] \subset \mathrm{End}(J(C_{f,p})),\ \zeta_p \mapsto \delta_p. \tag{4}$$

Notice also that the rings $\mathbb{Z}[\delta_p]$ and $\mathbb{Z}[\zeta_p]$ are isomorphic.

THEOREM 1.2. *Let $n \ge 5$ be an integer and p an odd prime such that K contains a primitive pth root of unity.*

Suppose that the Galois group $\mathrm{Gal}(f)$ of $f(x)$ contains a subgroup H that acts doubly transitively on the n-element set $\mathfrak{R}_f$ of roots of the polynomial $f(x)$ and enjoys the following properties.

(i) *The index of every maximal subgroup of H does not divide $n-1$.*

(ii) *p does not divide $|H|$. (E.g., $p>n$.)*

Then $\mathrm{End}^0(J(C_{f,p}))=\mathbb{Q}[\delta_p]$ *and* $\mathrm{End}(J^{(f,p)})=\mathbb{Z}[\delta_p]$.

THEOREM 1.3. *Let K be a subfield of $\mathbb{C}$ such that all the coefficients of $f(x)$ lie in K. Assume also that $f(x)$ is an irreducible polynomial in $K[x]$ of degree $n \ge 5$ and its Galois group over K is either the full symmetric group $\mathbf{S}_n$ or the alternating group $\mathbf{A}_n$. Then*

$$\mathrm{End}(J(C_{f,p}))=\mathbb{Z}[\delta_p]\cong\mathbb{Z}[\zeta_p].$$

REMARK 1.4. Theorem 1.3 was stated in [**37**, Th. 4.2]. Its proof was based on an assertion that a certain "permutational" representation $(\mathbb{F}_p^B)^{00}$ (that is called the *heart*[1]) of the alternating group $\mathrm{Alt}(B)=\mathbf{A}_n$ over $\mathbb{F}_p$ is *very simple*[2] [**36**, Th. 4.7]. Unfortunately, there is an error in the proof of [**36**, Th. 4.7] when $n=5, p>5$, caused by an improper use of [**36**, Cor. 4.4] (see [**36**, p. 108, lines 4-5]). So, the proof in [**37**] works only under an additional assumption that either $n>5$ or $p \le 5$.

---

[1]See [**13, 17**] and Section 3 below for the definition of the *heart*.

[2]See [**34, 38**] and Section 4 below for the definition and basic properties of very simple representations.

In this note we handle the remaining case when $n = 5$, $p > 5$. It turns out that if $p \not\equiv \pm 1 \mod 5$ then the representation of the group $\mathbf{A}_5$ is very simple, which allows us to salvage in this case the arguments of [**36**].

However, if $p \equiv \pm 1 \mod 5$ then the 4-dimensional representation $\left(\mathbb{F}_p^B\right)^{00}$ of $\mathbf{A}_5$, viewed as the representaton of the covering group $\mathrm{SL}(2, \mathbb{F}_5)$, splits into a tensor product of two 2-dimensional representations. In particular, $\left(\mathbb{F}_p^B\right)^{00}$ is *not* very simple, and we use a notion of a *central simple representation* (see Section 4 below), in order to prove Theorem 1.3 in this case.

REMARKS 1.5. If $f(x) \in K[x]$ then the curve $C_{f,p}$ and its Jacobian $J(C_{f,p})$ are defined over K. Let $K_a \subset \mathbb{C}$ be the algebraic closure of K. Then all endomorphisms of $J(C_{f,p})$ are defined over K_a. Hence, in order to prove Theorems 1.3 and 6.7, it suffices to check that the ring of all K_a-endomorphisms of $J(C_{f,p})$ coincides with $\mathbb{Z}[\delta_p]$.

Now let us try to relax the restrictions on p, keeping the double transitivity of $\mathrm{Gal}(f)$. Our next result deals with doubly transitive sporadic simple (Galois) groups, whose description may be found in [**22**], [**5**, Ch. 6 and Ch. 7, Sect. 7.7, pp. 252–253].

THEOREM 1.6. *Let p be an odd prime and*

$$\mathrm{Gal}(f) \subset \mathrm{Perm}(\mathfrak{R}_f) \cong \mathbf{S}_n$$

a permutation group that acts doubly transitively on the n-element set $\mathfrak{R}_f$. Suppose that $(n, \mathrm{Gal}(f))$ enjoys one of the following properties.

(**M**) *$n \in \{11, 12, 22, 23, 24\}$, and $\mathrm{Gal}(f)$ is isomorphic to the corresponding Mathieu group $\mathbf{M}_n$. If $n = 11$ then we assume additionally that $p > 3$.*

(**HS**) *$n = 176$, $p > 7$, and $\mathrm{Gal}(f))$ is isomorphic to the sporadic simple Higman-Sims group* HS.

(**CO3**) *$n = 276$, $p \notin \{3, 5, 11\}$, and $\mathrm{Gal}(f))$ is isomorphic to the sporadic simple Conway group* Co_3.

Then $\mathrm{End}^0(J(C_{f,p})) = \mathbb{Q}[\delta_p]$ *and* $\mathrm{End}(J(C_{f,p})) = \mathbb{Z}[\delta_p]$.

REMARK 1.7. The case (**M**) of Theorem 1.6 gives a (partial) answer to a question of Ravi Vakil that was asked during my talk at Simons Symposium "Geometry Over Non-closed Fields" (Puerto Rico, March 2015).

Now let us discuss the case when our Galois groups are doubly transitive finite simple *Chevalley groups* - they are classified in [**4**] and their action described in details in [**5**, Sect. 7.7]. (For general results about Chevalley groups see [**28**].)

THEOREM 1.8. *Suppose that p is an odd prime and $\mathrm{Gal}(f)$ contains a subgroup H that acts doubly transitively on the n-element set $\mathfrak{R}_f$ and is isomorphic to a finite Chevalley group $\mathfrak{G}(\mathfrak{q})$, and the corresponding stabilizers correspond to Borel subgroups of $\mathfrak{G}(\mathfrak{q})$, which are maximal subgroups of index n.*

Suppose that $(n, \mathfrak{G}(\mathfrak{q}), p)$ enjoys one of the following properties.

(**L2**) *Let ℓ be a prime and $\mathfrak{r}$ a positive integer. Then $n = \mathfrak{q}+1$ where $\mathfrak{q} = \ell^{\mathfrak{r}} > 11$ and $\mathfrak{G}(\mathfrak{q})$ is the projective special linear group*

$$\mathbf{L}_2(\mathfrak{q}) = \mathrm{PSL}(2, \mathbb{F}_{\mathfrak{q}})$$

where $\mathbb{F}_{\mathfrak{q}}$ is a finite $\mathfrak{q}$-element field. Assume additionally that either $p \neq \ell$ or $\mathfrak{q} = \ell = p$.

(**Lmq**) *Let* $m \geq 3$ *be an integer,* ℓ *a prime,* $\mathfrak{r}$ *a positive integer. Then* $n = (\mathfrak{q}^m - 1)/(\mathfrak{q} - 1)$ *where* $\mathfrak{q} = \ell^{\mathfrak{r}}$ *and* $\mathfrak{G}(\mathfrak{q})$ *is the projective special linear group*

$$\mathbf{L}_m(\mathfrak{q}) = \mathrm{PSL}(m, \mathbb{F}_{\mathfrak{q}})$$

where $\mathbb{F}_{\mathfrak{q}}$ *is a finite* $\mathfrak{q}$*-element field. Assume additionally that* $p \neq \ell$ *and*

$$(m, \mathfrak{q}) \neq (3,2), (3,4), (4,2), (4,3), (6,2), (6,3).$$

(**U3**) *Let* ℓ *be a prime and* $\mathfrak{r}$ *a positive integer. Then* $n = \mathfrak{q}^3 + 1$ *where* $\mathfrak{q} = \ell^{\mathfrak{r}}$ *is a power of a prime* ℓ,

$$\mathfrak{q} \neq 2, 5$$

and $\mathfrak{G}(\mathfrak{q})$ *is the projective special unitary group* $\mathbf{U}_3(q) = \mathrm{PSU}_3(\mathbb{F}_q)$.

(**Sz**) *Let* $\mathfrak{r}$ *be a positive integer,*

$$\mathfrak{q} = 2^{2\mathfrak{r}+1}, \quad n = q^2 + 1, \; m = 2^{\mathfrak{r}+1}.$$

Then H *is the simple Suzuki group* $\mathrm{Sz}(\mathfrak{q}) = {}^2\,\mathrm{B}_2(q)$. *In addition,* p *does not divide* $(q + 1 + m)$.

(**Ree**) *Let* $\mathfrak{r}$ *be a positive integer,*

$$\mathfrak{q} = 3^{2\mathfrak{r}+1}, \quad n = q^2 + 1, \; m = 3^{\mathfrak{r}+1}.$$

The group H *is the simple Ree group* $\mathrm{Ree}(\mathfrak{q}) = {}^2\,\mathrm{G}_2(q)$. *In addition,* p *does not divide* $3(q + 1)(q + m + 1)(q - m + 1)$.

Then $\mathrm{End}^0(J(C_{f,p})) = \mathbb{Q}[\delta_p]$ *and* $\mathrm{End}(J(C_{f,p})) = \mathbb{Z}[\delta_p]$.

Now let us assume that r is any positive integer (recall that $q = p^r$). In this case we obtain the results about the *endomorphism algebra* $\mathrm{End}^0(J(C_{f,q})) = \mathrm{End}(J(C_{f,q})) \otimes \mathbb{Q}$ of $J(C_{f,q})$ that may be viewed as analogues of Theorems 1.3 and Theorem 1.2 for the endomorphism algebra $\mathrm{End}^0(J(C_{f,q})$.

THEOREM 1.9. *Suppose that* $n \geq 5$ *is an integer,* p *an odd prime that does not divide* n, *and* K *contains a primitive* q*th root of unity.*

Let us assume that the Galois group $\mathrm{Gal}(f)$ *of* $f(x)$ *contains a subgroup* H *that acts doubly transitively on the* n*-element set* $\mathfrak{R}_f$ *of roots of the polynomial* $f(x)$ *and enjoys the following properties.*

(i) *The index of every maximal subgroup of* H *does not divide* $n - 1$.

(ii) p *does not divide* $|H|$. *(E.g.,* $p > n$.*)*

Then

$$\mathrm{End}^0(J(C_{f,q})) = \mathbb{Q}[\delta_q] \cong \prod_{i=1}^{r} \mathbb{Q}(\zeta_{p^i}).$$

THEOREM 1.10. *Let* K *be a subfield of* $\mathbb{C}$ *such that all the coefficients of* $f(x)$ *lie in* K. *Suppose that* $f(x)$ *is an irreducible polynomial in* $K[x]$ *of degree* $n \geq 5$ *and its Galois group over* K *is either the full symmetric group* $\mathbf{S}_n$ *or the alternating group* $\mathbf{A}_n$. *Assume also that either* p *does not divide* n *or* q *divides* n. *Then*

$$\mathrm{End}^0(J(C_{f,q})) = \mathbb{Q}[\delta_q] \cong \prod_{i=1}^{r} \mathbb{Q}(\zeta_{p^i}).$$

REMARK 1.11. Theorem 1.10 was stated in [**34**, Th. 1.1]. Similarly (see Remark 1.4), the proof in [**34**] works only under an additional assumption that either $n > 5$ or $p \leq 5$. In this paper we handle the remaining case $n = 5, p > 5$.

REMARK 1.12. An analogue of Theorem 1.10 when $p \mid n$ but q does *not* divide n was proven in [**33**].

The paper is organized as follows. In Section 2 we discuss complex abelian varieties Z with multiplications from cyclotomic fields, paying special attention to the centralizers of these fields in $\mathrm{End}^0(Z)$ and their action on the differentials of the first kind when Z is a superelliptic Jacobian In Section 3 we discuss modular representations of permutation groups, paying a special attention to the *hearts* of these representations. In Section 4 we introduce central simple representations and recall basic properties of very simple representations that first appeared in [**34, 38**], paying a special attention to the very simplicity and central simplicity of hearts of permutational representations in the case of doubly transitive permutation groups. In Section 5 we return to our discussion of abelian varieties Z with multiplications from cyclotomic fields $\mathbb{Q}(\zeta_q)$, paying a special attention to the Galois properties of the set of δ_q-invariants. In Section 6 we review results of [**37**] about endomorphism algebras of superelliptic Jacobians. In Section 7 we prove our main results that deal with the case $q = p$. The proofs for the case of arbitrary q are contained in Section 8.

2. Endomorphism fields of abelian varieties and their centralizers

In what follows E is a number field and O_E the ring of algebraic integers in E. It is well known that O_E is a Dedekind ring and therefore every finitely generated torsion-free O_E-module is projective/locally free and isomorphic to a direct sum of locally free O_E-modules of rank 1. In addition, the natural map

$$O_E \otimes \mathbb{Q} \to E, \ e \otimes c \mapsto c \cdot e \ \forall e \in E, \ c \in \mathbb{Q}$$

is an isomorphism of $\mathbb{Q}$-algebras.

Let Z be an abelian variety over $\mathbb{C}$ of positive dimension g, let $\mathrm{End}(Z)$ be the ring of its endomorphisms. If n is an integer then we we write n_Z for the endomorphism

$$n_Z : Z \to Z, \ z \mapsto nz.$$

Clearly, $n_Z \in \mathrm{End}(Z)$. By definition, 1_Z is the identity selfmap of Z. In addition, $n_Z : Z \to Z$ is an *isogeny* if and only if $n \neq 0$.

We write

$$\mathrm{End}^0(Z) = \mathrm{End}(Z) \otimes \mathbb{Q}$$

for the corresponding endomorphism algebra of Z, which is a finite-dimensional semisimple $\mathbb{Q}$-algebra. Identifying $= \mathrm{End}(Z)$ with

$$\mathrm{End}(Z) \otimes 1 \subset \mathrm{End}(Z) \otimes \mathbb{Q} = \mathrm{End}^0(Z),$$

we may view $\mathrm{End}(Z)$ as an **order** in the $\mathbb{Q}$-algebra $\mathrm{End}^0(Z)$.

The action of $\mathrm{End}(Z)$ by functoriality on the g-dimensional complex vector space $\Omega^1(Z)$ of differentials of the first kind on Z gives us the ring homomorphism $\mathrm{End}(Z) \to \mathrm{End}_{\mathbb{C}}(\Omega^1(Z))$ [**27**, Ch. 1, Sect. 2.8], which extends by $\mathbb{Q}$-linearity to the homomorphism of $\mathbb{Q}$-algebras

$$j_Z : \mathrm{End}^0(Z) \to \mathrm{End}_{\mathbb{C}}(\Omega^1(Z)), \tag{5}$$

which sends 1_Z to the identity automorphism of the $\mathbb{C}$-vector space $\Omega^1(Z)$. Let E be a number field that is a $\mathbb{Q}$-subalgebra of $\mathrm{End}^0(Z)$ with the same $1 = 1_Z$. Let

Σ_E be the $[E:\mathbb{Q}]$-element set of field embeddings $\sigma: E \hookrightarrow \mathbb{C}$. Let us define for each $\sigma \in \Sigma_E$ the corresponding *weight subspace*

$$\Omega^1(Z)_\sigma = \{\omega \in \Omega^1(Z) \mid j_Z(e)\omega = \sigma(e)\omega \ \forall e \in E\} \subset \Omega^1(Z).$$

The well known splitting

$$E \otimes_{\mathbb{Q}} \mathbb{C} = \oplus_{\sigma\in\Sigma_E} E \otimes_{E,\sigma} \mathbb{C} = \oplus_\sigma \mathbb{C}_\sigma \text{ where } \mathbb{C}_\sigma = E \otimes_{E,\sigma} \mathbb{C} = \mathbb{C}$$

implies that

$$\Omega^1(Z) = \oplus_{\sigma\in\Sigma_E}\Omega^1(Z)_\sigma.$$

Let us put

$$n_\sigma = \dim_{\mathbb{C}}(\Omega^1(Z)_\sigma).$$

Let D be the centralizer of E in $\mathrm{End}^0(Z)$. Clearly, E lies in the center of D, which makes D is a finite-dimensional E-algebra. It is also clear that each subspace $\Omega^1(Z)_\sigma$ is $j_Z(D)$-invariant, which gives us a $\mathbb{Q}$-algebra homomorphism

$$j_{Z,\sigma}: D \to \mathrm{End}_{\mathbb{C}}(\Omega^1(Z)_\sigma), \tag{6}$$

that sends $1 = 1_Z \in D$ to the identity automorphism of the $\mathbb{C}$-vector space $\Omega^1(Z)_\sigma$.

LEMMA 2.1. *Let D be as above. Suppose that D is a central simple E-algebra of dimension d^2 where d is a positive integer. Then:*

(i) *d divides all the multiplicities n_σ. In particular, if $n_\sigma = 1$ for some $\sigma \in \Sigma_E$ then $d = 1$ and $D = E$.*

(ii) *Let M be the the number of σ's with $n_\sigma \neq 0$. Then*

$$dM \leq \dim(Z).$$

In particular, if $d = 2\dim(Z)/[E:\mathbb{Q}]$ then $M \leq [E:\mathbb{Q}]/2$.

PROOF. Our condition on D implies that $D_\sigma = D \otimes_{E,\sigma} \mathbb{C}$ is isomorphic to the matrix algebra $\mathrm{Mat}_d(\mathbb{C})$ of size d over $\mathbb{C}$ for all $\sigma \in \Sigma_E$.

We may assume that $n_\sigma > 0$. Then $\Omega^1(Z)_\sigma \neq \{0\}$ and $j_{Z,\sigma}(D) \neq \{0\}$. Extending $j_{Z,\sigma}$ by $\mathbb{C}$-linearity, we get a $\mathbb{C}$-algebra homomorphism

$$D_\sigma \to \mathrm{End}_{\mathbb{C}}((\Omega^1(Z)_\sigma),$$

which provides $\Omega^1(Z)_\sigma$ with the structure of a $D_\sigma = \mathrm{Mat}_d(\mathbb{C})$-module. This implies that each n_σ is divisible by d. This proves (i). In order to prove (ii), it suffices to notice that

$$\dim(Z) = \sum_\sigma \dim_{\mathbb{C}}(\Omega^1(Z)_\sigma) = \sum_\sigma n_\sigma \geq dM. \qquad \square$$

REMARK 2.2.

(i) Let Λ be the centralizer of δ_p in $\mathrm{End}(J(C_{f,p}))$, which coincides with the centralizer of $\mathbb{Z}[\delta_p]$ in $\mathrm{End}(J(C_{f,p}))$. Let us consider the $\mathbb{Q}$-subalgebra

$$\Lambda_{\mathbb{Q}} = \Lambda \otimes \mathbb{Q} \subset \mathrm{End}(J(C_{f,p})) \otimes \mathbb{Q} = \mathrm{End}^0(J(C_{f,p})).$$

Clearly, $\Lambda_{\mathbb{Q}}$ coincides with the centralizer of $\mathbb{Q}[\delta_p]$ in $\mathrm{End}^0(J(C_{f,p}))$. Since δ_p respects the *theta divisor* on the Jacobian $J(C_{f,p})$, the algebra $\Lambda_{\mathbb{Q}}$ is stable under the corresponding *Rosati involution* and therefore is semisimple as a $\mathbb{Q}$-algebra. It is also clear that the number field $\mathbb{Q}[\delta_p] \cong \mathbb{Q}(\zeta_p)$ lies in the center of $\Lambda_{\mathbb{Q}}$. Hence, $\Lambda_{\mathbb{Q}}$ becomes a semisimple $\mathbb{Q}[\delta_p]$-algebra.

(ii) Let i be an integer such that $1 \le i \le p-1$. We write σ_i for the field embedding
$$\sigma_i : \mathbb{Q}[\delta_p] \hookrightarrow \mathbb{C}$$
sending δ_p to ζ_p^{-i}, and consider the corresponding subspace $\Omega^1(J(C_{f,p}))_{\sigma_i}$ of differentials of the first kind on $J(C_{f,p})$. It is known [**37**, Remark 3.7] that if p does *not* divide n then
$$n_{\sigma_i} = \dim_{\mathbb{C}}(\Omega^1(J(C_{f,p}))_{\sigma_i}) = \left[\frac{ni}{p}\right]. \tag{7}$$
(iii) It follows from Lemma 2.1 applied to $Z = J(C_{f,p})$ and $E = \mathbb{Q}[\delta_p]$ that if p does *not* divide n and $\Lambda_{\mathbb{Q}}$ is a central simple $\mathbb{Q}[\delta_p]$-algebra of dimension d^2 then d divides all $[ni/p]$ for all integers i with $1 \le i \le p-1$.

(iv) Suppose that either $n = p+1$, or $n-1$ is *not* divisible by p. Then the greatest common divisor of all n_{σ_i}'s is 1 [**39**, Lemma 8.1(D) on pp. 516–517]. It follows from Lemma 2.1 that if $\Lambda_{\mathbb{Q}}$ is a central simple $\mathbb{Q}[\delta_p]$-algebra then $\Lambda_{\mathbb{Q}} = \mathbb{Q}[\delta_p]$.

(v) Suppose that p divides $n-1$, say, $n = kp+1$ where k is a positive integer. Then the greatest common divisor of all n_{σ_i}'s is k. [**39**, Lemma 8.1(D) on pp. 516–517] It follows that if $\Lambda_{\mathbb{Q}}$ is a central simple $\mathbb{Q}[\delta_p]$-algebra of dimension d^2 then d divides k.

(vi) The number of i with $n_{\sigma_i} > 0$ is at least $(p+1)/2$ [**36**, p. 101]. It follows that if $\Lambda_{\mathbb{Q}}$ is a central simple $\mathbb{Q}[\delta_p]$-algebra of dimension d^2 then, in light of Proposition 2.1,
$$d \cdot \frac{p+1}{2} \le g$$
where $g = \dim(J(C_{f,p}))$ is the genus of $C_{f,p}$. This implies that
$$d \le \frac{2g}{p+1} < \frac{2g}{p-1}.$$

LEMMA 2.3. *Let $\mathcal{H}$ be a finite-dimensional E-algebra, and Λ an order in $\mathcal{H}$ that contains O_E. (In particular, Λ is a finitely generated torsion-free O_E-module and the natural map $\Lambda \otimes \mathbb{Q} \to \mathcal{H}$ is an isomorphism of finite-dimensional $\mathbb{Q}$-algebras.)*

Suppose that there are a positive integer d and a maximal ideal $\mathfrak{m}$ of O_E with residue field $k = O_E/\mathfrak{m}$ such that the k-algebra $\Lambda/\mathfrak{m}\Lambda$ is isomorphic to the matrix algebra $\mathrm{Mat}_d(k)$ of size d over k.

Then $\mathcal{H}$ is a central simple E-algebra of dimension d^2.

PROOF. Let $C_{\mathbb{Q}}$ the center of $\mathcal{H}$ that is a finite-dimensional commutative E-algebra. Then $C := C_{\mathbb{Q}} \cap \Lambda$ is the center of Λ.

Clearly, C contains O_E and is a saturated O_E-submodule of Λ. The latter means that if $eu \in C$ for some $u \in \Lambda$ and nonzero $e \in O_E$ then $u \in \Lambda$. This implies that the quotient Λ/C is a torsion-free (and finitely generated) O_E-module and therefore is projective. It follows that C is a direct summand of the O_E-module D and therefore there is an O_E-submodule $\mathfrak{P}$ of Λ such that
$$\Lambda = C \oplus \mathfrak{P}.$$
Similarly, O_E is a saturated O_E-submodule of C and, by the same token, there is a locally free O_E-submodule $\mathfrak{Q}$ of C such that
$$C = O_E \oplus \mathfrak{Q} \text{ and } \Lambda = C \oplus \mathfrak{P} = O_E \oplus \mathfrak{Q} \oplus \mathfrak{P}.$$

Then the the natural map of $O_E/\mathfrak{m} = k$-modules,

$$O_E/\mathfrak{m} \oplus \mathfrak{Q}/\mathfrak{m}\mathfrak{Q} = C/\mathfrak{m}C \to \Lambda/\mathfrak{m}\Lambda \cong \mathrm{Mat}_d(k)$$

is *injective* and its image lies in the center k of $\mathrm{Mat}_d(k)$. The k-dimension arguments imply that $\mathfrak{Q}/\mathfrak{m}\mathfrak{Q} = \{0\}$. Since $\mathfrak{Q}$ is finitely generated projective, $\mathfrak{Q} = \{0\}$, i.e., $C = O_E$ and the center of $\mathcal{H}$ is

$$C_{\mathbb{Q}} = C \otimes \mathbb{Q} = O_E \otimes \mathbb{Q} = E.$$

Hence, $\mathcal{H}$ is a finite-dimensional E-algebra with center E.

Let us check the simplicity of $\mathcal{H}$. Let $J_{\mathbb{Q}}$ be a proper two-sided ideal of $\mathcal{H}$. We need to check that $J_{\mathbb{Q}} = \{0\}$. In order to do that, let us consider the intersection $J := J_{\mathbb{Q}} \cap \Lambda$, which is obviously a two-sided ideal of Λ. It is also clear that J is a saturated O_E-submodule of Λ, i.e., the quotient Λ/J is a torsion free (finitely generated) O_E-module. Hence, Λ/J is a projective O_E-module. It follows that J is a direct summand of the O_E-module Λ, i.e., there exists an O_E-submodule $\mathfrak{Q}$ of Λ such that $\Lambda = J \oplus \mathfrak{Q}$. If $\mathfrak{Q} = \{0\}$ then $\Lambda = J$ and

$$\mathcal{H} = \Lambda \otimes \mathbb{Q} = J \otimes \mathbb{Q} = J_{\mathbb{Q}};$$

so $J_{\mathbb{Q}} = \mathcal{H}$, which is not true, because $J_{\mathbb{Q}}$ is a *proper* ideal of $\mathcal{H}$. This implies that $\mathfrak{Q} \neq \{0\}$ and therefore $\mathfrak{Q}/\mathfrak{m}\mathfrak{Q} \neq \{0\}$. We have

$$\Lambda/\mathfrak{m}\Lambda = J/\mathfrak{m}J \oplus \mathfrak{Q}/\mathfrak{m}\mathfrak{Q}.$$

Clearly, $J/\mathfrak{m}J$ is a proper two-sided ideal of the simple algebra $\Lambda/\mathfrak{m}\Lambda \cong \mathrm{Mat}_d(k)$. This implies that $J/\mathfrak{m}J = \{0\}$, which implies that $J = \{0\}$ and therefore $J_{\mathbb{Q}} = \{0\}$.

To summarize: $\mathcal{H}$ is a simple finite-dimensional E-algebra with center E, i.e., a finite-dimensional central simple E-algebra.

On the other hand, the E-dimension of $\mathcal{H}$ equals the rank of the locally free O_E-module Λ, which, in turn, equals the $k = O_E/\mathfrak{m}$-dimension of $\Lambda/\mathfrak{m}\Lambda$. Since $\Lambda/\mathfrak{m}\Lambda \cong \mathrm{Mat}_d(k)$ has k-dimension d^2, the E-dimension of H is also d^2. It follows that $\mathcal{H}$ is a central simple E-algebra of dimension d^2. □

3. Permutation groups and permutation modules

Our exposition in this section follows closely [**36**, Sect. 2], see also [**17**].

Let $n \geq 5$ be an integer, B a n-element set, and $\mathrm{Perm}(B)$ the group of permutations of B, which is isomorphic to the full symmetic group S_n. The group S_n has order $n!$ and contains precisely one (normal) subgroup of index 2 that we denote by $\mathrm{Alt}(B)$. Any isomorphism between $\mathrm{Perm}(B)$ and S_n induces an isomorphism between $\mathrm{Alt}(B)$ and the alternating group $\mathbf{A}_n$. Since $n \geq 5$, the group $\mathrm{Alt}(B)$ is simple non-abelian; its order is $n!/2$. Let G be a subgroup of $\mathrm{Perm}(B)$.

Let F be a field. We write F^B for the n-dimensional F-vector space of maps $h : B \to F$. The space F^B is endowed with a natural action of $\mathrm{Perm}(B)$ defined as follows. Each $s \in \mathrm{Perm}(B)$ sends a map $h : B \to F$ into $sh : b \mapsto h(s^{-1}(b))$. The permutation module F^B contains the $\mathrm{Perm}(B)$-stable hyperplane

$$(F^B)^0 = \{h : B \to F \mid \sum_{b \in B} h(b) = 0\}$$

and the $\mathrm{Perm}(B)$-invariant line $F \cdot 1_B$ where 1_B is the constant function 1. The quotient $F^B/(F^B)^0$ is a trivial 1-dimensional $\mathrm{Perm}(B)$-module.

Clearly, $(F^B)^0$ contains $F \cdot 1_B$ if and only if char(F) divides n. If this is *not* the case then there is a Perm(B)-invariant splitting

$$F^B = (F^B)^0 \oplus F \cdot 1_B.$$

Let G be a subgroup of Perm(B). Clearly, F^B and $(F^B)^0$ carry natural structures of G-modules or (which is the same) of $F[G]$-modules. (Hereafter $F[G]$ stands for the *group algebra* of G.)

If $F = \mathbb{Q}$ then the character of $\mathbb{Q}^B$ sends each $g \in G$ to the number of fixed points of g in B ([**26**], ex. 2.2, p.12); it takes on values in $\mathbb{Z}$ and called the *permutation character* of B. Let us denote by $\phi = \phi_B : G \to \mathbb{Q}$ the character of $(\mathbb{Q}^B)^0$.

If char$(F) = 0$ then the $F[G]$-module $(F^B)^0$ is absolutely simple[3] if and only if the action of G on B is doubly transitive ([**26**, ex. 2.6, p. 17], [**17**]). (Notice that $1 + \phi$ is the permutation character. This implies that the character ϕ also takes on values in $\mathbb{Z}$.) In particular, $\mathbb{Q}_p[G]$-module $(\mathbb{Q}_p^B)^0$ is absolutely simple if and only if the action of G on B is doubly transitive.

In what follows we concentrate on the case of $F = \mathbb{F}_p$.

REMARK 3.1.

- Let p be a prime that does *not* divide the order of G. This condition is automatically fulfilled if $p > n$, because G, being isomorphic to a subgroup of $\mathbf{S}_n$, has order that divides $n!$.
- Suppose that the action of G on B is *doubly transitive*. Taking into account that the representation theory of G over $\mathbb{Q}_p$ is "the same over $\mathbb{F}_p$ as over $\mathbb{Q}_p$" ([**26**, Sect. 15.5, Prop.43], [**17**]), we conclude that the $\mathbb{F}_p[G]$-module $(\mathbb{F}_p^B)^0$ is *absolutely simple* (see also [**39**, Cor. 7.5 on p. 513]).

DEFINITION 3.2. Let G be a subgroup of Perm(B).
If $p \mid n$ then let us define the G-module

$$\left(\mathbb{F}_p^B\right)^{00} := (\mathbb{F}_p^B)^0/(\mathbb{F}_p \cdot 1_B).$$

If p does not divide n then let us put

$$\left(\mathbb{F}_p^B\right)^{00} := (\mathbb{F}_p^B)^0.$$

The G-module $(\mathbb{F}^B)^0$ is called the *heart* of the permutation representation of G on B [**17**]. It follows from the definition that $\dim_{\mathbb{F}_p}((\mathbb{F}_p^B)^{00})) = n - 1$ if n is not divisible by p and $\dim_{\mathbb{F}_p}((\mathbb{F}_p^B)^{00}) = n - 2$ if $p \mid n$.

LEMMA 3.3. *Assume that* $G = \mathrm{Perm}(B)$ *or* $\mathrm{Alt}(B)$. *Then the G-module* $\left(\mathbb{F}_p^B\right)^{00}$ *is absolutely simple.*

PROOF. This result is well known (and goes back to Dickson). See [**3**, Th. 5.2 on p. 133], [**31**], [**17**], [**36**, Lemma 2.6]. □

[3] Recall that a simple $F[G]$-module V is called *absolutely simple* if the centralizer of G in $\mathrm{End}_F(V)$ coincides with F or equivalently the natural homomorphism $F[G] \to \mathrm{End}_F(V)$ of F-algebras is surjective.

REMARK 3.4. It turns out that the case of $n = 5$ and

$$G = \mathrm{Alt}(B) \cong \mathbf{A}_5 \cong \mathrm{PSL}(2, \mathbb{F}_5) = \mathrm{SL}(2, \mathbb{F}_5)/\{\pm 1\}$$

is rather special when

$$p \equiv \pm 1 \bmod 5. \tag{8}$$

Namely, in this case $p > 5$ and the $G = \mathrm{PSL}(2, \mathbb{F}_5)$-module $(\mathbb{F}_p^B)^{00} = (\mathbb{F}_p^B)^0$ viewed as the $\mathrm{SL}(2, \mathbb{F}_5)$-module splits into a nontrivial tensor product. In order to see this, recall [**7**, Sect. 38] that $\mathrm{SL}(2, \mathbb{F}_5)$ has the ordinary character θ_2 of degree 4 (which, is the lift of ϕ_5 from $\mathbf{A}_5$) and two *ordinary irreducible* characters η_1 and η_2 of degree 2 with

$$\mathbb{Q}(\eta_1) = \mathbb{Q}(\eta_2) = \mathbb{Q}(\sqrt{5}),$$

whose product $\eta_1\eta_2$ coincides with θ_2. By the quadratic reciprocity law, the congruence (8) implies that $\sqrt{5} \in \mathbb{F}_p$ and therefore $\sqrt{5}$ lies in the field $\mathbb{Q}_p$ of p-adic numbers, because $p \ne 2, 5$ is *odd*. This means that

$$\mathbb{Q}_p(\eta_1) = \mathbb{Q}_p(\eta_2) = \mathbb{Q}_p.$$

By a theorem of Janusz [**12**, Theorem (d) on pp. 3–4], characters of both η_1 and η_2 can be realized over $\mathbb{Q}_p$, i.e., there are two-dimensional $\mathbb{Q}_p$-vector spaces V_1 and V_2 and linear representations

$$\begin{aligned} \rho_1 &: \mathrm{SL}(2, \mathbb{F}_5) \to \mathrm{Aut}_{\mathbb{Q}_p}(V_1) \cong \mathrm{GL}(2, \mathbb{Q}_p), \\ \rho_2 &: \mathrm{SL}(2, \mathbb{F}_5) \to \mathrm{Aut}_{\mathbb{Q}_p}(V_2) \cong \mathrm{GL}(2, \mathbb{Q}_p), \end{aligned} \tag{9}$$

whose *characters* are η_1 and η_2 respectively. Let T_1 and T_2 be any $\mathrm{SL}(2, \mathbb{F}_5)$-invariant $\mathbb{Z}_p$-lattices of rank 2 in V_1 and V_2 respectively. Since the order 120 of the group $\mathrm{SL}(2, \mathbb{F}_5)$ is prime to p and the $\mathbb{Q}_p[\mathrm{SL}(2, \mathbb{F}_5)]$-modules V_1 and V_2 are simple, it follows from [**26**, Sect. 15.5, Prop. 43]) that their *reductions* modulo p

$$\bar{V_1} = T_1/pT_1, \ \bar{V_2} = T_2/pT_2$$

are simple $\mathbb{F}_p[\mathrm{SL}(2, \mathbb{F}_5)]$-modules. On the other hand, the tensor product

$$T := T_1 \otimes_{\mathbb{Z}_p} T_2 \subset V_1 \otimes_{\mathbb{Q}_p} V_2$$

is a $\mathrm{SL}(2, \mathbb{F}_5)$ -invariant $\mathbb{Z}_p$-lattice of rank 4 in $V_1 \otimes_{\mathbb{Q}_p} V_2 =: V$. Taking into account that ϕ_B is irreducible, the equality

$$\eta_1\eta_2 = \theta_2 \tag{10}$$

of the corresponding class functions on $\mathrm{SL}(2, \mathbb{F}_5)$ implies that the $\mathbb{Q}_p[\mathrm{SL}(2, \mathbb{F}_5)]$-module V is simple and the $\mathbb{F}_p[\mathrm{SL}(2, \mathbb{F}_5)]$-module

$$T/pT = (T_1 \otimes_{\mathbb{Z}_p} T_2)/p = (T_1/pT_1) \otimes_{\mathbb{F}_p} (T_1/pT_1) = \bar{V_1} \otimes_{\mathbb{F}_p} \bar{V_2} \tag{11}$$

is simple. On the other hand, the equality (10) implies the existence of an isomorphism

$$u : V = V_1 \otimes_{\mathbb{Q}_p} V_2 \cong (\mathbb{Q}_p^B)^0$$

of the $\mathbb{Q}_p[\mathrm{SL}(2, \mathbb{F}_5)]$-modules.

Obviously,

$$(\mathbb{Z}_p^B)^0 := \{h : B \to \mathbb{Z}_p \mid \sum_{b \in B} h(b) = 0\}$$

is a $\mathrm{SL}(2, \mathbb{F}_5)$-invariant $\mathbb{Z}_p$-lattice of rank 4 in $(\mathbb{Q}_p^B)^0$. (Here $\mathrm{SL}(2, \mathbb{F}_5)$ acts on $(\mathbb{Q}_p^B)^0$ through the quotient $\mathrm{SL}(2, \mathbb{F}_5)/\{\pm 1\} = \mathbf{A}_5$.) Notice that $u(T)$ is a (may be, another) $\mathrm{SL}(2,\mathbb{F}_5)$-invariant $\mathbb{Z}_p$-lattice of rank 4 in $(\mathbb{Q}_p^B)^0$ and the $\mathbb{F}_p[\mathrm{SL}(2,\mathbb{F}_5)]$-module $u(T)/p\ u(T)$ is obviously isomorphic to T/pT. In light of [**26**, Sect. 15.1, Th. 32], the *simplicity* of the $\mathbb{F}_p[\mathrm{SL}(2,\mathbb{F}_5)]$-modules T/pT (and, hence, of $u(T)/p\ u(T)$) implies that the $\mathbb{F}_p[\mathrm{SL}(2,\mathbb{F}_5)]$-modules T/pT and $(\mathbb{Z}_p^B)^0/p(\mathbb{Z}_p^B)^0$ are isomorphic. Taking into account (11) and that $(\mathbb{Z}_p^B)^0/p\cdot(\mathbb{Z}_p^B)^0=(\mathbb{F}_p^B)^0$, we conclude that that the $\mathrm{SL}(2,\mathbb{F}_5)$-modules $\bar{V}_1\otimes_{\mathbb{F}_p}\bar{V}_2$ and $(\mathbb{F}_p^B)^0$ are isomorphic.

REMARK 3.5. One may find an explicit construction of the group embeddings $\mathrm{SL}(2,\mathbb{F}_5)\to\mathrm{GL}(2,\mathbb{F}_p)$ (when p satisfies (8)) in the book of M. Suzuki [**29**, Ch. 3, Sect. 6].

4. Very simple and central simple representations

DEFINITION 4.1. Let V be a vector space of over a field F, let G be a group and $\rho: G\to\mathrm{Aut}_F(V)$ a linear representation of G in V. Let $R\subset\mathrm{End}_F(V)$ be a F-subalgebra containing the identity map

$$\mathrm{Id}: V\to V.$$

(i) We say that R is G-normal if

$$\rho(\sigma)R\rho(\sigma)^{-1}\subset R\quad\forall\sigma\in G.$$

(ii) We say that a normal G-subalgebra is *obvious* if it coincides either with $F\cdot\mathrm{Id}$ or with $\mathrm{End}_F(V)$.
(iii) We say that the G-module V is *very simple* if every G-normal subalgebra of $\mathrm{End}_F(V)$ is obvious.
(iv) We say that the G-module V is *cental simple* if every G-normal subalgebra of $\mathrm{End}_F(V)$ is a central simple F-algebra.
(v) We say that the G-module V is *strongly simple* if every G-normal subalgebra of $\mathrm{End}_F(V)$ is a simple F-algebra.

REMARK 4.2.

(i) Clearly, a very simple G-module is central simple and strongly simple. It is also clear that a central simple G-module is strongly simple.
(ii) Clearly, a subalgebra $R\subset\mathrm{End}_F(V)$ is G-normal if and only if it is $\rho(G)$-normal. It follows readily that the G-module V is very simple (resp. central simple) (resp. strongly simple) if and only if the corresponding $\rho(G)$-module V is very simple (resp. central simple) (resp. strongly simple). It is known [**34**, Rem. 2.2(ii)] that a very simple module is absolutely simple.
(iii) If R is a G-normal subalgebra of $\mathrm{End}_F(V)$ then

$$\rho(\sigma)R\rho(\sigma)^{-1}=R\quad\forall\sigma\in G.$$

Indeed, suppose that there is $u\in R$ such that for some $\sigma\in G$

$$u\notin\rho(\sigma)R\rho(\sigma)^{-1}.$$

This implies that

$$\rho(\sigma^{-1})u\rho(\sigma^{-1})^{-1}=\rho(\sigma)^{-1}u\rho(\sigma)\notin\rho(\sigma)^{-1}\big(\rho(\sigma)R\rho(\sigma)^{-1}\big)\rho(\sigma)=R.$$

It follows that

$$\rho(\sigma^{-1})R\rho(\sigma^{-1})^{-1}\not\subset R,$$

which contradicts the normality of R, because $\sigma^{-1} \in G$. (Of course, if $\dim_F(V)$ is finite, the desired equality follows readily from the coincidence of F-dimensions of R and $\rho(\sigma)R\rho(\sigma)^{-1}$.)

(iv) If G' is a subgroup of G then every G-normal subalgebra is also a normal G'-subalgebra. It follows that if the G'-module V is very simple then the G-module V is also very simple.

(v) Let us check that a strongly simple G-module V is simple. Indeed, if it is not then there is a *proper* G-invariant F-vector subspace W of V. Then the F-subalgebra

$$R := \{u \in \operatorname{End}_F(V) \mid u(W) \subset W\}$$

is G-normal but even *not semisimple*, because it contains a proper two-sided ideal

$$I(W,V) := \{u \in \operatorname{End}_F(V) \mid u(V) \subset W\}.$$

This proves the simplicity of V.

The centralizer $\operatorname{End}_G(V)$ is obviously G-normal. This implies that it is a division algebra over F. (Actually, it follows from the simplicity of the G-module V.

If the G-module V is central simple (resp. very simple) then normal $\operatorname{End}_G(V)$ is a central division F-algebra (resp. coincides with $F \cdot \mathrm{Id}$).

(vi) If R is a G-normal subalgebra of $\operatorname{End}_F(V)$ then for each $\sigma \in G$ the map

$$R \to R, \ u \mapsto \rho(\sigma)u\rho(\sigma)^{-1}$$

is an *automorphism* of the F-algebra R (in light of (iii)). This implies that if C is the center of R then

$$\rho(\sigma)C\rho(\sigma)^{-1} = C$$

for all $\sigma \in G$. This means that C is a G-normal subalgebra of $\operatorname{End}_F(V)$.

Recall that a module V over a ring R is called *isotypic* if either V is simple or is isomorphic to direct sum of finitely many copies of a simple R-module W. The following assertion is contained in [**34**, Lemma 7.4]

LEMMA 4.3. *Let H be a group, F a field, V a vector space of finite positive dimension N over F. Let $\rho : H \to \operatorname{Aut}_F(V)$ be an irreducible linear representation of H. Let R be a H-normal subalgebra of* $\operatorname{End}_F(V)$. *Then:*

(i) *The faithful R-module V is semisimple.*

(ii) *Either the R-module V is isotypic or there is a subgroup H' of finite index r in H such that $r > 1$ and r divides N.*

PROPOSITION 4.4. *Let F be a field, whose Brauer group* $\mathrm{Br}(F) = \{0\}$. *(E.g., F is either finite or an algebraically closed field.) Let V be a vector space of finite positive dimension N over F. Let H be a group and $\rho : H \to \operatorname{Aut}_F(V)$ a linear absolutely irreducible representation of H in V. Suppose that every maximal subgroup of H has index that does not divide N.*

Then the H-module V is central simple.

PROOF. Slightly abusing the notation, we write F instead of $F \cdot \mathrm{Id}$.

Let R be a H-normal subalgebra of $\operatorname{End}_F(V)$. It follows from Lemma 4.3 that the faithful R-module V is *isotypic*, i.e., there is a simple faithful R-module W and a positive integer a such that the R-modules V and W^a are isomorphic. The existence of a faithful simple R-module implies that R is a simple F-algebra. In particular, the center k of R is a field. We have

$$F = F \cdot \mathrm{Id} \subset k \subset R \subset \mathrm{End}_F(V).$$

Then V carries the natural structure of a F-vector space. This implies that the degree $[k : F]$ divides $\dim_F(V) = N$.

The center k of H-normal R is also H-normal (see Remark 4.2(v)). This gives rise to the group homomorphism

$$H \to \mathrm{Aut}(k/F), \ \sigma \mapsto \{c \mapsto \rho(\sigma)u\rho(\sigma)^{-1}\}. \tag{12}$$

Here $\mathrm{Aut}(k/F)$ is the automorphism group of the field extension k/F. By Galois theory, the order of $\mathrm{Aut}(k/F)$ divides $[k : F]$, which in turn, divides N. This implies that the kernel of the homomorphism (12) is a subgroup of H, whose index divides N. Our condition on indices of subgroups of H implies that the kernel coincides with the whole H, i.e., the homomorphism (12) is trivial. This means that all elements of k commute with $\rho(\sigma)$ for all $\sigma \in H$. The absolute irreducibility of ρ implies that $k \subset F$ and therefore

$$k = F = F \cdot \mathrm{Id}.$$

So, R is a simple F-algebra with center $F \cdot \mathrm{Id}$, i.e., is a central simple F-algebra. This ends the proof. □

THEOREM 4.5. *Let F be a field, whose Brauer group $\mathrm{Br}(F) = \{0\}$. (E.g., F is either finite or an algebraically closed field.) Let V be an F-vector space of finite dimension $N > 1$. Let G be a group and*

$$\rho : G \to \mathrm{Aut}_F(V)$$

be a group homomorphism. Let H be a normal subgroup of G that enjoys the following properties.

(i) *If H' is a subgroup of H of finite index N' and N' divides N then $H' = H$.*
(ii) *H is a simple non-abelian group. Assume additionally that either $H = G$, or H is the only proper normal subgroup of G.*
(iii) *The H-module V is absolutely simple, i.e., the representation of H in V is irreducible and the centralizer $\mathrm{End}_H(V) = F \cdot \mathrm{Id}$.*

Let $R \subset \mathrm{End}_F(V)$ be a G-normal subalgebra. Then there are positive integers a and b that enjoy the following properties.

(a) *$N = ab$;*
(b) *The F-algebra R is isomorphic to the matrix algebra $\mathrm{Mat}_a(F)$ of size a over F. In particular, the G-module V is central simple.*
(c) *The R-module V is semisimple, isotypic and isomorphic to R^b. In addition, the centralizer $\tilde{R} = \mathrm{End}_R(V)$ is a normal G-subalgebra that is isomorphic to the matrix algebra $\mathrm{Mat}_b(F)$ of size b over F.*
(d) *Suppose that $a \neq 1, b \neq 1$ (i.e., R is not obvious). Then both homomorphisms*

$$\mathrm{Ad}_R : G \to \mathrm{Aut}(R) = R^*/F^* \,\mathrm{Id} \cong \mathrm{GL}(a, F)/F^* = \mathrm{PGL}(a, F),$$
$$\mathrm{Ad}_R(\sigma)(u) = \rho(\sigma)u\rho(\sigma)^{-1} \ \forall u \in R$$

and

$$\mathrm{Ad}_{\tilde{R}} : G \to \mathrm{Aut}(\tilde{R}) = \tilde{R}^*/F^*\,\mathrm{Id} \cong \mathrm{GL}(b,F)/F^* = \mathrm{PGL}(b,F),$$
$$\mathrm{Ad}_{\tilde{R}}(\sigma)(u) = \rho(\sigma)u\rho(\sigma)^{-1}\ \forall u \in \tilde{R}$$

(with $\sigma \in G$) are injective. In addition,

$$\mathrm{Ad}_R(H) \subset \mathrm{PSL}(a,F),\ \mathrm{Ad}_{\tilde{R}}(H) \subset \mathrm{PSL}(b,F).$$

(e) *The H-module V is central simple.*

PROOF.

STEP 0. Since H is a simple group, V is a faithful H-module. In light of (ii), V is a faithful G-module.

Clearly, V is a faithful R-module. Since R is G-normal,

$$\rho(\sigma)R\rho(\sigma)^{-1} = R \quad \forall \sigma \in G. \tag{13}$$

It follows from (ii) that

$$H = [H,H] \subset [G,G] \subset G$$

and either $G = H$ or G/H is a finite simple group (e.g., a cyclic group of prime order).

STEP 1. By Lemma 4.3(i), V is a *semisimple* R-module.

STEP 2. In light of Lemma 4.3(ii), property (i) implies that the R-module V is *isotypic*.

STEP 3. Since the faithful R-module V is an isotypic, there exist a faithful simple R-module W and a positive integer b such that $V \cong W^b$. If we put $a = \dim_F(W)$ then we get

$$ba = b \cdot \dim_F(W) = \dim_F(V) = N.$$

Clearly, $\mathrm{End}_R(V)$ is isomorphic to the matrix algebra $\mathrm{Mat}_b(\mathrm{End}_R(W))$ of size b over $\mathrm{End}_R(W)$.

Consider the centralizer

$$k := \mathrm{End}_R(W)$$

of R in $\mathrm{End}_F(W)$. Since W is a simple R-module, k is a finite-dimensional division algebra over F. Since $\mathrm{Br}(F) = \{0\}$, k must be a field. Hence, the automorphism group $\mathrm{Aut}_F(k)$ of the F-algebra k is actually the automorphism group $\mathrm{Aut}(k/F)$ of the field extension k/F. It follows that $\mathrm{Aut}_F(k) = \mathrm{Aut}(k/F)$ is finite and its order divides the degree $[k : F]$. We have

$$\tilde{R} = \mathrm{End}_R(V) \cong \mathrm{Mat}_b(k).$$

Clearly, the F-subalgebra $\tilde{R} = \mathrm{End}_R(V) \subset \mathrm{End}_F(V)$ is stable under the "adjoint action" of G, which gives rise to the group homomorphism

$$\mathrm{Ad}_{\tilde{R}} : G \to \mathrm{Aut}(\tilde{R}).$$

Since k is the center of $\mathrm{Mat}_b(k)$, it is stable under the action of G and of its subgroup H. This gives rise to the group homomorphism

$$H \to \mathrm{Aut}(k/F),\ h \mapsto \{\lambda \mapsto \rho(h)\lambda\rho(h)^{-1}\ \forall \lambda \in k\}\ \forall h \in H,$$

whose kernel H' has index $[H : H']$ dividing $[k : F]$. Since V carries the natural structure of a k-vector space, $[k : F]$ divides $\dim_F(V) = N$, the index $[H : H']$

divides N. In light of (i), $H' = H$, i.e., the homomorphism is trivial. This means that center k of $\mathrm{End}_R(V)$ commutes with $\rho(H)$. Since $\mathrm{End}_H(V) = F$, we have $k = F$. This implies that $\mathrm{End}_R(V) \cong \mathrm{Mat}_b(F)$ and

$$\mathrm{Ad}_{\tilde{R}} : G \to \mathrm{Aut}_F(\tilde{R}) = \tilde{R}^*/F^* \,\mathrm{Id} \cong \mathrm{GL}(b,F)/F^* = \mathrm{PGL}(b,F)$$

kills H if and only if $\tilde{R} = \mathrm{End}_R(V) \subset \mathrm{End}_H(V) = F \cdot \mathrm{Id}$. Since $\tilde{R} = \mathrm{End}_R(V) \cong \mathrm{Mat}_b(F)$, the homomorphism $\mathrm{Ad}_{\tilde{R}}$ kills H if and only if $b = 1$, i.e., V is an absolutely simple (faithful) R-module. This means that if $b > 1$ then $\mathrm{Ad}_{\tilde{R}}$ does *not* kill H, i.e., the normal subgroup $\ker(\mathrm{Ad}_{\tilde{R}})$ of G does *not* contain H. In light of (ii), this implies that the group homomorphism

$$\mathrm{Ad}_{\tilde{R}} : G \to \mathrm{Aut}(\tilde{R}) \cong \mathrm{PGL}(b,F)$$

is *injective* if $b > 1$.

Since V is a semisimple module over the subalgebra R of $\mathrm{End}_F(V)$ and $\tilde{R}$ is the centralizer of R in $\mathrm{End}_F(V)$, it follows from the Jacobson density theorem that

$$R = \mathrm{End}_{\tilde{R}}(V) \cong \mathrm{End}_F(W) \cong \mathrm{Mat}_a(F).$$

The "adjoint action" of G on R gives rise to the homomorphism

$$\mathrm{Ad}_R : G \to \mathrm{Aut}(R) = R^*/F^* \,\mathrm{Id} \cong \mathrm{PGL}(a,F).$$

Clearly, Ad_R kills H if and only if R commutes with $\rho(H)$, i.e., $R = F \cdot \mathrm{Id}$, which is equivalent to the equality $a = 1$. This means that if $a > 1$ then Ad_R does *not* kill H, i.e., the normal subgroup $\ker(\mathrm{Ad}_R)$ of G does *not* contain H. In light of (ii), this implies that the group homomorphism

$$\mathrm{Ad}_R : G \to \mathrm{Aut}(\tilde{R}) \cong \mathrm{PGL}(a,F)$$

is *injective* if $a > 1$.

To summarize: a normal G-subalgebra R is *not* obvious if and only if

$$a > 1, b > 1.$$

If this is the case then both group homomorphisms

$$\mathrm{Ad}_R : G \to \mathrm{PGL}(a,F), \ \mathrm{Ad}_{\tilde{R}} : G \to \mathrm{PGL}(b,F)$$

are *injective*.

The last assertions of Theorem 4.5(d) about the images of H follow from the equality $H = [H,H]$ and the inclusions

$$[\mathrm{GL}(a,F), \mathrm{GL}(a,F)] \subset \mathrm{SL}(a,F), \ [\mathrm{GL}(b,F), \mathrm{GL}(b,F)] \subset \mathrm{SL}(b,F).$$

The assertion (e) follows readily from the second assertion of (b) (if we replace G by H). This ends the proof. □

COROLLARY 4.6. *Keeping the assumption and notation of Theorem 4.5, assume additionally that $N = 2\ell$ where ℓ is a prime. If the H-module V is not very simple then there exist group embeddings*

$$G \hookrightarrow \mathrm{PGL}(2,F), \ H \hookrightarrow \mathrm{PSL}(2,F).$$

PROOF. Let R be a H-normal non-obvious H-subalgebra and $\tilde{R} = \mathrm{End}_R(V)$. By Theorem 4.6, there are positive integers a and b such that

$$ab = N, a > 1, b > 1; R \cong \mathrm{Mat}_a(F), \tilde{R} \cong \mathrm{Mat}_b(F).$$

Our conditions on N imply that either $a = 2, b = \ell$ or $a = \ell, b = 2$. By Theorem 4.5, there are group embeddings

$$G \hookrightarrow \mathrm{PGL}(a, F),\ H \hookrightarrow \mathrm{PSL}(a, F)$$

and

$$G \hookrightarrow \mathrm{PGL}(b, F),\ H \hookrightarrow \mathrm{PSL}(b, F).$$

Since either a or b is 2, there are group embeddings

$$G \hookrightarrow \mathrm{PGL}(2, F),\ H \hookrightarrow \mathrm{PSL}(2, F). \qquad \square$$

THEOREM 4.7. *Suppose that $n \geq 5$ is an integer, B is an n-element set, and p is a prime. Let us consider the vector space $\left(\mathbb{F}_p^B\right)^{00}$ over the field $\mathbb{F}_p$ endowed with the natural structure of a* $\mathrm{Perm}(B)$*-module (see Definition* 3.2*), and let*

$$\rho : \mathrm{Perm}(B) \to \mathrm{Aut}_{\mathbb{F}_p}\left(\left(\mathbb{F}_p^B\right)^{00}\right)$$

be the corresponding structure homomorphism.

Then:

(i) *The* $\mathrm{Perm}(B)$*-module* $\left(\mathbb{F}_p^B\right)^{00}$ *is very simple.*

(ii) *The* $\mathrm{Alt}(B)$*-module* $\left(\mathbb{F}_p^B\right)^{00}$ *is very simple if and only if either $n > 5$ or*

$$n = 5,\ p \not\equiv \pm 1 \mod 5.$$

(iii) *Suppose that*

$$n = 5,\ p \equiv \pm 1 \mod 5.$$

and $R \subset \mathrm{End}_{\mathbb{F}_p}(\left(\mathbb{F}_p^B\right)^{00})$ is a $\mathrm{Alt}(B)$*-normal subalgebra.*

Then either $R = \mathbb{F}_p \cdot \mathrm{Id}$, or $R = \mathrm{End}_{F_p}(\left(\mathbb{F}_p^B\right)^{00})$, or the $\mathbb{F}_p$-algebra R is isomorphic to the matrix algebra $\mathrm{Mat}_2(\mathbb{F}_p)$ of size 2 over $\mathbb{F}_p$.

(iv) *The* $\mathrm{Alt}(B)$*-module* $\left(\mathbb{F}_p^B\right)^{00}$ *is central simple*

REMARK 4.8. The assertion of Theorem 4.7 was earlier proven in the following cases.

(A) $p \in \{2, 3\}$, see [**34**, Ex. 7.2] and [**36**, Cor. 4.3].

(B) $p > 3$ and $n \geq 8$, see [**36**, Cor. 4.6].

(C) $N = \dim_{\mathbb{F}_p}(\left(\mathbb{F}_p^B\right)^{00})$ is a prime. It follows readily from [**36**, Cor. 4.4(i)] applied to $H = \mathrm{Alt}(B)$ and $V = (\mathbb{F}_p^B)^{00}$. (We may apply this result from [**36**], because $\mathrm{Alt}(B)$ is a simple non-abelian group of order $n!/2$ and therefore its order is bigger that the order of $\mathbf{S}_N$, since $N \leq n - 1$.)

So, In the course of the proof we may assume that

$$p > 3;\ n \in \{5, 6, 7\}. \tag{14}$$

PROOF OF THEOREM 4.7. We assume that (14) holds.

STEP 1. First assume that $p \mid n$. Then either $n = p = 5$ or $n = p = 7$. In both cases

$$N = \dim_{\mathbb{F}_p}((\mathbb{F}_p^B)^{00}) = n - 2$$

is a prime. Now the very simplicity of the $\mathrm{Alt}(B)$-module $(\mathbb{F}_p^B)^{00}$ follows from Remark 4.8(A). So, we may assume that p does *not* divide n and therefore

$$N = n - 1.$$

STEP 2. If $n = 6$ then $N = 5$ and the very simplicity of the $\mathrm{Alt}(B)$-module $(\mathbb{F}_p^B)^{00}$ follows from Remark 4.8(C). So, we may assume that

$$n \in \{5, 7\}.$$

STEP 3. Suppose that $n = 7$. Then $N = 6 = 2 \times 3$ where 3 is a prime. It follows from Corollary 4.6 that if the $\mathrm{Alt}(B)$-module $(\mathbb{F}_p^B)^{00}$ is *not* very simple then there is a group homomorphism

$$\mathrm{Ad}_R : \mathrm{Alt}(B) \hookrightarrow \mathrm{PSL}(2, \mathbb{F}_p).$$

However, it is known [**29**, Th. 6.25 on p. 412 and Th. 626 on p. 414] that $\mathrm{PSL}(2, \mathbb{F}_p)$ does *not* contain a subgroup isomorphic to $\mathbf{A}_7$. Since $\mathrm{Alt}(B) \cong \mathbf{A}_7$, we get a contradiction, which implies that the $\mathrm{Alt}(B)$-module $(\mathbb{F}_p^B)^{00}$ is very simple if $n = 7$.

STEP 4. Suppose that $n = 5$. We are going to apply Corollary 4.6 to $H = \mathrm{Alt}(B)$, $G = \mathrm{Perm}(B)$ or $\mathrm{Alt}(B)$, and $V = (\mathbb{F}_p^B)^{00}$.

Since p does *not* divide $n = 5$, we get $p > 5$, and $n - 1 = 4 = 2 \times 2$ where 2 is a prime.

- Suppose that the $\mathrm{Perm}(B)$-module $(\mathbb{F}_p^B)^{00}$ is *not* very simple. It follows from Corollary 4.6 (applied to $G = \mathrm{Perm}(B), H = \mathrm{Alt}(B)$) that there is a group embedding

$$\mathrm{Ad}_R : \mathrm{Perm}(B) \hookrightarrow \mathrm{PGL}(2, \mathbb{F}_p).$$

  This implies that $\mathrm{PGL}(2, \mathbb{F}_p)$ contains a subgroup isomorphic to S_5, because $\mathrm{Perm}(B) \cong \mathbf{S}_5$. Since $p > 5$, the order 120 of $\mathbf{S}_5$ is *not* divisible by p. However, there are no finite subgroups of $\mathrm{PGL}(2, \mathbb{F}_p)$ that are isomorphic to $\mathbf{S}_5$ [**29**, Th. 6.25 on p. 412 and Th. 626 on p. 414]; see also [**24**, Sect. 2.5]. The obtained contradiction proves that the $\mathrm{Perm}(B)$-module $(\mathbb{F}_p^B)^{00}$ is very simple if $n = 5$.
- It follows from Corollary 4.6 (applied to $G = H = \mathrm{Alt}(B)$ and $V = (\mathbb{F}_p^B)^{00}$) that if the $\mathrm{Alt}(B)$-module $(\mathbb{F}_p^B)^{00}$ is *not* very simple then there is an injective group homomorphism

$$\mathrm{Ad}_R : \mathrm{Alt}(B) \hookrightarrow \mathrm{PSL}(2, \mathbb{F}_p).$$

  Then the order 60 of the group $\mathrm{Alt}(B)$ divides the order $(p^2 - 1)p/2$ of the group $\mathrm{PSL}(2, \mathbb{F}_p)$. This implies that 5 divides $p^2 - 1 = (p + 1)(p - 1)$, i.e., $p \equiv \pm 1 \mod 5$. This implies that if $n = 5$ and $p \not\equiv \pm 1 \mod 5$ then $\mathrm{Alt}(B)$-module $(\mathbb{F}_p^B)^{00}$ is very simple.

STEP 5. Suppose that $n = 5$ and $p \equiv \pm 1 \mod 5$. Let us prove that the $\mathrm{Alt}(B)$-module $(\mathbb{F}_p^B)^{00} = (\mathbb{F}_p^B)^0$ is *not* very simple. Recall (Remark 3.3) that there is a surjective homomorphism $\mathrm{SL}(2, \mathbb{F}_5) \twoheadrightarrow \mathbf{A}_5$, and there are $\mathrm{SL}(2, \mathbb{F}_5)$-modules $\bar{V}_1$ and $\bar{V}_2$ with

$$\dim_{\mathbb{F}_p}(\bar{V}_1) = \dim_{\mathbb{F}_p}(\bar{V}_2) = 2,$$

and an isomorphism of $\mathrm{SL}(2, \mathbb{F}_5)$-modules $(\mathbb{F}_p^B)^0 \cong \bar{V}_1 \otimes_{\mathbb{F}_p} \bar{V}_2$. This isomorphism induces an isomorphism of $\mathbb{F}_p$-algebras

$$\mathrm{End}_{\mathbb{F}_p}((\mathbb{F}_p^B)^0) = \mathrm{End}_{\mathbb{F}_p}(\bar{V}_1) \otimes_{\mathbb{F}_p} \mathrm{End}_{\mathbb{F}_p}(\bar{V}_2),$$

under which (the images of) the subalgebras

$$R = \mathrm{End}_{\mathbb{F}_p}(\bar{V}_1) \otimes 1, \ \tilde{R} = 1 \otimes \mathrm{End}_{\mathbb{F}_p}(\bar{V}_2)$$

are $\mathrm{Alt}(B)$-normal subalgebras of $\mathrm{End}_{\mathbb{F}_p}((\mathbb{F}_p^B)^0)$, see [**38**, Example 3.1(ii)]. In particular, the $\mathrm{Alt}(B)$-module $(\mathbb{F}_p^B)^{00} = (\mathbb{F}_p^B)^0$ is *not* very simple.

On the other hand, it follows from Theorem 4.5 that if R is a *non*-obvious $\mathrm{Alt}(B)$-normal subalgebra of $\mathrm{End}_{\mathbb{F}_p}((\mathbb{F}_p^B)^0)$ then $R \cong \mathrm{Mat}_a(\mathbb{F}_p)$ where a positive integer a is a *proper* divisor of

$$\dim_{\mathbb{F}_p}((\mathbb{F}_p^B)^0) = 4 = 2^2.$$

This implies that $a = 2$ and $R \cong \mathrm{Mat}_2(\mathbb{F}_p)$.

The assertion (iv) of Theorem 4.7 follows readily from already proven (ii) and (iii). □

THEOREM 4.9. *Let $n \in \{11, 12, 22, 23, 24\}$. Let B be an n-element set B and $G \subset \mathrm{Perm}(B)$ the corresponding Mathieu group $\mathbf{M}_n$, which acts doubly transitively on B. Let p be an odd prime. If $n = 11$, then we assume additionally that $p > 3$.*

Then

- *the G-module $(\mathbb{F}_p^B)^{00}$ is central simple;*
- *if $n \neq p+1$ and $n-1$ is divisible by p, then the G-module $(\mathbb{F}_p^B)^{00}$ is very simple.*

PROOF. It follows from ([**13**], [**17**, Table 1]) that the absolutely simple $G = \mathbf{M}_n$-module $(\mathbb{F}_p^B)^{00}$ is absolutely simple. By [**1**] the index of every maximal subgroup of $\mathbf{M}_n$ is at least

$$n > n-1 \geq N = \dim_{\mathbb{F}_p}\left((\mathbb{F}_p^B)^{00}\right)$$

(recall that N is either $n-1$ or $n-2$).

In light of Proposition 4.4, the $G = \mathbf{M}_n$-module $(\mathbb{F}_p^B)^{00}$ is central simple. It remains to prove the very simplicity in the "exceptional" cases when $n \neq p+1$ and $n-1$ is divisible by p. We prove that in all the exceptional cases the $\mathbf{M}_n$-module $(\mathbb{F}_p^B)^{00}$ is *very* simple. After that the desired result will follow from Theorem 6.6.

- $n = 11$. Then $p = 5$ and $n - 1 = 2 \times 5$ where both 2 and 5 are primes. If the $\mathbf{M}_{11}$-module $(\mathbb{F}_5^B)^{00}$ is *not* very simple then it follows from Theorem 4.5(iii-d) that there is a group embedding $\mathbf{M}_{11} \hookrightarrow \mathrm{PSL}(2, \mathbb{F}_5)$, which is not true. Hence, the $\mathbf{M}_{11}$-module $(\mathbb{F}_5^B)^{00}$ is very simple.
- $n = 12$. Then $n - 1$ is a prime and there are no exceptional cases.
- $n = 22$. Then $n - 1 = 22 - 1 = 3 \cdot 7$ where both 3 and 7 are primes. Then $p = 3$ or 7. If the $\mathbf{M}_{22}$-module $(\mathbb{F}_3^B)^{00}$ is *not* very simple then it follows from Theorem 4.5(iii-d) that there is a group embedding $\mathbf{M}_{22} \hookrightarrow \mathrm{PSL}(3, \mathbb{F}_p)$. Such an embedding does not exist if $p = 3$, because the order of $\mathrm{PSL}(3, \mathbb{F}_3)$ is *not* divisible by 11 while 11 divides the order of $\mathbf{M}_{22}$. Hence, the $\mathbf{M}_{22}$-module $(\mathbb{F}_3^B)^{00}$ is very simple.

  Such an embedding does not exist if $p = 7$ as well, because the order of $\mathrm{PSL}(3, \mathbb{F}_7)$ is *not* divisible by 11, which divides the order of $\mathbf{M}_{22}$. Hence, the $\mathbf{M}_{22}$-module $(\mathbb{F}_7^B)^{00}$ is also very simple.
- $n = 23$. Then $n - 1 = 22 = 2 \cdot 11$ where both 2 and 11 are primes. Then $p = 11$. If the $\mathbf{M}_{23}$-module $(\mathbb{F}_{11}^B)^{00}$ is *not* very simple then it follows from Theorem 4.5(iii-d) that there is a group embedding $\mathbf{M}_{23} \hookrightarrow \mathrm{PSL}(2, \mathbb{F}_{11})$. Such an embedding does not exist. Hence, the $\mathbf{M}_{23}$-module $(\mathbb{F}_{11}^B)^{00}$ is very simple.
- $n = 24$. Then $n-1 = 23$ is a prime and there are no exceptional cases. □

PROPOSITION 4.10. *Let G be a doubly transitive permutation subgroup of a n-element set B. Let $p > 3$ be a prime. Suppose that (n, G) enjoys one of the following properties.*

(1) *$n = 176$ and G is isomorphic to* HS;
(2) *$n = 276$ and G is isomorphic to* Co_3.

Then the G-module $(\mathbb{F}_p^B)^{00}$ is central simple.

PROOF. It follows from [**17**, Tables] that in both cases the G-module $(\mathbb{F}_p^B)^{00}$ is absolutely simple.

CASE 1. According to the Atlas [**1**], if H is a maximal subgroup of HS with index $[\mathrm{HS} : H] < 176$ then $[\mathrm{HS} : H] = 100$, which divides neither $176-1$ nor $176-2$. By Proposition 4.4, the G-module $(\mathbb{F}_p^B)^{00}$ is central simple.

CASE 2. According to the Atlas [**1**], if H is a maximal subgroup subgroup of Co_3 then its index $m = [\mathrm{Co}_3 : H]$ is greater or equal than 276 [**1**]; in particular, it divides neither $276 - 1$ nor $276 - 2$. By Proposition 4.4, the G-module $(\mathbb{F}_p^B)^{00}$ is central simple as well. □

THEOREM 4.11.

(i) *Let ℓ be a prime, $\mathfrak{r}$ a positive integer, and $n = \mathfrak{q} + 1$ where $\mathfrak{q} = \ell^{\mathfrak{r}} > 11$.*
(ii) *Let G be a subgroup of $\mathrm{Perm}(B)$. Suppose that G contains a subgroup H that is isomorphic to $\mathbf{L}_2(\mathfrak{q}) = \mathrm{PSL}(2, \mathbb{F}_{\mathfrak{q}})$ where $\mathbb{F}_{\mathfrak{q}}$ is a $\mathfrak{q}$-element field.*

If p is an odd prime then the G-module $(\mathbb{F}_p^B)^{00}$ is central simple.

PROOF. It suffices to check that the $H \cong \mathbf{L}_2(q)$-module $(\mathbb{F}_p^B)^{00}$ is central simple. First, our conditions on q imply that each subgroup of $\mathbf{L}_2(\mathfrak{q})$ (except $\mathbf{L}_2(\mathfrak{q})$ itself) has index $\geq \mathfrak{q} + 1 = n$ [**29**, p. 414, (6.27)]. This implies that H acts transitively on the $(\mathfrak{q}+1)$-element set B and the stabilizer H_b of any $b \in B$ has index $\mathfrak{q}+1$. It follows from [**29**, Th. 6.25 on p. 412] that $H_b \subset \mathbf{L}_2(\mathfrak{q})$ is conjugate to the (Borel) subgroup of upper-triangular matrices modulo $\{\pm 1\}$. It follows that the $\mathbf{L}_2(\mathfrak{q})$-set B is isomorphic to the *projective line* $\mathbb{P}^1(\mathbb{F}_{\mathfrak{q}})$ with the standard fractional-linear action of $\mathbf{L}_2(\mathfrak{q})$, which is doubly transitive.

Notice that

$$q + 1 = n > N = \dim_{\mathbb{F}_p}\left((\mathbb{F}_p^B)^{00})\right),$$

because N is either $n-1$ or $n-2$. It follows that the index of any maximal subgroup of $\mathbf{L}_2(\mathfrak{q})$ does *not* divide N. On the other hand, according to [**17**, Table 1], the $H = \mathbf{L}_2(\mathfrak{q})$-module is absolutely simple. It follows now from Proposition 4.4 that the H-module V is central simple. □

THEOREM 4.12. *Let O be a Dedekind ring, T a locally free/projective O-module of finite positive rank r. Let E be the field of fractions of O, $\mathfrak{m}$ a maximal ideal in O and $k = O/\mathfrak{m}$ its residue field. Let us consider the r-dimensional E-vector space $T_E = T \otimes_O E$ and the r-dimensional k-vector space $T_k = T/\mathfrak{m}T = T \otimes_O k$.*

Let G be a group and $\rho : G \to \mathrm{Aut}_O(T)$ be a group homomorphism (O-linear representation). Let us consider the corresponding E-linear representation of G

$$\rho_E : G \to \mathrm{Aut}_E(T_E),\ \sigma \mapsto \{t \otimes e \mapsto \rho(\sigma)(t) \otimes e\ \forall t \in T, e \in E\}\ \forall \sigma \in G$$

and the corresponding k-linear representation of G

$$\rho_k : G \to \mathrm{Aut}_k(T_k),\ \sigma \mapsto \{t \otimes c \mapsto \rho(\sigma)(t) \otimes c\ \forall t \in T, c \in k\}\ \forall \sigma \in G.$$

If ρ_k is central simple (resp. very simple) then ρ_E is central simple (resp. very simple).

PROOF. We view $T = T \otimes 1$ as a certain G-invariant lattice in $T \otimes_O E = T_E$ and $\mathrm{End}_O(T) = \mathrm{End}_O(T) \otimes 1$ as a certain G-invariant lattice (subalgebra) in $\mathrm{End}_O(T) \otimes_O E = \mathrm{End}_E(T_E)$.

Let R_E be a G-normal E-subalgebra of $\mathrm{End}_E(T_E)$. Let C_E be the center of R_E and J_E a *proper* ideal of R_E. We have

$$E \subset C, \ J \subset R_E, \ J \neq R_E.$$

Let us consider the O-subalgebra $R := R_E \cap \mathrm{End}_O(T)$. Clearly, R is a saturated O-submodule of $\mathrm{End}_O(T)$, i.e., the quotient $\mathrm{End}_O(T)/R$ is torsion free (finitely generated) O-module.

The natural map

$$R \otimes_O E \to R_E, \ u \otimes e \mapsto e \cdot u$$

is an isomorphism of E-algebras. This implies that:

(i) $C := C \cap \mathrm{End}_O(T)$ is the center of R that contains O as a saturated O-submodule, i.e., the quotient C/O is a torsion free (finitely generated) O-module. In addition, C is a saturated O-submodule of R, i.e., R/C is a torsion free finitely generated O-module.

(ii) The intersection $J := J_E \cap \mathrm{End}_O(T)$ is a *proper* ideal of R that is a saturated O-submodule of R, i.e., the quotient R/J is a torsion free (finitely generated) O-module.

Since the O-modules $\mathrm{End}_O(T)/R$, C/O, R/C and R/J finitely generated torsion free, they are *projective*, because the ring O is *Dedekind*. This implies that there are locally free submodules $R_1 \subset \mathrm{End}_O(T)$, $C_1 \subset C$, $D \subset R$, and $I \subset R$ such that

$$\mathrm{End}_O(T) = R \oplus R_1, \ C = O \oplus C_1, \ R = C \oplus D, \ I \oplus J = R. \tag{15}$$

Since J is a proper ideal, $I \neq 0$. Since C_1 and J are torsion-free finitely generated O-modules they are also locally free/projective. Now let us consider the $k = O/\mathfrak{m}$-subalgebra

$$R_k = R \otimes_O k \subset \mathrm{End}_O(T) \otimes_O k = \mathrm{End}_k(T_k)$$

where $T_k := T \otimes_O k$. Clearly,

(1) R_k is a G-normal subalgebra of $\mathrm{End}_k(T_k)$;
(2) $k \oplus (C_1 \otimes_O k)$ lies in the center of R_k.
(3) $R_k = (J \otimes_O k) \oplus (I \otimes_O k) \neq \{0\}$. This implies that $J_k = J \otimes_O k$ is a *proper* two-sided ideal of R_k, because $I \neq \{0\}$ and therefore $I \otimes_O k \neq \{0\}$.

Suppose that ρ_k is central simple. Then R_k is a simple k-algebra with center k. It follows that

$$C_1 \otimes_O k = \{0\}, \ J_k = J \otimes_O k = \{0\}.$$

Since C_1 and J are locally free, we conclude that

$$C_1 = \{0\}, \ J = \{0\},$$

which implies that

$$J_E = \{0\}, C = O \oplus C_1 = O \oplus \{0\} = O$$

and therefore $C_E = E$. This means that T_E is a central simple E-algebra, which proves that ρ_E is also central simple.

Assume now that ρ_k is very simple. Then either $R_k = k$ or $R_k = \mathrm{End}_k(T_k)$. In the latter case, applying (15), we get

$$\mathrm{End}_k(T_k) = (R \otimes_O k) \oplus (R_1 \otimes_O k) = R_k \oplus (R_1 \otimes_O k) = \mathrm{End}_k(T_k) \oplus (R_1 \otimes_O k).$$

Now k-dimension arguments imply that $R_1 \otimes_O k = \{0\}$ and therefore $R_1 = \{0\}$. This implies that $\mathrm{End}_O(T) = R \oplus R_1 = R$ and therefore $R_E = \mathrm{End}_E(T_E)$.

Assume now that $R_k = k$. It follows from (15) that

$$R = O \oplus (C_1 \oplus D)$$

and therefore

$$k = R \otimes_O k = k \oplus \ (C_1 \oplus D) \otimes_O k.$$

Again, k-dimension arguments imply that $(C_1 \oplus D) \otimes_O k = \{0\}$ and therefore $C_1 \oplus D = \{0\}$. It follows that $R = O$ and therefore $R_E = E$. This proves that ρ_E is semisimple. □

5. Abelian varieties and cyclotomic fields

Let p be a prime, r a positive integer, and $q = p^r$. Let $E = \mathbb{Q}(\zeta_q)$ be the qth cyclotomic field and $O_E = \mathbb{Z}[\zeta_q]$ its ring of integers.

Let us put

$$\eta = \eta_q := 1 - \zeta_q \in \mathbb{Z}[\zeta_q].$$

It is well known [**32**] that the principal ideal $\eta_q\mathbb{Z}[\zeta_q]$ of $\mathbb{Z}[\zeta_q]$ is maximal and contains $p\mathbb{Z}[\zeta_q]$. Actually,

$$p\mathbb{Z}[\zeta_q] = \eta_q^{\phi(q)}\mathbb{Z}[\zeta_q].$$

It follows that there is $\eta' \in \mathbb{Z}[\zeta_q]$ such that

$$\eta'\mathbb{Z}[\zeta_q] = \eta_q^{\phi(q)-1}\mathbb{Z}[\zeta_q], \ \eta_q\eta' = \eta'\eta_q = p. \tag{16}$$

The residue field $\mathbb{Z}[\zeta_q]/\eta\mathbb{Z}[\zeta_q]$ coincides with $\mathbb{F}_p$. It is also well known [**32**] that

$$\mathbb{Z}_p[\zeta_q] = \mathbb{Z}[\zeta_q] \otimes \mathbb{Z}_p$$

is the ring of integers in the p-adic qth cyclotomic field $\mathbb{Q}_p(\zeta_q)$ and $\eta_q\mathbb{Z}_p[\zeta_q]$ is the maximal ideal of $\mathbb{Z}_p[\zeta_q]$ with residue field

$$\mathbb{Z}_p[\zeta_q]/\eta_q\mathbb{Z}_p[\zeta_q] = \mathbb{Z}[\zeta_q]/\eta_q\mathbb{Z}[\zeta_q] = \mathbb{F}_p.$$

Let K be a field of characteristic different from p. Let K_a be the algebraic closure of K and $K_s \subset K_a$ the separable algebraic closure of K. We write $\mathrm{Gal}(K)$ for the automorphism group $\mathrm{Aut}(K_a/K) = \mathrm{Gal}(K_s/K)$ of the corresponding field extension.

Let Z be an abelian variety of positive dimension g over K, and $\mathrm{End}_K(Z)$ (resp. $\mathrm{End}(Z)$) the ring of its K-endomorphisms (resp. the ring of all K_a-endomorphisms). By a theorem of Chow, all endomorphisms of Z are defined over K_s. In addition, $Z[p] \subset Z(K_s)$ where $Z[p]$ is the kernel of multiplication by p in $Z(K_a)$. If m is an integer then we write $m_Z \in \mathrm{End}_K(Z)$ for multiplication by m in Z.

Suppose that we are given the ring embedding

$$\mathbf{i} : O_E \hookrightarrow \mathrm{End}_K(Z) \subset \mathrm{End}(Z)$$

such that $1 \in O_E$ goes to the identity automorphism 1_Z of Z. In light of (16), $\mathbf{i}(\eta_q) : Z \to Z$ and $\mathbf{i}(\eta') : Z \to Z$ are isogenies and the kernel $\ker(\mathbf{i}(\eta_q))$ of $\mathbf{i}(\eta_q)$ lies in $Z[p]$. In addition,

$$\ker(\mathbf{i}(\eta_q)) = \mathbf{i}(\eta')(Z[p]) \subset Z[p]. \tag{17}$$

Indeed, since $\eta_q\eta' = p$, we have $\mathbf{i}(\eta_q)\mathbf{i}(\eta') = p_Z$ and

$$\mathbf{i}(\eta')(Z[p]) \subset \ker(\mathbf{i}(\eta_q)).$$

Conversely, suppose that $z \in \ker(\mathbf{i}(\eta_q))$. Since $\mathbf{i}(\eta')$ is an isogeny, it is surjective and therefore there is $\tilde{z} \in Z(K_a)$ such that $\mathbf{i}(\eta')(\tilde{z}) = z$. This implies that

$$0 = \mathbf{i}(\eta_q)(z) = \mathbf{i}(\eta_q)\mathbf{i}(\eta')\tilde{z} = p_Z\tilde{z} = p\tilde{z}.$$

It follows that $\tilde{z} \in Z[p]$ and therefore $\ker(\mathbf{i}(\eta_q)) \subset \mathbf{i}(\eta')(Z[p])$, which ends the proof of (17).

Let us put

$$\boldsymbol{\delta} := \mathbf{i}(\zeta_q) \in \mathrm{End}_K(Z) \subset \mathrm{End}(Z). \tag{18}$$

REMARK 5.1.

(i) Since $\eta_q = 1 - \zeta_q$, we get $\mathbf{i}(\eta_q) = 1_Z - \boldsymbol{\delta}$ and therefore

$$\ker(\mathbf{i}(\eta_q)) = \{z \in Z(K_a) \mid \boldsymbol{\delta}(z) = z\} =: Z^{\boldsymbol{\delta}}. \tag{19}$$

(ii) Since $\ker(\mathbf{i}(\eta_q))$ is a subgroup of $Z[p]$, it carries the natural structure of a $\mathbb{F}_p$-vector space. In other words, $\ker(\mathbf{i}(\eta_q))$ is a $\mathbb{F}_p$-vector subspace of $Z[p]$.

(iii) Since the endomorphism $\mathbf{i}(\eta_q)$ is defined over K, $\ker(\mathbf{i}(\eta_q))$ is a $\mathrm{Gal}(K)$-invariant subspace of $Z[p]$. The action of $\mathrm{Gal}(K)$ on $\ker(\mathbf{i}(\eta_q))$ gives rise to the natural linear representation

$$\rho_\eta = \rho_{\eta,Z} : \mathrm{Gal}(K) \to \mathrm{Aut}_{\mathbb{F}_p}(\ker(\mathbf{i}(\eta_q))), \tag{20}$$
$$\sigma \mapsto \{z \mapsto \sigma(z)\ \forall z \in \ker(\mathbf{i}(\eta_q)) \subset Z[p] \subset Z(K_s)\}\ \forall \sigma \in \mathrm{Gal}(K).$$

LEMMA 5.2. *$\phi(q) = [E : \mathbb{Q}]$ divides $2\dim(Z) = 2g$ and $\ker(\mathbf{i}(\eta_q))$ is a $\mathbb{F}_p$-vector space of dimension*

$$h_E := \frac{2\dim(Z)}{[E : \mathbb{Q}]} = \frac{2g}{\phi(q)}.$$

PROOF. By a result of Ribet [19, Prop. 2.2.1 on p. 769], the $\mathbb{Z}_p$-*Tate module* $T_p(Z)$ of Z is a free module over the ring

$$\mathbb{Z}_p[\boldsymbol{\delta}] = \mathbf{i}(O_E) \otimes \mathbb{Z}_p \cong \mathbb{Z}[\zeta_q] \otimes \mathbb{Z}_p = \mathbb{Z}_p[\zeta_q]$$

of rank $h_E = 2g/\phi(q)$. (In particular, h_E is an integer.) This implies that the $\mathbb{Z}_p[\delta]$-module $Z[p] = T_p(Z)/pT_p(Z)$ is isomorphic to $(\mathbb{Z}_p[\boldsymbol{\delta}]/p)^{h_E}$. It follows from (17) that the $\mathbb{F}_p$-vector space

$$\ker(\mathbf{i}(\eta_q)) \cong (\eta'\mathbb{Z}_p[\zeta_q]/p)^{h_E} =$$
$$(\eta'\mathbb{Z}_p[\zeta_q]/\eta'\eta_q\mathbb{Z}_p[\zeta_q])^{h_E} = (\mathbb{Z}_p[\zeta_q]/\eta_q\mathbb{Z}_p[\zeta_q])^{h_E} = \mathbb{F}_p^{h_E}.$$

This proves that $\ker(\mathbf{i}(\eta_q))$ is a $\mathbb{F}_p$-vector space of dimension h_E. □

Let Λ be the centralizer of $\mathbf{i}(O_E)$ in $\mathrm{End}(Z)$. Clearly, $\mathbf{i}(O_E)$ lies in the center of Λ. It is also clear that

$$\Lambda(\ker \mathbf{i}(\eta_q)) \subset \ker(\mathbf{i}(\eta_q)),$$

which gives rise to the natural homomorphism of $O/\eta_q O = \mathbb{F}_p$-algebras

$$\kappa : \Lambda/\mathbf{i}(\eta_q)\Lambda \to \mathrm{End}_{\mathbb{F}_p}(\ker \mathbf{i}(\eta_q)),\ u + \mathbf{i}(\eta_q)\Lambda \mapsto \{z \mapsto u(z)\}\ \forall z \in \ker \mathbf{i}(\eta_q). \tag{21}$$

PROPOSITION 5.3. *The homomorphism κ defined in* (21) *is injective.*

PROOF. Suppose that $u \in \Lambda$ and $u(\ker \mathbf{i}(\eta_q)) = \{0\}$. We need to prove that $u \in \mathbf{i}(\eta_q)\Lambda$. In order to do that, notice that the endomorphism of Z

$$v := \mathbf{i}(\eta')\ u = u\ \mathbf{i}(\eta') \in \Lambda \subset \mathrm{End}(Z)$$

kills $Z[p]$, because

$$v(Z[p]) = u\ \mathbf{i}\,(\eta')(Z[p]) = u\ (\mathbf{i}\,(\eta')Z[p]) = u\big(\ker \mathbf{i}(\eta_q)\big) = \{0\}.$$

This implies that there is $\tilde{v} \in \mathrm{End}(Z)$ such that $v = p\tilde{v}$. Since v commutes with $\mathbf{i}(O_E)$, $\tilde{v}$ also commutes with $\mathbf{i}(O_E)$, i.e., $\tilde{v} \in \Lambda$. We have

$$\mathbf{i}(\eta')\mathbf{i}(\eta_q)\tilde{v} = p\tilde{v} = v = \mathbf{i}(\eta')u.$$

This implies that in $\mathrm{End}(Z)$

$$\mathbf{i}(\eta')\big(\mathbf{i}(\eta_q)\tilde{v} - u\big) = 0.$$

Multiplying it by $\mathbf{i}(\eta_q)$ from the left and taking into account that $\mathbf{i}(\eta_q)\mathbf{i}(\eta') = \mathbf{i}(p)$, we get

$$p\big(\mathbf{i}(\eta_q)\tilde{v} - u\big) = 0$$

in $\mathrm{End}(Z)$. It follows that $\mathbf{i}(\eta_q)\tilde{v} = u$. Since $\tilde{v} \in \Lambda$, we are done. □

REMARK 5.4.

(i) Since Z is defined over K, one may associate with every $u \in \mathrm{End}(Z)$ and $\sigma \in \mathrm{Gal}(K)$ an endomorphism ${}^{\sigma}u \in \mathrm{End}(Z)$ such that

$$(22) \qquad {}^{\sigma}u(z) = \sigma u(\sigma^{-1}z) \quad \forall z \in Z(K_a).$$

(ii) Recall that $\mathbf{i}(O_E) \subset \mathrm{End}_K(Z)$ consists of K-endomorphisms of Z. It follows that if $u \in \mathrm{End}(Z)$ commutes with $\mathbf{i}(O_E)$ then ${}^{\sigma}u$ commutes with $\mathbf{i}(O_E)$ for all $\sigma \in \mathrm{Gal}(K)$. In other words, if $u \in \Lambda$ then ${}^{\sigma}u \in \Lambda$ for all $\sigma \in \mathrm{Gal}(K)$.

(iii) Since $O_E = \mathbb{Z}[\zeta_q]$, we have $\mathbf{i}(O_E) = \mathbb{Z}[\boldsymbol{\delta}]$. It follows that Λ coincides with the centralizer of $\boldsymbol{\delta}$ in $\mathrm{End}(Z)$.

PROPOSITION 5.5. *The image $R := \kappa(\Lambda/\eta_q\Lambda)$ is a* $\mathrm{Gal}(K)$*-normal subalgebra of* $\mathrm{End}_{\mathbb{F}_p}(\ker \mathbf{i}(\eta_q))$.

PROOF. Let $u \in \Lambda$. Then

$$\bar{u} := \kappa(u + \eta_q\Lambda) \in R \subset \mathrm{End}_{\mathbb{F}_p}(\ker \mathbf{i}(\eta_q)),$$
$$\bar{u} : z \mapsto u(z)\ \forall z \in \ker \mathbf{i}(\eta_q).$$

Then ${}^{\sigma}u \in \Lambda$ for all $\sigma \in \mathrm{Gal}(K)$ and

$$\overline{{}^{\sigma}u} = \kappa({}^{\sigma}u + \eta_q\Lambda) \in \mathrm{End}_{\mathbb{F}_p}(\ker \mathbf{i}(\eta_q)),$$
$$\overline{{}^{\sigma}u} : z \mapsto \sigma u \sigma^{-1}(z) = \rho_{\eta}(\sigma)u\rho_{\eta}(\sigma)^{-1}(z) = \rho_{\eta}(\sigma)\bar{u}\rho_{\eta}(\sigma)^{-1}(z).$$

In other words, for each $\bar{u} \in R$

$$\rho_{\eta}(\sigma)\bar{u}\rho_{\eta}(\sigma)^{-1} \in R\ \forall \sigma \in \mathrm{Gal}(K).$$

This proves that R is $\mathrm{Gal}(K)$-normal. □

REMARK 5.6.

(i) Extending $\mathbf{i}$ by $\mathbb{Q}$-linearity, we get a $\mathbb{Q}$-algebra embedding

$$E = O_E \otimes \mathbb{Q} \to \mathrm{End}(Z) \otimes \mathbb{Q} =: \mathrm{End}^0(Z), \quad u \otimes c \mapsto cu \ \forall u \in O, c \in \mathbb{Q}$$

that we continue to denote by $\mathbf{i}$. Clearly, $\mathbf{i}(E)$ coincides with the $\mathbb{Q}$-subalgebra $\mathbb{Q}[\boldsymbol{\delta}]$ of $\mathrm{End}^0(Z)$ generated by δ_q. Clearly, $\mathbf{i} : E \to \mathbb{Q}[\boldsymbol{\delta}]$ is a field isomorphism of number fields, and $\mathbf{i}(O_E)$ is the ring of integers in the number field $\mathbb{Q}[\delta]$.

(ii) Let us consider the $\mathbb{Q}$-subalgebra

$$\mathcal{H} = \Lambda \otimes \mathbb{Q} \subset \mathrm{End}(Z) \otimes \mathbb{Q} = \mathrm{End}^0(Z).$$

Then the center of $\mathcal{H}$ contains $\mathbf{i}(O) \otimes \mathbb{Q} = \mathbb{Q}[\boldsymbol{\delta}]$. In other words, $\mathcal{H}$ is a $\mathbb{Q}[\boldsymbol{\delta}]$-algebra of finite dimension.

(iii) We have

(23)
$$\Lambda = \mathcal{H} \cap \mathrm{End}(Z).$$

where the intersection is taken in $\mathrm{End}^0(Z)$. (Here we identify $\mathrm{End}(Z)$ with $\mathrm{End}(Z) \otimes 1$ in $\mathrm{End}^0(Z)$.) Indeed, the inclusion $\Lambda \subset \mathcal{H} \cap \mathrm{End}(Z)$ is obvious. Conversely, suppose that $u \in \mathcal{H} \cap \mathrm{End}(Z)$. Then $u \in \mathrm{End}(Z)$ and $mu \in \Lambda$ for some positive integer m. This means that

$$(mu)\boldsymbol{\delta} = \boldsymbol{\delta}(mu),$$

which means that $m(u\boldsymbol{\delta} - \boldsymbol{\delta}u) = 0$ in $\mathrm{End}(Z)$. It follows that $u\boldsymbol{\delta} - \boldsymbol{\delta}u = 0$, i.e., $u \in \Lambda$. It follows that $\mathcal{H} \cap \mathrm{End}(Z) \subset \Lambda$, which ends the proof of (23).

PROPOSITION 5.7.

(i) *If the* $\mathrm{Gal}(K)$*-module* $\ker \mathbf{i}(\eta_q)$ *is central simple then* $\mathcal{H}$ *is a central simple* $\mathbb{Q}[\boldsymbol{\delta}]$*-algebra.*

(ii) *If the* $\mathrm{Gal}(K)$*-module* $\ker \mathbf{i}(\eta_q)$ *is very simple then either*

$$\mathcal{H} = \Lambda \otimes \mathbb{Q} = \mathbf{i}(E) = \mathbb{Q}[\boldsymbol{\delta}], \ \Lambda = \mathbf{i}(O_E) = \mathbb{Z}[\boldsymbol{\delta}]$$

or $\mathcal{H} = \Lambda \otimes \mathbb{Q}$ *is a central simple* $\mathbb{Q}[\boldsymbol{\delta}]$*-algebra, whose dimension is the square of* $2\dim(Z)/[E : \mathbb{Q}] = 2g/\phi(q)$.

PROOF.

(i) The central simplicity implies that the $\mathrm{Gal}(K)$-normal subalgebra

$$R = \kappa(\Lambda/\eta\Lambda) \cong \Lambda/\eta\Lambda$$

is a central simple $\mathbb{F}_p$-algebra and therefore is isomorphic to the matrix algebra $\mathrm{Mat}_d(\mathbb{F}_p)$ of a certain size d. Applying Lemma 2.3 to $O = \mathbf{i}(O_E)$, the maximal ideal $\mathfrak{m} = \mathbf{i}(\eta_q O_E)$ and the residue field $k = \mathbb{F}_p$, we conclude that $\mathcal{H}$ is a central simple $\mathbb{Q}[\delta_q]$-algebra.

(ii) The very simplicity implies that either $\Lambda/\eta_q\Lambda = \mathbb{F}_p$ or

$$\Lambda/\eta\Lambda \cong \mathrm{End}_{\mathbb{F}_p}(\ker \mathbf{i}(\eta_q)) \cong \mathrm{Mat}_{h_E}(\mathbb{F}_p).$$

In the latter case, Lemma 2.3 tells us that $\mathcal{H}$ is a central simple $\mathbb{Q}[\boldsymbol{\delta}]$-algebra of dimension h_E^2.

In the former case, Lemma 2.3 tells us that $\mathcal{H}$ is a central simple $\mathbb{Q}[\boldsymbol{\delta}]$-algebra of dimension 1, i.e., $\mathcal{H} = \mathbb{Q}[\boldsymbol{\delta}]$. Hence,

$$\mathbb{Z}[\boldsymbol{\delta}] \subset \Lambda \subset \mathbb{Q}[\boldsymbol{\delta}].$$

Since $\mathbb{Z}[\boldsymbol{\delta}] \cong \mathbb{Z}[\zeta_q]$ is integrally closed and Λ is a free $\mathbb{Z}$-module of finite rank, $\mathbb{Z}[\boldsymbol{\delta}] = \Lambda$. □

6. Cyclic covers and Jacobians

Hereafter we fix an odd prime p.

Let us assume that K is a subfield of $\mathbb{C}$. We write K_a for the algebraic closure of K in $\mathbb{C}$ and write $\mathrm{Gal}(K)$ for the absolute Galois group $\mathrm{Aut}(K_a/K)$. We also fix in K_a a primitive pth root of unity $\zeta = \zeta_p$.

Let $f(x) \in K[x]$ be a separable polynomial of degree $n \geq 4$. We write $\mathfrak{R}_f$ for the n-element set of its roots and denote by $L = L_f = K(\mathfrak{R}_f) \subset K_a$ the corresponding splitting field of $f(x)$. As usual, the Galois group $\mathrm{Gal}(L/K)$ is called the Galois group of f and denoted by $\mathrm{Gal}(f)$. Clearly, $\mathrm{Gal}(f)$ permutes elements of $\mathfrak{R}_f$ and the natural map of $\mathrm{Gal}(f)$ into the group $\mathrm{Perm}(\mathfrak{R}_f)$ of all permutations of $\mathfrak{R}_f$ is an embedding. We will identify $\mathrm{Gal}(f)$ with its image and consider it as the certain permutation group of $\mathfrak{R}_f$. Clearly, $\mathrm{Gal}(f)$ is transitive if and only if f is irreducible in $K[x]$. Therefore the $\mathrm{Gal}(f)$-module $(\mathbb{F}_p^{\mathfrak{R}_f})^{00}$ is defined. The canonical surjection

$$\mathrm{Gal}(K) \twoheadrightarrow \mathrm{Gal}(f)$$

provides $(\mathbb{F}_p^{\mathfrak{R}_f})^{00}$ with the canonical structure of the $\mathrm{Gal}(K)$-module via the composition

$$\mathrm{Gal}(K) \twoheadrightarrow \mathrm{Gal}(f) \subset \mathrm{Perm}(\mathfrak{R}_f) \subset \mathrm{Aut}((\mathbb{F}_p^{\mathfrak{R}_f})^{00}).$$

Let us put

$$V_{f,p} := (\mathbb{F}_p^{\mathfrak{R}_f})^{00}. \tag{24}$$

Let $C = C_{f,p}$ be the smooth projective model of the smooth affine K-curve

$$y^p = f(x).$$

The genus

$$g = g(C) = g(C_{f,p})$$

of C is $(p-1)(n-1)/2$ if p does *not* divide p and $(p-1)(n-2)/2$ if it does ([**16**], pp. 401–402, [**30**], Prop. 1 on p. 3359, [**21**], p. 148).

Assume that K contains ζ. There is a non-trivial biregular automorphism of C

$$\delta_p : (x, y) \mapsto (x, \zeta y).$$

Clearly, δ_p^p is the identity selfmap of C.

Let

$$J^{(f,p)} := J(C) = J(C_{f,p})$$

be the Jacobian of C. It is a g-dimensional abelian variety defined over K and one may view δ_p as an element of

$$\mathrm{Aut}(C) \subset \mathrm{Aut}(J(C)) \subset \mathrm{End}(J(C))$$

such that

$$\delta_p \neq \mathrm{Id}, \quad \delta_p^p = \mathrm{Id}$$

where Id is the identity endomorphism of $J(C)$. Here $\mathrm{End}(J(C))$ stands for the ring of all K_a-endomorphisms of $J(C)$. As usual, we write $\mathrm{End}^0(J(C)) = \mathrm{End}^0(J^{(f,p)})$ for the corresponding $\mathbb{Q}$-algebra $\mathrm{End}(J(C)) \otimes \mathbb{Q}$.

Recall (4) that there is a ring embedding

$$\mathbf{i}_{p,f} : \mathbb{Z}[\zeta_p] \cong \mathbb{Z}[\delta_p] \subset \mathrm{End}(J^{(f,p)}), \ \zeta_p \mapsto \delta_p.$$

Let us put

$$J^{(f,p)}(\eta_p) =: \ker(\mathbf{i}_{p,f}(\eta_p)) \subset J^{(f,p)}(K_a) \tag{25}$$

where $\eta_p = 1 - \zeta_p \in \mathbb{Z}[\delta_p]$ (Section 5).

REMARK 6.1. Let

$$\Lambda := \mathrm{End}_{\delta_p}(J^{(f,p)})$$

be the centralizer of δ_p in $\mathrm{End}(J^{(f,p)})$. Clearly,

$$\mathcal{H} := \Lambda \otimes \mathbb{Q} \subset \mathrm{End}(J^{(f,p)}) \otimes \mathbb{Q} \subset \mathrm{End}^0(J^{(f,p)})$$

is the centralizer of $\mathbb{Q}[\delta_p]$ in $\mathrm{End}^0(J^{(f,p)})$.

THEOREM 6.2 (Prop. 6.2 in [**21**], Prop. 3.2 in [**23**]). *There is a canonical isomorphism of the* $\mathrm{Gal}(K)$*-modules*

$$J^{(f,p)}(\eta_p) \cong V_{f,p}.$$

REMARK 6.3. Clearly, the natural homomorphism $\mathrm{Gal}(K) \to \mathrm{Aut}_{\mathbb{F}_p}(V_{f,p})$ coincides with the composition

$$\mathrm{Gal}(K) \twoheadrightarrow \mathrm{Gal}(f) \subset \mathrm{Perm}(\mathfrak{R}_f) \subset \mathrm{Aut}\left((\mathbb{F}_p^{\mathfrak{R}_f})^{00}\right) = \mathrm{Aut}_{\mathbb{F}_p}(V_{f,p}).$$

COROLLARY 6.4.

(i) *If the* $\mathrm{Gal}(f)$*-module* $V_{f,p}$ *is central simple then the* $\mathrm{Gal}(K)$*-module* $J^{(f,p)}(\eta_p)$ *is central simple and* $\mathcal{H}=\Lambda\otimes\mathbb{Q}$ *is a central simple* $\mathbb{Q}[\delta_p]$*-algebra.*

(ii) *If the* $\mathrm{Gal}(f)$*-module* $V_{f,p}$ *is very simple then the* $\mathrm{Gal}(K)$*-module* $J^{(f,p)}(\eta_p)$ *is very simple and either* $\Lambda = \mathbf{i}(O)$ *or* $\mathcal{H} = \Lambda\otimes\mathbb{Q}$ *is a central simple* $\mathbb{Q}[\delta_p]$*-algebra, whose dimension is the square of* $2\dim(J^{(f,p)})/(p-1)$.

PROOF. It follows from Remark 4.2(ii) combined with Theorem 6.2 that if the $\mathrm{Gal}(f)$-module $V_{f,p}$ is central simple (resp. very simple) then the $\mathrm{Gal}(K)$-module $J^{(f,p)}(\eta_p)$ (defined in (25)) is central simple (resp. very simple). Now the desired result follows readily from Proposition 5.7 applied to $Z = J^{(f,p)}$, $q = p$, and $\mathbf{i} = \mathbf{i}_{p,f}$. □

The following assertion was proven in [**37**, Th. 3.6].

THEOREM 6.5. *Suppose that* $n \geq 4$. *Assume that* $\mathbb{Q}[\delta_p]$ *is a maximal commutative subalgebra of* $\mathrm{End}^0(J^{(f,p)})$.

Then $\mathrm{End}^0(J^{(f,p)}) = \mathbb{Q}[\delta_p] \cong \mathbb{Q}(\zeta_p)$ *and therefore* $\mathrm{End}(J^{(f,p)}) = \mathbb{Z}[\delta_p] \cong \mathbb{Z}[\zeta_p]$.

THEOREM 6.6. *Let* p *be an odd prime and* $\zeta \in K$. *Suppose that the* $\mathrm{Gal}(f)$*-module* $(\mathbb{F}_p^{\mathfrak{R}_f})^{00}$ *enjoys one of the following properties.*

(i) *The* $\mathrm{Gal}(f)$*-module* $V_{f,p} = (\mathbb{F}_p^{\mathfrak{R}_f})^{00}$ *is very simple.*

(ii) *The* $\mathrm{Gal}(f)$*-module* $V_{f,p} = (\mathbb{F}_p^{\mathfrak{R}_f})^{00}$ *is central simple. In addition, either* $n = p+1$, *or* $n-1$ *is not divisible by* p.

Then $\mathrm{End}^0(J^{(f,p)}) = \mathbb{Q}[\delta_p]$ *and* $\mathrm{End}(J^{(f,p)}) = \mathbb{Z}[\delta_p]$.

PROOF OF THEOREM 6.6. In light of Theorem 6.5, it suffices to check that $\mathbb{Q}[\delta_p]$ coincides with its own centralizer in $\mathrm{End}^0(J^{(f,p)})$. Recall that $J^{(f,p)}$ is a g-dimensional abelian variety defined over K.

The properties of the $\mathrm{Gal}(f)$-module $(\mathbb{F}_p^{\mathfrak{R}_f})^{00}$ and the integers n and p imply (thanks to Remark 4.2(ii) and Theorem 6.2) that either the $\mathrm{Gal}(K)$-module $J^{(f,p)}(\eta_p)$ is very simple, or the following conditions hold.

(a) The $\mathrm{Gal}(K)$-module $J^{(f,p)}(\eta_p)$ is central simple.
(b) Either $n = p+1$, or $n-1$ is not divisible by p.

In all the cases the normal $\mathbb{F}_p$-subalgebra $R \cong \Lambda/\eta_p\Lambda$ is isomorphic to the matrix algebra $\mathrm{Mat}_d(\mathbb{F}_p)$ for some positive integer d.

Applying Corollary 5.7, we conclude that $\mathcal{H} = \Lambda_{\mathbb{Q}} = \Lambda \otimes \mathbb{Q}$ is a central simple $\mathbb{Q}[\delta_p]$-algebra of dimension d^2 for some positive integer d. In addition, if the $\mathrm{Gal}(K)$-module $J^{(f,p)}(\eta_p)$ is very simple, then either

$$d = 1, \ \mathcal{H} = \mathbb{Q}[\delta_p], \Lambda = \mathbb{Z}[\delta_p]$$

or

$$d = 2g/(p-1).$$

According to Remark 2.2(vi), $d \neq 2g/(p-1)$. So, in the very simple case $\mathcal{H} = \mathbb{Q}[\delta_p], \Lambda = \mathbb{Z}[\delta_p]$.

Now suppose that $J^{(f,p)}(\eta_p)$ is *not* very simple. Then either $n = p+1$ or $n-1$ is not divisible by p. It follows from Remark 2.2(iv) that $d = 1$. This implies that $H = \mathbb{Q}[\delta_p]$. Therefore

$$\mathbb{Z}[\delta_p] \subset \Lambda \subset \mathbb{Q}[\delta_p].$$

This implies that $\Lambda = \mathbb{Z}[\delta_p]$ and therefore the centralizer of $\mathbb{Q}[\delta_p]$ in $\mathrm{End}^0(J^{(f,p)})$ coincides with $\Lambda \otimes \mathbb{Q} = \mathbb{Q}[\delta_p]$. □

THEOREM 6.7. *Let $n \geq 5$ be an integer, p an odd prime, and K contains a primitive pth root of unity. Let us put $N := n-1$ if p does not divide n and $N := n-2$ if $p \mid n$. Suppose that the Galois group* $\mathrm{Gal}(f)$ *of $f(x)$ contains a subgroup H such that the representation of H in $(\mathbb{F}_p^{\mathfrak{R}_f})^{00}$ is absolutely irreducible. Assume additionally that*

(i) *the index of every maximal subgroup of H does not divide N.*
(ii) *Either $n = p+1$, or $n-1$ is not divisible by p.*

Then $\mathrm{End}^0(J^{(f,p)}) = \mathbb{Q}[\delta_p]$ *and* $\mathrm{End}(J^{(f,p)}) = \mathbb{Z}[\delta_p]$.

PROOF OF THEOREM 6.7. Enlarging K if necessary, we may and will assume that $H = \mathrm{Gal}(f)$. It follows from Proposition 4.4 that the absolutely simple H-module $(\mathbb{F}_p^{\mathfrak{R}_f})^{00}$ is central simple. Applying Theorem 6.6, we conclude that $\mathrm{End}^0(J^{(f,p)}) = \mathbb{Q}[\delta_p]$ and $\mathrm{End}(J^{(f,p)}) = \mathbb{Z}[\delta_p]$. □

REMARK 6.8. See [**5**, Sect. 7.7] and [**17**] for the list of doubly transitive permutation groups $H \subset \mathrm{Perm}(B)$ and primes p such that the H-module $(\mathbb{F}_p^B)^{00}$ is (absolutely) simple. (See also [**22**, Sect. 4], [**4**, Main Theorem] and [**15**].)

7. Jacobians of cyclic covers of prime degree p

PROOF OF THEOREM 1.2. Enlarging K if necessary, we may and will assume that

$$H = \mathrm{Gal}(f) \subset \mathrm{Perm}(\mathfrak{R}_f).$$

Since $p > n$, the prime p divides neither n nor $n-1$. In particular,

$$(\mathbb{F}_p^{\mathfrak{R}_f})^{00} = (\mathbb{F}_p^{\mathfrak{R}_f})^{0}.$$

In light of Remark 3.1 applied to $B = \mathfrak{R}_f$ and $G = H$, the double transitivity of H implies that the H-module $(\mathbb{F}_p^{\mathfrak{R}_f})^0$ is absolutely simple. Now the desired result follows readily from Theorem 6.7. □

Proof of Theorem 1.3. Assume that $n \geq 5$ and $\mathrm{Gal}(f) = \mathrm{Perm}(\mathfrak{R}_f)$ or $\mathrm{Alt}(\mathfrak{R}_f)$. Enlarging K if necessary, we may assume that $\mathrm{Gal}(f) = \mathrm{Alt}(\mathfrak{R}_f)$. Taking into account that $\mathrm{Alt}(\mathfrak{R}_f)$ is non-abelian simple while the field extension $K(\zeta)/K$ is abelian, we conclude that the Galois group of f over $K(\zeta)$ is also $\mathrm{Alt}(\mathfrak{R}_f)$. (In particular, $f(x)$ remains irreducible over $K(\zeta)$.) So, in the course of the proof of Theorem 1.3, we may assume that $\zeta \in K$ and $\mathrm{Gal}(f) = \mathrm{Alt}(\mathfrak{R}_f)$.

It is well known that the index of every maximal subgroup of $\mathrm{Alt}(\mathfrak{R}_f) \cong \mathbf{A}_n$ is at least n; notice that

$$n > N = \dim_{\mathbb{F}_p}(V_{f,p}) = \dim_{\mathbb{F}_p}\left((\mathbb{F}_p^B)^{00}\right).$$

(Recall that $N = n-1$ or $n-2$.) By Theorem 4.7(iv), the $\mathrm{Gal}(f)$-module $V_{f,p} = (\mathbb{F}_p^B)^{00}$ is *central simple*. It is *very simple* if either $n > 5$ or $p \leq 5$, thanks to Theorem 4.7(ii). On the other hand, if $n = 5$ and $p > 5$ then $n-1$ is *not* divisible by p. Now the desired result follows readily from Theorem 6.6. □

Proof of Theorem 1.6. Since the groups $\mathbf{M}_n$, HS and Co_3 are simple non-abelian groups, replacing K by $K(\zeta)$, we may and will assume that $\zeta \in K$. Now the desired result follows readily from Theorem 6.7 combined with Theorem 4.9 and Proposition 4.10. □

Proof of Theorem 1.8. Enlarging K, we may assume that $\mathrm{Gal}(f) = H \cong \mathfrak{G}(\mathfrak{q})$. Since $\mathfrak{G}(\mathfrak{q})$ is a simple nonabelian group, replacing K by $K(\zeta)$, we may and will assume that $\zeta \in K$. In light of [**17**, Table 1], our conditions on H and p imply that the H-module $V_{f,p}$ is *absolutely simple.*

L2. It follows from Theorem 4.11 that the $\mathrm{Gal}(f)$-module $(\mathbb{F}_p^{\mathfrak{R}_f})^{00} = V_{f,p}$ is central simple.

On the other hand, if $n-1$ is divisible by p then $p = \ell$, because $n-1 = (\mathfrak{q}+1)-1 = \mathfrak{q}$ which is a power of the prime number ℓ. Hence, our assumptions imply that $n = \mathfrak{q}+1 = p+1$. So, either $n-1$ is *not* divisible by p or $n = p+1$. Now we may apply Theorem 6.6, which gives us $\mathrm{End}^0(J^{(f,p)}) = \mathbb{Q}[\delta_p]$ and $\mathrm{End}(J^{(f,p)}) = \mathbb{Z}[\delta_p]$.

Lmq. It follows from a result of Guralnick and Tiep [**9**, Th. 1.1] that every nontrivial projective representation of $\mathrm{Gal}(f) = H = \mathrm{L}_m(\mathfrak{q})$ in characteristic p has dimension $\geq \dim_{\mathbb{F}_p}(V_{f,p})$. In light of [**35**, Cor. 5.4], the $\mathrm{Gal}(f)$-module $V_{f,p}$ is very simple. Now the desired result follows readily from Theorem 6.7.

U3. It follows readily from the Mitchell's list of maximal subgroups of $\mathrm{U}_3(\mathfrak{q})$ ([**10**, pp. 212–213], [**8**, Th. 6.5.3 and its proof, pp. 329–332] that the index of every maximal subgroup of $\mathrm{U}_3(\mathfrak{q})$ is greater or equal than

$$\mathfrak{q}^3 + 1 = n > N$$

where $N = \dim_{\mathbb{F}_p}(V_{f,p})$ is either $n-1 = \mathfrak{q}^3$ or $n-2 = \mathfrak{q}^3 - 1$. On the other hand, $n-1 = \mathfrak{q}^3$ is a power of the prime ℓ and therefore is *not* divisible by the prime p, since $\ell \neq p$. Now the desired result follows readily from Theorem 6.7.

Sz. It follows from the classification of subgroups of $\mathrm{Sz}(\mathfrak{q})$ [**11**, Remark 3.12(e) on p. 194] that every maximal subgroup of $\mathrm{Sz}(\mathfrak{q})$ has index $\geq \mathfrak{q}^2 + 1 = n$. Since

$n-1=\mathfrak{q}^2$ is a power of 2, the odd prime p does *not* divide $n-1$. Now the desired result follows readily from Theorem 6.7.

REE. Our conditions on p imply that $p \neq 3$. Since $n-1=\mathfrak{q}^2$ is a power of 3, the prime p does *not* divide $n-1$. It follows from the classification of subgroups of $\mathrm{Ree}(\mathfrak{q})$ [**14**, Th. C] that every maximal subgroup of $\mathrm{Sz}(\mathfrak{q})$ has index $\geq \mathfrak{q}^3+1=n$. (See also [**6**, Remark 5.4].) Now the desired result follows readily from Theorem 6.7. □

8. Jacobians of cyclic covers of degree q

In this section we discuss the case when $q=p^r>2$ where r is any positive integer, K is a subfield of $\mathbb{C}$ and $f(x) \in K[x]$ a degree n polynomial without repeated roots. We assume that $n \geq 5$ and either $q \mid n$ or p does *not* divide n. Let $J(C_{f,q})$ be the Jacobian of the curve $C_{f,q}$ and δ_q the automorphism of $J(C_{f,q})$, whicch are defined in the beginning of Section 1.

REMARK 8.1. One may define a positive-dimensional abelian subvariety

$$J^{(f,q)} := \mathcal{P}_{q/p}(\delta_q)(J(C_{f,q}))$$

of $J(C_{f,q})$ [**34**, p. 355] that is defined over $K(\zeta_q)$ and enjoys the following properties [**34**] (see also [**39**]).

(i) If $q=p$ then $J^{(f,p)}=J(C_{f,p})$ (as above).
(ii) $J^{(f,q)}$ is defined over $K(\zeta_q)$.
(iii) $J^{(f,q)}$ is a δ_q-invariant abelian subvariety of $J(C_{f,q})$. If

$$\Phi_q(t)=\sum_{i=0}^{p-1} t^{ip^{r-1}} \in \mathbb{Z}[t]$$

is the qth cyclotomic polynomial, then $\Phi_q(\delta_q)(J^{(f,q)})=0$. This gives rise to the ring embedding

$$\mathbf{j}_{q,f} : \mathbb{Z}[\zeta_q] \hookrightarrow \mathrm{End}(J^{(f,q)})$$

under which ζ_q goes to to the restriction of δ_q to $J^{(f,q)}$, which we denote by $\boldsymbol{\delta}_q \in \mathrm{End}(J^{(f,q)})$. Then the subring $\mathbb{Z}[\boldsymbol{\delta}_q]$ of $\mathrm{End}(J^{(f,q)})$ is isomorphic to $\mathbb{Z}[\zeta_q]$ (via $\mathbf{j}_{q,f}$), and the $\mathbb{Q}$-subalgebra $\mathbb{Q}[\boldsymbol{\delta}_q]$ of $\mathrm{End}^0(J^{(f,q)})$ is isomorphic to $\mathbb{Q}(\zeta_q)$.
(iv) If p does *not* divide n then there is an isogeny of abelian varieties

$$J(C_{f,q}) \to J(C_{f,q/p}) \times J^{(f,q)}$$

that is defined over $K(\zeta_q)$. (Notice that $q/p=p^{r-1}$, so $J(C_{f,q/p})=J(C_{f,p^{r-1}})$.) By induction, this gives us an isogeny of abelian varieties

$$J(C_{f,q}) \to J(C_{f,p}) \times \prod_{i=2}^{r} J^{(f,r^i)} = \prod_{i=1}^{r} J^{(f,r^i)}$$

that is also defined over $K(\zeta_q)$ [**34**, Cor. 4.12].

(v) If $\zeta_q \in K$, then the $\mathrm{Gal}(K)$-submodule $J^{(f,q)}(\eta_q) := \ker(\mathbf{j}_{q,f}(\eta_q))$ of $J^{(f,q)}(K_a)$ is isomorphic to $V_{f,p}$. (See [**34**, Lemma 4.11], [**39**, Th. 9.1].) In particular, $J^{(f,q)}(\eta_q)$ is an N-dimensional vector space over $\mathbb{F}_p$ where
- $N = n-1$ if p does *not* divide n;
- $N = n-2$ if p divides n and q divides n.

(vi) Consider the action of the subfield $\mathbb{Q}[\boldsymbol{\delta}_q]$ of $\mathrm{End}^0(J^{(f,q)})$ on $\Omega^1(J^{(f,q)})$. Let $i < q$ be a positive integer that is *not* divisible by p and $\sigma_i : \mathbb{Q}[\boldsymbol{\delta}_q] \hookrightarrow \mathbb{C}$ be the field embedding that sends $\boldsymbol{\delta}_q$ to ζ_q^{-i}. Clearly,

$$\Omega^1(J^{(f,q)}) = \oplus_i \Omega^1(J^{(f,q)})_{\sigma_i}$$

where $\Omega^1(J^{(f,q)})_{\sigma_i}$ are the corresponding weight subspaces (see Section 2). Let us consider the nonnegative integers

$$n_{\sigma_i} := \dim_{\mathbb{C}}(\Omega^1(J^{(f,q)})_{\sigma_i}).$$

(1) If p does *not* divide n then

$$n_{\sigma_i} = \left[\frac{ni}{q}\right]$$

[**34**, Remark 4.13]. In addition, the number of i with $n_{\sigma_i} \neq 0$ is strictly greater than

$$\frac{(p-1)p^{r-1}}{2} = \frac{\phi(q)}{2} = \frac{[\mathbb{Q}[\boldsymbol{\delta}_q] : \mathbb{Q}]}{2}$$

[**34**, pp. 357–358]).

(2) If p is odd and q divides n then the GCD of all n_{σ_i}'s is 1 [**39**, Lemma 8.1(D)].

(3) If p is an odd prime that does *not* divide n, and either $n = q+1$ or $n-1$ is *not* divisible by q, then the GCD of all n_{σ_i}'s is 1 [**39**, Lemma 8.1(D)].

(vii) If p is odd and $\mathbb{Q}[\boldsymbol{\delta}_q]$ is a maximal commutative subalgebra of $\mathrm{End}^0(J^{(f,q)})$ then

$$\mathrm{End}^0(J^{(f,q)}) = \mathbb{Q}[\boldsymbol{\delta}_q],\ \mathrm{End}(J^{(f,q)}) = \mathbb{Z}[\boldsymbol{\delta}_q]$$

([**34**, Th. 4.16], [**39**, Th. 8.3]).

THEOREM 8.2. *Let $n \geq 5$ be an integer, p an odd prime, and K contains a primitive qth root of unity. Suppose that either p does not divide n or q divides n.*

Let us put $N := n-1$ if p does not divide n and $N := n-2$ if $q \mid n$. Suppose that the Galois group $\mathrm{Gal}(f)$ of $f(x)$ contains a subgroup H such that the representation of H in $\left(\mathbb{F}_p^{\mathfrak{R}_f}\right)^{00} = V_{f,p}$ is absolutely irreducible. Assume additionally that the index of every maximal subgroup of H does not divide N and one of the following conditions holds.

(i) *The representation of H in $\left(\mathbb{F}_p^{\mathfrak{R}_f}\right)^{00} = V_{f,p}$ is very simple.*

(ii) *Either p does not divide n and $n-1$ is not divisible by q, or $n = q+1$, or $q \mid n$.*

Then $\mathrm{End}^0(J^{(f,q)}) = \mathbb{Q}[\boldsymbol{\delta}_q]$ and $\mathrm{End}(J^{(f,q)}) = \mathbb{Z}[\boldsymbol{\delta}_q]$. In particular, $J^{(f,q)}$ is an absolutely simple abelian variety.

PROOF. Enlarging K if necessary, we may and will assume that $\mathrm{Gal}(f) = H$. It follows from Proposition 4.4 that the absolutely simple H-module $(\mathbb{F}_p^{\mathfrak{R}_f})^{00} = V_{f,p}$ is central simple.

Recall (Remark 8.1(v)) that the $\mathrm{Gal}(K)$-module $J^{(f,q)}(\eta_q)$ is isomorphic to $V_{f,p}$ and therefore is also central simple. In addition, it is very simple if and only if the H-module $V_{f,p}$ is very simple.

Let Λ be the centralizer of $\mathbb{Z}[\boldsymbol{\delta}_q]$ in $\mathrm{End}(J^{(f,q)})$ and

$$\mathcal{H} = \Lambda_{\mathbb{Q}} := \Lambda \otimes \mathbb{Q}$$

the centralizer of $\mathbb{Q}[\boldsymbol{\delta}_q]$ in $\mathrm{End}^0(J^{(f,q)})$. Applying Proposition 5.7 to

$$Z = J^{(f,q)}, \ E = \mathbb{Q}[\delta_q], \ \mathbf{i} = \mathbf{j}_{q,f}, \ O_E = \mathbb{Z}[\boldsymbol{\delta}_q],$$

we conclude that $\mathcal{H} = \Lambda_{\mathbb{Q}} = \Lambda \otimes \mathbb{Q}$ is a central simple $\mathbb{Q}[\boldsymbol{\delta}_q]$-algebra of dimension d^2 for some positive integer d.

In addition, if the $\mathrm{Gal}(K)$-module $J^{(f,q)}(\eta_q)$ is very simple, then either

$$d = 1, \ \mathcal{H} = \mathbb{Q}[\boldsymbol{\delta}_q], \Lambda = \mathbb{Z}[\boldsymbol{\delta}_q]$$

or

$$d = N = \frac{2g}{\phi(q)} = \dim_{\mathbb{F}_p}(V_{f,p})$$

where

$$g = \dim(J^{(f,q)}), \ \phi(q) = [\mathbb{Q}(\zeta_q) : \mathbb{Q}] = [\mathbb{Q}[\boldsymbol{\delta}_q] : \mathbb{Q}].$$

In light of Remark 8.1(i-ii) combined with Proposition 5.7(ii),

$$d \neq \frac{2g}{[\mathbb{Q}[\boldsymbol{\delta}_q] : \mathbb{Q}]}.$$

So, in the very simple case $\mathcal{H} = \mathbb{Q}[\boldsymbol{\delta}_q], \quad \Lambda = \mathbb{Z}[\boldsymbol{\delta}_q]$.

Now suppose that $J^{(f,q)}(\eta_q)$ is *not very* simple. Then $V_{f,p}$ is not very simple. This implies that either $n = q + 1$, or p does not divide n and $n - 1$ is not divisible by q, or $q \mid n$. It follows from Proposition 5.7(i) combined with Remark 8.1(vii) that $d = 1$. This implies that $\mathcal{H} = \mathbb{Q}[\boldsymbol{\delta}_q]$. Therefore

$$\mathbb{Z}[\boldsymbol{\delta}_q] \subset \Lambda \subset \mathbb{Q}[\boldsymbol{\delta}_q].$$

This implies that $\Lambda = \mathbb{Z}[\boldsymbol{\delta}_q]$ and therefore $\mathcal{H} = \mathbb{Q}[\boldsymbol{\delta}_q]$ is a maximal commutative subalgebra of $\mathrm{End}^0(J^{(f,q)})$. Now the desired result follows from Remark 2.2(viii). $\square$

PROOF OF THEOREMS 1.10 AND 1.9. Enlarging K if necessary, we may assume that K contains a primitive qth root of unity, and

- $\mathrm{Gal}(f) = \mathrm{Alt}(\mathfrak{R}_f) =: H$ in the case of Theorem 1.10;
- $\mathrm{Gal}(f) = H$ in the case of Theorem 1.9.

It follows from Theorem 8.2 that $\mathrm{End}^0(J^{(f,q)}) = \mathbb{Q}[\boldsymbol{\delta}_q] \cong \mathbb{Q}(\zeta_q)$ and $J^{(f,q)}$ is an absolutely simple abelian variety for $q = p^r$ when r is any positive integer. This implies that for distinct positive integers i and j there are no nonzero homomorphisms between $J^{(f,p^i)}$ and $J^{(f,p^j)}$, because they are absolutely simple abelian varieties with non-isomorphic endomorphism algebras. This implies that the endomorphism algebra $\mathrm{End}^0(Y)$ of the product $Y := \prod_{i=1}^r J^{(f,r^i)}$ is $\prod_{i=1}^r \mathbb{Q}(\zeta_{p^i})$, whose $\mathbb{Q}$-dimension is $q - 1$.

In light of Remark 8.1(iv), if p does *not* divide n then Y is isogenous to $J(C_{f,q})$. It follows that $\mathrm{End}^0(J(C_{f,q}))$ also has $\mathbb{Q}$-dimension $(q-1)$. However, we know that the $\mathbb{Q}$-algebra $\mathrm{End}^0(J(C_{f,q}))$ contains the $\mathbb{Q}$-subalgebra $\mathbb{Q}[\delta_q]$ of $\mathbb{Q}$-dimension $(q-1)$, thanks to (3). This implies that

$$\mathrm{End}^0(J(C_{f,q})) = \mathbb{Q}[\delta_q] \cong \prod_{i=1}^{r} \mathbb{Q}(\zeta_{p^i}).$$

This ends the proof of Theorem 1.9.

Let us finish the proof of Theorem 1.10, following [**34**, Remark 4.3 and Proof of Th. 5.2 on p. 360]. It remains to do the case when $q \mid n$, say, $n = qm$ for some positive integer m. Since the case $q = p$ was already covered by already proven Theorem 1.3, we may assume that $q \geq p^2 \geq 9$ and therefore $n \geq 9$. Recall that $\mathrm{Gal}(f) = \mathrm{Alt}(\mathfrak{R}_f) \cong \mathbf{A}_n$. Let $\alpha \in K_a$ be a root of $f(x)$. Let us consider the overfield $K_1 = K(\alpha)$ of K. We have $f(x) = (x-\alpha)f_1(x) \in K_1[x]$ where $f_1(x)$ is a degree $(n-1)$ irreducible polynomial over K_1 with Galois group $\mathbf{A}_{n-1}$. Let us consider the polynomials

$$h(x) = f_1(x+\alpha),\ h_1(x) = x^{n-1} \in K_1[x]$$

of degree $n-1 \geq 9-1 = 8$. Notice that $n-1$ is not divisible by p and the Galois group of $h_1(x)$ over K_1 is still $\mathbf{A}_{n-1}$. The standard substitution

$$x_1 = \frac{1}{x-\alpha},\ y_1 = \frac{y}{(x-\alpha)^m}$$

establishes a birational isomorphism between the curves $C_{f,q}$ and $C_{h_1,q}$ [**30**, p. 3359]. This implies that the Jacobians $J(C_{f,q})$ and $J(C_{h_1,q})$ are isomorphic and therefore their endomorphism algebras are also isomorphic. Applying to the jacobian $J(C_{h_1,q})$ the already proven part of Theorem 1.10, we conclude that the $\mathbb{Q}$-algebra $\mathrm{End}^0(J(C_{h_1,q}))$ has $\mathbb{Q}$-dimension $(q-1)$. This implies that $\mathrm{End}^0(J(C_{f,q}))$ also has $\mathbb{Q}$-dimension $q-1$. However, we know that $\mathrm{End}^0(J(C_{f,q}))$ contains the $\mathbb{Q}$-subalgebra $\mathbb{Q}[\delta_q]$ of $\mathbb{Q}$-dimension $(q-1)$ (see (3) in Section 1). This implies that

$$\mathrm{End}^0(J(C_{f,q})) = \mathbb{Q}[\delta_q] \cong \prod_{i=1}^{r} \mathbb{Q}(\zeta_{p^i}).$$

This ends the proof of Theorem 1.10. □

Acknowledgment

I am grateful to the referee, whose comments helped to improve the exposition.

References

[1] J. H. Conway, R. T. Curtis, S. P. Norton, R. A. Parker, and R. A. Wilson, $\mathbb{ATLAS}$ *of finite groups: Maximal subgroups and ordinary characters for simple groups*, Oxford University Press, Eynsham, 1985. With computational assistance from J. G. Thackray. MR827219

[2] Charles W. Curtis and Irving Reiner, *Representation theory of finite groups and associative algebras*, Pure and Applied Mathematics, Vol. XI, Interscience Publishers (a division of John Wiley & Sons, Inc.), New York-London, 1962. MR144979

[3] H. K. Farahat, *On the natural representation of the symmetric groups*, Proc. Glasgow Math. Assoc. **5** (1962), 121–136 (1962). MR139666

[4] Charles W. Curtis, William M. Kantor, and Gary M. Seitz, *The 2-transitive permutation representations of the finite Chevalley groups*, Trans. Amer. Math. Soc. **218** (1976), 1–59, DOI 10.2307/1997427. MR422440

[5] John D. Dixon and Brian Mortimer, *Permutation groups*, Graduate Texts in Mathematics, vol. 163, Springer-Verlag, New York, 1996, DOI 10.1007/978-1-4612-0731-3. MR1409812

[6] T. Eritsyan, *Endomorphism rings and algebras of Jacobians of certain superelliptic curves*, Ph.D. Thesis, Penn State, 2022.

[7] Larry Dornhoff, *Group representation theory. Part A: Ordinary representation theory*, Pure and Applied Mathematics, vol. 7, Marcel Dekker, Inc., New York, 1971. MR347959

[8] D. Gorenstein, R. Lyons, R. Solomon, *The classification of the finite simple groups, Number 3*. Mathematical Surveys and Monographs **40.3**, American Mathematical Society, Providence, RI, 1998.

[9] Robert M. Guralnick and Pham Huu Tiep, *Low-dimensional representations of special linear groups in cross characteristics*, Proc. London Math. Soc. (3) **78** (1999), no. 1, 116–138, DOI 10.1112/S0024611599001720. MR1658160

[10] Alan R. Hoffer, *On unitary collineation groups*, J. Algebra **22** (1972), 211–218, DOI 10.1016/0021-8693(72)90141-X. MR301624

[11] Bertram Huppert and Norman Blackburn, *Finite groups. III*, Grundlehren der Mathematischen Wissenschaften, vol. 243, Springer-Verlag, Berlin-New York, 1982. MR662826

[12] G. J. Janusz, *Simple components of $Q[\mathrm{SL}(2,q)]$*, Comm. Algebra **1** (1974), 1–22, DOI 10.1080/00927877408548606. MR344323

[13] Michael Klemm, *Über die Reduktion von Permutationsmoduln* (German), Math. Z. **143** (1975), no. 2, 113–117, DOI 10.1007/BF01187052. MR379639

[14] Peter B. Kleidman, *The maximal subgroups of the Chevalley groups $G_2(q)$ with q odd, the Ree groups ${}^2G_2(q)$, and their automorphism groups*, J. Algebra **117** (1988), no. 1, 30–71, DOI 10.1016/0021-8693(88)90239-6. MR955589

[15] Alexander Kleshchev, Lucia Morotti, and Pham Huu Tiep, *Irreducible restrictions of representations of symmetric and alternating groups in small characteristics*, Adv. Math. **369** (2020), 107184, 66, DOI 10.1016/j.aim.2020.107184. MR4097880

[16] Ja Kyung Koo, *On holomorphic differentials of some algebraic function field of one variable over* **C**, Bull. Austral. Math. Soc. **43** (1991), no. 3, 399–405, DOI 10.1017/S0004972700029245. MR1107394

[17] Brian Mortimer, *The modular permutation representations of the known doubly transitive groups*, Proc. London Math. Soc. (3) **41** (1980), no. 1, 1–20, DOI 10.1112/plms/s3-41.1.1. MR579714

[18] David Mumford, *Abelian varieties*, Tata Institute of Fundamental Research Studies in Mathematics, vol. 5, Published for the Tata Institute of Fundamental Research, Bombay; by Oxford University Press, London, 1970. MR282985

[19] Kenneth A. Ribet, *Galois action on division points of Abelian varieties with real multiplications*, Amer. J. Math. **98** (1976), no. 3, 751–804, DOI 10.2307/2373815. MR457455

[20] Donald Passman, *Permutation groups*, W. A. Benjamin, Inc., New York-Amsterdam, 1968. MR237627

[21] Bjorn Poonen and Edward F. Schaefer, *Explicit descent for Jacobians of cyclic covers of the projective line*, J. Reine Angew. Math. **488** (1997), 141–188, DOI 10.1515/crll.1997.488.141. MR1465369

[22] Cheryl E. Praeger and Leonard H. Soicher, *Low rank representations and graphs for sporadic groups*, Australian Mathematical Society Lecture Series, vol. 8, Cambridge University Press, Cambridge, 1997. MR1430574

[23] Edward F. Schaefer, *Computing a Selmer group of a Jacobian using functions on the curve*, Math. Ann. **310** (1998), no. 3, 447–471, DOI 10.1007/s002080050156. MR1612262

[24] Jean-Pierre Serre, *Propriétés galoisiennes des points d'ordre fini des courbes elliptiques* (French), Invent. Math. **15** (1972), no. 4, 259–331, DOI 10.1007/BF01405086. MR387283

[25] Jean-Pierre Serre, *Topics in Galois theory*, Research Notes in Mathematics, vol. 1, Jones and Bartlett Publishers, Boston, MA, 1992. Lecture notes prepared by Henri Damon [Henri Darmon]; With a foreword by Darmon and the author. MR1162313

[26] Jean-Pierre Serre, *Linear representations of finite groups*, Graduate Texts in Mathematics, Vol. 42, Springer-Verlag, New York-Heidelberg, 1977. Translated from the second French edition by Leonard L. Scott. MR450380

[27] Goro Shimura, *Abelian varieties with complex multiplication and modular functions*, Princeton Mathematical Series, vol. 46, Princeton University Press, Princeton, NJ, 1998, DOI 10.1515/9781400883943. MR1492449

[28] Robert Steinberg, *Lectures on Chevalley groups*, University Lecture Series, vol. 66, American Mathematical Society, Providence, RI, 2016. Notes prepared by John Faulkner and Robert Wilson; Revised and corrected edition of the 1968 original [MR0466335]; With a foreword by Robert R. Snapp, DOI 10.1090/ulect/066. MR3616493

[29] Michio Suzuki, *Group theory. I*, Grundlehren der Mathematischen Wissenschaften, vol. 247, Springer-Verlag, Berlin-New York, 1982. Translated from the Japanese by the author. MR648772

[30] Christopher Towse, *Weierstrass points on cyclic covers of the projective line*, Trans. Amer. Math. Soc. **348** (1996), no. 8, 3355–3378, DOI 10.1090/S0002-9947-96-01649-2. MR1357406

[31] Ascher Wagner, *The faithful linear representations of least degree of S_n and A_n over a field of odd characteristic*, Math. Z. **154** (1977), no. 2, 103–114, DOI 10.1007/BF01241824. MR437631

[32] Lawrence C. Washington, *Introduction to cyclotomic fields*, 2nd ed., Graduate Texts in Mathematics, vol. 83, Springer-Verlag, New York, 1997, DOI 10.1007/978-1-4612-1934-7. MR1421575

[33] Jiangwei Xue, *Endomorphism algebras of Jacobians of certain superelliptic curves*, J. Number Theory **131** (2011), no. 2, 332–342, DOI 10.1016/j.jnt.2010.07.013. MR2910215

[34] Yuri G. Zarhin, *Hyperelliptic Jacobians and modular representations*, Moduli of abelian varieties (Texel Island, 1999), Progr. Math., vol. 195, Birkhäuser, Basel, 2001, pp. 473–490. MR1827030

[35] Yuri G. Zarhin, *Very simple 2-adic representations and hyperelliptic Jacobians*, Mosc. Math. J. **2** (2002), no. 2, 403–431, DOI 10.17323/1609-4514-2002-2-2-403-431. Dedicated to Yuri I. Manin on the occasion of his 65th birthday. MR1944511

[36] Yuri G. Zarhin, *Cyclic covers of the projective line, their Jacobians and endomorphisms*, J. Reine Angew. Math. **544** (2002), 91–110, DOI 10.1515/crll.2002.030. MR1887891

[37] Yuri G. Zarhin, *The endomorphism rings of Jacobians of cyclic covers of the projective line*, Math. Proc. Cambridge Philos. Soc. **136** (2004), no. 2, 257–267, DOI 10.1017/S0305004103007102. MR2040573

[38] Yuri G. Zarhin, *Very simple representations: variations on a theme of Clifford*, Progress in Galois theory, Dev. Math., vol. 12, Springer, New York, 2005, pp. 151–168, DOI 10.1007/0-387-23534-5_9. MR2148465

[39] Yuri G. Zarhin, *Endomorphism algebras of abelian varieties with special reference to superelliptic Jacobians*, Geometry, algebra, number theory, and their information technology applications, Springer Proc. Math. Stat., vol. 251, Springer, Cham, 2018, pp. 477–528, DOI 10.1007/978-3-319-97379-1_22. MR3880401

Department of Mathematics, Pennsylvania State University, University Park, Pennsylvania 16802

Email address: zarhin@math.psu.edu

SELECTED PUBLISHED TITLES IN THIS SERIES

832 **Nils Bruin, David Kohel, and Chloe Martindale, Editors,** Arithmetic, Geometry, Cryptography and Coding Theory, 2026

831 **Irena Swanson and Kei-ichi Watanabe, Editors,** Lectures on Tight Closure and Its Applications, 2025

830 **O. Bezushchak, Yu. Drozd, P. Etingof, V. Futorny, R. Grigorchuk, and E. Zelmanov, Editors,** Modern Algebra Volume 2, 2025

829 **O. Bezushchak, Yu. Drozd, P. Etingof, V. Futorny, R. Grigorchuk, and E. Zelmanov, Editors,** Modern Algebra Volume 1, 2025

828 **Hafedh Herichi, Maria Rosaria Lancia, Therese Basa Landry, Anna Rozanova-Pierrat, and Steffen Winter, Editors,** Fractal Geometry in Pure and Applied Mathematics, 2025

827 **Carmen Caprau, J. Scott Carter, Neslihan Gügümcü, and Sam Nelson, Editors,** Algebraic Structures in Knot Theory, 2025

826 **Ahmed Laghribi and André Leroy, Editors,** Algebraic Structures and Applications, 2025

825 **Saurabh Verma, María A. Navascués, and Amit Priyadarshi, Editors,** Advances in Dimension Theory, Fractal Functions and Measures, 2025

824 **Hanna E. Makaruk and Robert M. Owczarek, Editors,** Inverse Problems: In Memory of Professor Zbigniew Oziewicz, 2025

823 **Mikhail Bershtein, Anton Dzhamay, and Andrei Okounkov, Editors,** The Versatility of Integrability, 2025

822 **Ahmad Barhoumi, Roozbeh Gharakhloo, and Andrei Martinez-Finkelshtein, Editors,** Recent Developments in Orthogonal Polynomials, 2025

821 **Bogdan D. Suceavă, Adara M. Blaga, Marian Ioan Munteanu, Cezar Oniciuc, Shoo Seto, Gabriel-Eduard Vîlcu, and Lihan Wang, Editors,** Recent Advances in Differential Geometry and Related Areas, 2025

820 **Sudeb Mitra, Rakesh Kumar Parmar, Saminathan Ponnusamy, and Toshiyuki Sugawa, Editors,** Geometric Function Theory and Related Topics, 2025

819 **Howard S. Cohl, Roberto S. Costas-Santos, and Robert S. Maier, Editors,** Applications and q-Extensions of Hypergeometric Functions, 2025

818 **Howard S. Cohl, Roberto S. Costas-Santos, and Robert S. Maier, Editors,** Classical Hypergeometric Functions and Generalizations, 2025

817 **Mikhail Khovanov, Joshua Sussan, and Anton Zeitlin, Editors,** From Representation Theory to Mathematical Physics and Back, 2025

816 **Luca F. Di Cerbo and Laurenţiu G. Maxim, Editors,** Geometry and Topology of Aspherical Manifolds, 2025

815 **Daniel Orr and Joshua Jeishing Wen, Editors,** Macdonald Theory and Beyond, 2025

814 **Susan Montgomery and Siu-Hung Ng, Editors,** Quantum Groups, Hopf Algebras, and Applications, 2025

813 **Thomas Creutzig, Julia Plavnik, and David Ridout, Editors,** Quantum Symmetries, 2025

812 **Ross Geoghegan, Craig R. Guilbault, and Kim Ruane, Editors,** Topology at Infinity of Discrete Groups, 2025

811 **Krishnendu Gongopadhyay and Shashikant A. Katre, Editors,** Geometry, Groups and Mathematical Philosophy, 2025

810 **Robert J. Berman and Yanir A. Rubinstein, Editors,** Convex and Complex: Perspectives on Positivity in Geometry, 2025

809 **Fabrice Baudoin and Luca Rizzi, Editors,** New Trends in Sub-Riemannian Geometry, 2025

For a complete list of titles in this series, visit the
AMS Bookstore at **www.ams.org/bookstore/conmseries/**.

FSC
www.fsc.org
MIX
Papier | Fördert
gute Waldnutzung
FSC® C083411

Kolibri 360 GmbH
Ferdinand-Jühlke-Straße 7
99095 Erfurt, Deutschland
produktsicherheit@kolibri360.de